Introduction to Ship Engine Room Systems

Introduction to Ship Engine Room Systems outlines the key systems, machinery, and equipment found in a ship's engine room. It explores the basics of their function with overall practical guidance for engine room operation and maintenance, recognising emerging environmental challenges. It covers the following topics:

- The role and function of the steering and propulsion systems
- Power generation
- The heating, ventilation, and air-conditioning systems
- The water management system
- Engine room fires and emergency response systems
- Engine room watch procedures and checklists

This book serves as an accessible introductory text for engineering students at Higher National Certificate (HNC), Higher National diploma (HND), and foundation degree (Fd) levels; marine engineering cadets; and non-engineering marine professionals such as deck officers and cadets who want a general guide to how the engine room functions.

Introduction to Ship Engine Room Systems

Alexander Arnfinn Olsen

LONDON AND NEW YORK

Cover image: Shutterstock

First published 2023
by Routledge
4 Park Square, Milton Park, Abingdon, Oxon OX14 4RN

and by Routledge
605 Third Avenue, New York, NY 10158

Routledge is an imprint of the Taylor & Francis Group, an informa business

© 2023 Alexander Arnfinn Olsen

The right of Alexander Arnfinn Olsen to be identified as author of this work has been asserted in accordance with sections 77 and 78 of the Copyright, Designs and Patents Act 1988.

All rights reserved. No part of this book may be reprinted or reproduced or utilised in any form or by any electronic, mechanical, or other means, now known or hereafter invented, including photocopying and recording, or in any information storage or retrieval system, without permission in writing from the publishers.

Trademark notice: Product or corporate names may be trademarks or registered trademarks, and are used only for identification and explanation without intent to infringe.

British Library Cataloguing-in-Publication Data
A catalogue record for this book is available from the British Library

ISBN: 978-1-032-34228-3 (hbk)
ISBN: 978-1-032-34227-6 (pbk)
ISBN: 978-1-003-32109-5 (ebk)

DOI: 10.1201/9781003321095

Typeset in Sabon
by SPi Technologies India Pvt Ltd (Straive)

Dedicated to James Leslie McLeod

"The sea! the sea! the open sea!
The blue, the fresh, the ever free!"

26.01.1919 – 01.12.2003

Contents

Preface	xvii
Acknowledgements	xix
Abbreviations, glossary, and terms used	xxi
Temperature conversion table	xxv
Introduction	xxvii

PART I
Steering and propulsion systems — 1

1 Rudder and steering gear	**3**
How the rudder helps the ship to turn	7
The steering gear	12
Evaluating the steering gear	17
Steering gear system failure	18
Emergency steering system	20
Notes	21
2 Propeller design and function	**23**
Principles of propeller design and function	24
Classifications of propeller	29
Propeller dimensions	30
Propulsion system assembly	31
Propeller shaft	31
Marine gearbox	33
Stern tube and propeller shafts	33
Propeller boss	34
Boss control on CPP	36
Propeller maintenance and overhauling	36
Boring and sighting	36
Renewing the propeller	38
Side thrusters	39
Design and function of the thrusters	40
Side thruster maintenance	42
Notes	42

3 Introduction to the main engine — 45

- Types of main engines — 46
  - Steam engines — 46
    - Reciprocating steam engine — 47
    - Steam turbine engines — 47
    - Nuclear powered — 47
  - Diesel engines — 48
  - Gas turbines — 49
  - LNG engines — 49
  - Other types of engines — 49
    - Electric only — 50
    - Diesel-electric — 50
    - Turbo-electric — 50
    - Stirling engine — 50
- How main engines are installed inside the ship's engine room — 51

4 Key components of the main engine — 55

- Main bearings — 55
  - Types of main bearings — 56
  - Common bearing defects — 57
- Engine brace and strut — 58
- Common rail system — 59
- Turbocharger — 61
  - Surge line — 61
  - Terms associated with turbocharger surge — 62
  - Categories of turbocharger surge — 62
  - Causes of turbocharger surge — 63
  - Preventing turbocharger surge — 63
- Entablature — 64
- Air bottle — 65
- Jack bolts — 66
- Piston, piston skirt, and piston rod — 67
  - Piston skirt — 67
  - Piston rods — 67
  - Trunk piston — 68
- Tie rods — 68
- Dampers and de-tuners — 69
  - Dampers — 69
  - De-tuners — 70

5 Main engine pre-start checks and monitoring — 73

- Engine monitoring — 74
  - Engine monitoring systems — 74
  - Understanding the indicator diagram — 75
- Power balancing — 78

6 Slow steaming and economic fuel consumption   81

   Preparing to slow steam   83
   Optimising the main engine for slow steaming   85
      Optimisation of the ship's main engine   85
   Defining economical fuel consumption   86
      Performance curves   86
      Economical fuel consumption   88

7 Exhaust gas system and scrubbers   91

   Exhaust gas piping   91
   EGB   93
   Silencer   93
   Spark arrestor   93
   Expansion joints   94
   Exhaust gas scrubbers   94
      Wet scrubbers   95
         Open-loop scrubber system   99
         Closed-loop scrubber system   100
         Hybrid scrubber system   101
      Dry scrubbers   101
   Notes   102

8 Engine room lubrication systems   103

   Lube oil systems   103
      Main engine lubricating system   103
      Turbocharger lubricating oil system   105
      Cylinder lubrication system   105
      Piston rod stuffing, the box, and scavenge space drainage system   106
   Lube oil properties   106
   Assessing the lube oil   107

9 Basic engine room machinery maintenance and troubleshooting   111

   Principles of troubleshooting   111
   Common faults and malfunctions   112
   Crankshaft faults and malfunctions   115
      Crankcase inspections   115
   Main bearing faults and malfunctions   117
      Bearing inspections and surveys   119
   Fuel valve overhauling   120
   Fuel changeover procedures   120
   Overhauling the cylinder liner   122
   Timing chain tightening and adjustments   124
   Excessive water loss from the freshwater expansion tank   128
   Overspeeding and prevention   129
      Preventing overspeeding of engine   129

10 Mechanical measuring tools and gauges — 131
   Workshop processes — 135

PART II
Power generation — 139

11 Marine diesel generators — 141
   Summary of the marine electrical system on ships — 141
   Working principles of the marine generator — 144
      Marine power distribution — 144
      Marine emergency power — 144
   Estimating the power requirement for the ship — 145
   Starting and stopping the generator — 149
      Generator starting procedure – automatic start — 150
      Generator starting procedure – manual start — 150
      Generator stopping procedure – automatic stop — 151
      Generator stopping procedure – manual stop — 151
      Situations where the generator must be stopped immediately — 151
   Generator synchronisation — 153
      Emergency synchronising lamps (three-bulb method) — 154
   Maintenance and overhauling of the main generator — 155
   Decarbonisation (d'carbing) — 155
   Emergency generator — 158
      Emergency generator maintenance — 159

12 Marine electrical systems — 161
   Main switchboard — 163
   Busbar — 164
   Governor — 166
      Classification of governors based on their operating principles — 168
      Maintenance of the governor — 170
      Droop — 170
      Speeder spring — 171
      Deadband — 171
      Hunting — 171
   Safety principles of the marine electrical system — 171
      Avoiding electrocution — 171
      Main switchboard safety devices — 172
      Air circuit breakers — 173
      Preferential trips — 174
      Blackout conditions — 174

13 Electrical distribution systems and redundancy — 177

14 Air compressor — 185
   Main components of the air compressor system — 190

Safety features of the compressor   *191*
Air compressor maintenance   *191*
Air compressor starting and stopping procedures   *194*
Air compressor troubleshooting   *195*

PART III
Heating, ventilation, and air conditioning   197

15 Marine boiler   199

Requirement 1 – compensating for heat loss   *199*
Requirement 2 – raising the fuel oil temperature   *200*
Requirement 3 – all other services   *201*
Boiler ratings   *201*
Procedures for starting and stopping the boiler   *203*
Boiler misfires and malfunctions   *204*
Marine boiler failures   *205*
Boiler feedwater contamination   *206*
Boiler blowdown procedures   *207*
    Procedure for scumming and bottom blowdown   *209*
    Advantages and disadvantages of boiler blowdown   *209*
Cleaning the gauge glass   *210*

16 Central cooling system   211

Heat exchangers   *213*
    Shell and tube-type heat exchanger   *213*
    Plate-type heat exchanger   *214*
    Plate-fin heat exchanger   *215*
    Dynamic scraped surface heat exchanger   *215*
    Phase change heat exchanger   *216*
    Spiral heat exchanger   *216*
    Direct contact heat exchanger   *216*
    Charge air cooler   *216*
    Inter- and aftercoolers (for air compressors)   *216*

17 Refrigeration and air conditioning   219

PART IV
Water management systems   227

18 Ballast water management   229

Ballast water exchange methods   *230*
BWM plan and implementation   *232*
Record keeping   *234*
Ballast tanks on ships   *236*
Ballast tank monitoring   *238*
Ballast tank protection   *238*

Ballast tank inspections	239
Performing ballasting operations	239
Ballast water treatment	241
Ozone generator for ballast water treatment	244
Ballast water system maintenance	245
MARPOL annex I	247
Control of the discharge of oil under MARPOL annex I, regulation 4	248
Discharges in special areas	248
Discharges from oil tankers	248
Discharges from oil tankers in special areas	249
Complying with MARPOL annex I	249
Certificates, plans, and records under MARPOL annex I	249
Other essential requirements of MARPOL annex I	251
Roles and responsibilities under MARPOL annex I	252
Complying with the ballast water convention	253
Notes	255

19 Oily water separator — 257

Design, construction, and working principles of the OWS	258
Operating the OWS	259
OWS maintenance	261
Oil discharge and monitoring and control system on oil tankers	264
Sludge production and management	265
Sludge tanks	266
Oily water evaporation and sludge incineration	266
Engine room bilge water generation	267
Cargo-hold bilge water generation	268
Good bilge management practices help improve OWS performance	268
Construction and working principles of the waste incinerator	270
Good practices for the incinerator	272
Ship oil pollution emergency plan (SOPEP)	273
Notes	274

20 Wastewater management — 275

Design and construction of the sewage treatment plant	275
Special area regulations	277
Maintenance and checks	278
Reducing marine pollution from ships	279
Notes	280

21 Freshwater generation — 281

Operating the freshwater generator	282
Reverse osmosis	282

22 Pipes, tubes, bends, and valves — 285

Bends and elbows	286

Short-radius and long-radius elbows	*286*
Mitre bends	*288*
Pipe fittings	*288*
Valves	*293*
Gate valve	*293*
Globe valve	*295*

PART V
Engine room tanks and bunkering operations — 301

23 Main fuel, diesel, and lube oil tanks on ships — 303

Fuel and diesel oil tanks	*304*
Lubricating oil tanks	*304*
Tank inspections	*305*
Tank cleaning	*308*
Preparations to be done before cleaning the ship's tanks	*309*
During the tank cleaning	*309*
After the tank cleaning	*309*
Tank maintenance	*309*

24 Bunkering operations — 311

Oil bunkerage procedures	*312*
LNG bunkering procedures	*314*
Bunkerage disputes	*317*
Coriolis flow metre	*317*
Fuel oil bunkering malpractices	*319*
Tricks of the trade	*320*
Fuel oil storage on ships	*327*
Fuel oil consumption calculations	*328*
Measuring and reporting fuel oil consumption	*328*
Position, arrival, and departure reports	*330*
Determining the fuel bunkered and fuel in tanks	*330*
Measuring and reporting distance travelled	*331*
Method to measure hours underway	*332*
Emission factors	*332*
Information to be submitted to the IMO ship fuel oil consumption database	*332*
Notes	*333*

PART VI
Engine room fires and emergency response — 335

25 General emergency drills, alarms, and emergency systems — 337

General alarms and emergencies	*337*
Engineer's call and alarms	*338*
Engine room drills and training procedures	*338*

26 Engine room explosions and fires — 341

- FCP — 341
- Crankcase explosions — 342
- Starter air-line explosion — 343
- Purifier room fires — 343
- Scavenge fires — 344
- Exhaust gas boiler fires — 345
- Incinerator fires — 347
- Electrical insulation fires — 347
- Battery room fires — 348
- Bacteria fires — 348

27 Engine room drills, firefighting procedures and apparatus — 351

- Fire drills — 351
- Preventative measures and firefighting appliances — 352
- Sprinkler system: automatic fire detection, alarm, and extinguishing system — 354
- CO_2 firefighting systems — 355
- ISC — 356
- Firefighting on oil tankers — 358
- Notes — 360

28 Engine room flooding — 361

- Leaks from machinery and equipment — 361
- Leaks from the overboard valve — 361
- Flooding caused by cracks and fissures in the ship's hull or sideboard — 362
- Watertight bulkheads and flooding prevention — 362
- Watertight doors — 366
    - SOLAS rules relating to watertight bulkheads — 367

PART VII
Engine room watch procedures — 369

29 Engine room watch procedures — 371

- Engine room watchkeeping — 373

30 Engine room logbook entries and checklists — 375

- Engine room logbooks and records — 375
- Manned engine room checklists — 377
    - Engine department departure checklist — 377
        - 24 hours prior to departure — 377
        - Six hours prior to departure — 377
        - One hour prior to departure — 378
        - 15 minutes prior to departure — 378
        - Checks to be made when the main engine is running — 378

Checks to be made when the vessel is full away	*379*
Engine department arrival checklist	*379*
One hour before the estimated time of arrival	*379*
On arrival at port	*380*
UMS checklists	*380*
Notes	*384*
Appendix: Recommended reading for marine engineers	*385*
Index	*389*

Preface

The modern ship's engine room is a vast and complex collection of systems, machineries and equipment. This book is not intended to be used as a technical reference, but rather as an introduction to the engine room for readers with a general interest in how ships operate, and for students and cadets starting out on a new and exciting career at sea. For ease of use, this book has been divided into seven parts:

- Part 1 of this book will examine the role and function of the steering and propulsion systems, including the rudder and propeller, propeller design and function, the main engine, exhaust gas system and scrubbers, the lubrication systems, basic engine room maintenance and troubleshooting, and mechanical measuring tools and gauges.
- Part 2 will look at power generation (generators, electrical systems, distribution systems and redundancy, and air compressors).
- Part 3 will explore the heating, ventilation, and air-conditioning systems (HVAC).
- Part 4 examines the water management system, such as ballast water, oily water, wastewater, and freshwater, as well as the myriad pipes, tubes, and bends that these systems use.
- Part 5 looks at engine room fires and emergency response systems.
- Part 6 will cover some of the main engine room watch procedures and checklists.

For further information on marine engineering, the appendix recommends several excellent books across a wide range of marine engineering subject areas. For a general introduction to the merchant navy and the types of vessels of the merchant navy, the following three texts are recommended reading:

- Olsen, Alexander Arnfinn. 2022. *Introduction to Container Ship Operations and Onboard Safety*. Routledge, London.
- Olsen, Alexander Arnfinn. 2022. *Practical Guide to the International Convention for the Prevention of Pollution from Ships (MARPOL 73/78)*. Magellan Maritime Press, Southampton.
- Olsen, Alexander Arnfinn. 2023. *Merchant Ship Types*. Routledge, London.

Acknowledgements

I would like to take this opportunity to thank everyone involved in the development of this book. With special gratitude to Tony Moore and Aimee Wragg at Routledge for their guidance and support, Vijay Kumar for his guidance and engineering expertise, Raveena Withanage for providing the technical and line drawings, and Melissa Brown Levine at Brown Levine Productions for their patience and support.

To you all, my grateful thanks.

Alexander Arnfinn Olsen

Abbreviations, glossary, and terms used

$	US Dollar
£	British Pound Sterling
€	Euro
°C	Degrees Centigrade
°F	Degrees Fahrenheit
ABS	American Bureau of Shipping
AC	Alternating Current
ACB	Air Circuit Breaker
ACT	Actuator
AFAM	Automatic Fresh Air Management Container
AHU	Air Handling Unit
AWG	American Wire Gauge
BDC	Bottom Dead Centre
BDC	Bunkerage Delivery Note
BHP	Break Horsepower
BLEVE	Boiling Liquid Expanding Vapour Explosion
BOD	Biological Oxygen Demand
BOGMS	Boil-Off Gas Management System
BOH	Boil-Off Hydrogen
BN	Base Number
BTM	Bearing Temperature Monitor
BDN	Bunker Delivery Note
BWM	Ballast Water Management
C/E, CHENG	Chief Engineer
CAS	Condition Assessment Scheme
CBT	Clean Ballast Tank
CFU	Colony-Forming Unit
COD	Chemical Oxygen Demand
COW	Crude Oil Washing
CPP	Controllable Pitch Propeller
DB	Decibel
DB	Double Bottom
DC	Direct Current
DCP	Dry Chemical Powder
DG	Diesel Generator

DOL	Direct Online Motor
DP	Diesel Particulate
ECA	Emission Control Area
ECDIS	Electronic Chart Display and Information System
ECR	Engine Control Room
EEBD	Emergency Escape Breathing Device
EEDA	Emergency Escape Breathing Apparatus
EEDI	Energy Efficiency Design Index
EGB	Exhaust Gas Boiler
EGCS	Exhaust Gas Cleaning Systems
EMF	Electromagnetic Force
EOW	Engineer Officer of the Watch
ETD	Estimated Time of Departure
ETO	Electrotechnical Officer
EX	Explosive Classified
FAD	Free Air Delivery
FCP	Fire Control Plan
FDF	Forced Draft Fans
FFA	Firefighting Appliances
FFU	Full Follow-Up Steering
FPP	Fixed Pitch Propeller
FSSC	Fire Safety System Code
FT	Feet
FW	Freshwater
GM	Metacentric Height
GMP	Garbage Management Plan
GMT	Greenwich Mean Time
GPS	Global Positioning System
HCI	Hydrochloride
HMI	Human Machine Interface
HSFO	High Sulphur Fuel Oil
HV	High Voltage
HVAC	Heating, Ventilation and Air Conditioning
HZ	Hertz
ICU	Injection Control Unit
IGC	International Code of the Construction and Equipment of Ships Carrying Liquefied Gases in Bulk
IGF	International Code of Safety for Ship Using Gases or Other Low-flashpoint Fuels
IMO	International Maritime Organisation
IOPP	International Oil Pollution Prevention Certificate
IN	Inches
ISC	International Shore Connection
ITTC	International Towing Tank Conference
KHZ	Kilo Hertz
KT	Knots
KV	Kilovolts
KVA	Kilovolts and Amps

KW	Kilowatts
LDO	Light Diesel Oil
LEL	Lower Explosive Limit
LNG	Liquified Natural Gas
LSA	Lifesaving Appliances
LSFO	Sulphur Fuel Oil
LSMGO	Low Sulphur Marine Gas Oil
LV	Low Voltage
M	Metres
MA/CA	Modified or Controlled Atmosphere
MARPOL	International Convention for the Prevention of Pollution from Ships (1973/1978)
MAWP	Maximum Allowable Working Pressure
MCR	Maximum Continuous Rating
MECC	Main Engine Crankcase
MEPC	Marine Environment Protection Committee
MI	Miles
MLC	Maritime Labour Convention
MPU	Electromagnetic Pickup
MRP	Maximum Rated Power
MSB	Main Switchboard
NDT	Non-destructive Testing
NOX	Nitrogen Oxide
ODM	Oil Discharge Monitor
ODMCS	Oil Discharge Monitoring and Control System
OOW	Officer of the Watch
ORB	Oil Record Book
OWS	Oily Water Separator
PCB	Polychlorinated Biphenyls
PD	Power to the Propeller
PE	Effective Power
PID	Proportional-Integral-Differential
PMS	Preventative Maintenance Schedule/System
PMS	Power Management System
PPM	Parts per Million
PSC	Port State Control
PTW	Permit to Work
PUMS	Periodically Unattended Machinery Spaces
PVC	Polyvinylchloride
QCDC	Quick Connect/Quick Disconnect [Coupling]
QPC	Quasi-Propulsive Coefficient
RMS	Royal Mail Ship
RPM	Revolutions (revs) per Minute
RPT	Rapid Phase Transition
SBT	Segregated Ballast Tank
SCBA	Self-Contained Breathing Apparatus
SFOC	Specific Fuel Oil Consumption
SMS	Safety Management System

SMT	Ship's Mean Time
SOC	Statement of Compliance
SOLAS	International Convention for the Safety of Life at Sea (1965, 1974, 1980)
SOPEP	Shipboard Oil Pollution Emergency Plan
SOX	Sulphur Oxide
SW	Seawater
TBN	Total Base Number
TDC	Top Dead Centre
TDS	Total Dissolved Solids
UK	United Kingdom
UMS	Unmanned Machinery Space
US	United States
UV	Ultra-Violet
V	Volts
V/A	Volts and Amperes
VCU	Valve Control Unit
VGP	Vessel General Permit
VGT	Variable Geometry Turbocharger
VHF	Very High Frequency
VPP	Variable Pitch Propeller
WIOM	Water in Oil Monitor

Temperature conversion table

°C = (°F − 32) ÷ 1.8 °F = (°C × 1.8) + 32

Celsius	Fahrenheit	Celsius	Fahrenheit	Celsius	Fahrenheit	Celsius	Fahrenheit
−50	−58	−10	14	26	78.8	66	150.8
−49	−56.2	−9	15.8	27	80.6	67	152.6
−48	−54.4	−8	17.6	28	82.4	68	154.4
−47	−52.6	−7	19.4	29	84.2	69	156.2
−46	−50.8	−6	21.2	30	86	70	158
−45	−49	−5	23	31	87.8	71	159.8
−44	−47.2	−4	24.8	32	89.6	72	161.6
−43	−45.4	−3	26.8	33	91.4	73	163.4
−42	−43.6	−2	28.4	34	93.2	74	165.2
−41	−41.8	−1	30.2	35	95	75	167
−40	−40	FREEZING POINT OF WATER		36	96.8	76	168.8
−39	−38.2			37	98.6	77	170.6
−38	−36.4			38	100.4	78	172.4
−37	−34.6	0	32	39	102.2	79	174.2
−36	−32.8			40	104	80	176
−35	−31	+1	33.8	41	105.8	81	177.8
−34	−29.2	2	35.6	42	107.6	82	179.6
−33	−27.4	3	37.4	43	109.4	83	171.4
−32	−25.6	4	39.2	44	111.2	84	183.2
−31	−23.8	5	41	45	113	85	185
−30	−22	6	42.8	46	114.8	86	186.8
−29	−20.2	7	44.6	47	116.8	87	188.6
−28	−18.4	8	46.4	48	118.4	88	190.4

(*Continued*)

(*Continued*)

Celsius	Fahrenheit	Celsius	Fahrenheit	Celsius	Fahrenheit	Celsius	Fahrenheit
−27	−16.6	9	48.2	49	120.2	89	192.2
−26	−14.8	10	50	50	122	90	194
−25	−13	11	51.8	51	123.8	91	1495.8
−24	−11.2	12	53.6	52	125.6	92	197.6
−23	−9.4	13	55.4	53	127.4	93	199.4
−22	−7.6	14	57.2	54	129.2	94	201.2
−21	−5.8	15	59	55	131	95	203
−20	−4	16	60.8	56	132.8	96	204.8
−19	−2.2	17	62.6	57	134.6	97	206.6
−18	−0.4	18	64.4	58	136.4	98	208.4
−17	1.4	19	66.2	59	138.2	99	210.2
−16	3.2	20	68	60	140		
−15	5	21	69.8	61	141.8	BOILING POINT OF WATER	
−14	6.8	22	71.6	62	143.6		
−13	8.6	23	73.4	63	145.4		
−12	10.4	24	75.2	64	147.2	100	212
−11	12.2	25	77	65	149		

Introduction

On modern merchant and naval vessels, the engine room is the compartment where the machinery for marine propulsion is situated. To increase a vessel's safety and scope for surviving critical damage, the machinery necessary for the vessel's operation is often segregated into separated compartments or spaces. Of these compartments or spaces, the engine room is the largest. On some vessels, there may be more than one engine room, such as the forward and aft, or port and starboard engine rooms, or they may simply be numbered (engine room 1, engine room 2, etc.). On all vessels, the engine room is located towards the bottom of the ship's hull, either at the stern or as near to the aft as possible. This design helps maximise the amount of cargo-carrying capacity of the vessel and ensures the prime mover is positioned close to the ship's propeller(s). This reduces equipment and maintenance costs and lessens the potential for problems associated with long shaft lines. Although modern engine rooms are complex assemblies of various machines and systems, the basic principles of ship propulsion and power generation have remained much the same today as they were when ships were powered by coal and steam. At the heart of the engine room is the ship's propulsion system, which consists of the main engine. This is usually some variation of marine combustion engine, be it diesel, gas-powered, or steam turbine. The engine room typically contains several engines, each serving different purposes. The main, or propulsion, engine is used to turn the vessel's propeller, which in turn propels or forces the vessel through the water. Main engines typically burn diesel oil or heavy fuel oil and in some cases can switch from one fuel to the other. Smaller, but still considerably large, sized engines drive electrical generators which provide the power for the ship's electrical systems. Large ships typically have three or more synchronised generators. This ensures a smooth and uninterrupted supply of electrical power. Even though each generator can supply sufficient power for the vessel under normal load, it is important the vessel has ample redundant power in the event one generator fails. By comparison, on steamships, power for both electricity and propulsion is supplied by one or more large boilers. As steamships are powered by steam, rather than by an engine, this compartment is called the boiler room. High-pressure steam from the boiler is used to drive reciprocating engines or turbines for propulsion, in conjunction with a turbogenerator for electricity.

Besides the propulsion and auxiliary engines, a typical engine room will also contain an assortment of many smaller engines and machineries, including air compressors, feed pumps, and fuel pumps. On modern ships, these machineries are usually powered by small diesel engines or electric motors, though some may operate off the low-pressure steam which is generated by the ship's boiler. The engines are cooled using liquid-to-liquid heat exchangers that are connected to the seawater tank or freshwater which is diverted to recirculate through the freshwater tanks. Both supplies draw heat from the engines via coolant and cooling oil lines. Heat exchangers which use oil are represented by a yellow mark on the pipe flange and rely on paper-type gaskets to seal the mating faces of each pipe. Seawater or brine cooling systems are represented by a green mark on the flange, and freshwater cooling systems are represented by blue marks on the flange.

Most vessels of a certain size are fitted with some form of bow (and in some cases, stern) thruster arrangement. Thrusters are laterally mounted propellers that suck or blow water from one side of the ship to the other side (i.e., port to starboard or left to right and vice versa). These thrusters are typically operated by small electric motors controlled from the bridge. The thrusters are normally only used during tight manoeuvring, such as berthing. Use of thrusters is almost always prohibited in tight confines such as drydocks and in canals. Thrusters, like the main propellers, are reversible by hydraulic operation. Small embedded hydraulic motors rotate the blades up to 180 degrees to reverse the direction of the thrust. A relatively recent development in thruster technology is the azipod. This is a fully rotatable propeller mounted on a swivelling pod that can rotate 360 degrees to direct thrust in any direction. This makes fine steering easier and often allows the ship to move sideways into a berth when used in conjunction with the ship's bow thruster.

In addition to the main engine compartment are several other machinery spaces. These may be incorporated into the main engine compartment (where the engine room is 'open plan') or else separated by bulkheads. The largest of these compartments is the mechanical room or plant room and usually holds the ship's auxiliary equipment, such as the air handlers, boilers, chillers, heat exchangers, water heaters and tanks, pumps, main distribution valves and pipes, backup generators, batteries, and other heating, ventilation, and air-conditioning (HVAC) plants. Ships with large and complex electrical systems (such as cruise ships) tend to have a separate electrical room. The electrical room is a space or compartment dedicated to electrical equipment such as the power distribution equipment and communications equipment. A typical electrical room will contain a variety of switchboards, distribution boards, circuit breakers and disconnectors, transformers, busbars, fire alarm control panels, and distribution frames. Given the complexity of these electrical systems, it is not uncommon for ships fitted with electrical rooms to have a dedicated electrotechnical officer on board.

Hopefully, we should be starting to build an understanding and appreciation of the size and complexity of the ship's engine room. Before we move onto the more intricate details of how these systems work, it is worth noting some of the main safety issues associated with the ship's engine room. As we can imagine, engine rooms are extremely noisy and hot, usually dirty, and are extremely dangerous environments. The presence of flammable fuels, lubricating oils and chemicals, high-voltage electrical equipment, and moving systems means that serious fire hazards are always present. These hazards must be monitored closely and continuously by the engine room department. Fortunately, there are various manual and automatic monitoring systems available. Just as important is the need for adequate ventilation. If the ship is powered by internal combustion or turbine engines, the engine room must have available some means of providing fresh air for the operation of the engines and associated ventilation. If crew members are normally present in these compartments (as opposed to unmanned machinery spaces (UMS), something we will cover later), additional ventilation must be made available to maintain the engine room ambient temperature within acceptable safe limits. On ships with UMS, or small pleasure boats where crew members are not required in the machinery spaces, it is often necessary to only provide intake air with minimal ventilation.

The department responsible for managing the engine room and for overseeing the ship's equipment and machinery is the engineering department. This is one of the two main departments found on most merchant vessels, with the other being the deck department. It is worth noting that some vessels (such as cruise ships and passenger ferries) have additional departments for housekeeping, guest services, and entertainment. The engineering department is an organisational unit that is responsible for the operation, maintenance, and repair of the ship's propulsion systems and the support systems for the crew, passengers, and cargo. These typically include the ship's engine, fuel oil, lubrication, water distillation, separation processes, lighting, air-conditioning, and refrigeration systems. The engineering department emerged with the introduction of marine engines for propulsion during the latter part of the nineteenth century. Due to advances in ship technology throughout the twentieth century, the engineering department is now recognised as being equally important as the deck department. Trained marine engineering officers are required to oversee the ship's machinery much in the same way as trained deck officers are responsible for safe navigation and vessel handling. Like deck officers, engineering officers must stand watch on a rotating basis.[1] The specific requirement for each watch differs between ships and companies. When on watch, the marine engineering officer is responsible for ensuring the safe and smooth operation of the ship's machinery. The most senior marine engineer is the chief engineer (CE, C/E, or ChEng). Subordinate to the chief engineer is the second engineer, followed by the third engineer, and then the fourth engineer. The most junior marine engineer is the fifth engineer, who is typically freshly qualified and has yet to serve sufficient sea time to qualify for the fourth engineer's rank. Some companies carry a trainee engine cadet on board. In terms of actual responsibilities, the chief engineer oversees the engineering department and reports only to the ship's master.[2] The second engineer is typically responsible for overseeing the day-to-day running of the engine department, which includes the operation and maintenance of the main engines. The second engineer stands the 0400–0800 and 1600–2000 watches. The third engineer is usually responsible for the auxiliary engines and boilers and stands the 0000–0400 and 1200–1600 watches. The fourth engineer, who stands the 0800–1200 and 0000–0400 watches, oversees the air compressors, purifiers, pumps, and other auxiliary machineries and equipment. Depending on the vessel, the fourth engineer may also be given responsibility for the boilers. Though nominally a member of the engineering department, the electrotechnical officer is not usually licensed and is therefore not permitted to stand watch. On larger cruise ships, reefer ships, and other specialist vessels such as drillships, an additional electrical officer may report to the electrotechnical officer. To become a licensed watchkeeper, there are three basic requirements that must be met. These are (1) age, (2) education and training, and (3) demonstrable seagoing experience.

Sitting below the engineering officers in the rank hierarchy are the engineering department ratings. Unlike officers, who must be licensed, ratings are unlicensed members of the crew and usually conduct the more mundane duties required to keep the engine room operating smoothly. Though the number of ratings will vary according to the type, size, and complexity of the engine room arrangement, most container ships will have between seven and ten engine ratings. Each rating is assigned a specific tasking commensurate with their training and experience. The most senior engine rating is the motorman, who supports and assists the second engineer in overseeing the operation of the main engine. The motorman may be a qualified engineering technician and can, upon completing the required courses, seek promotion to marine engineer rank. This experience allows the motorman to stand watch with the engineering officers, in addition to performing routine duties such as supporting the officers during maintenance. There may be as many as four motormen on board. Beneath the motorman is the oiler. The oiler is responsible for ensuring the machinery is always lubricated, as well as performing other menial tasks, such as cleaning and sounding the tanks. The most junior rating is the wiper. The wiper is responsible for keeping the engine room clean, for wiping away any oil and fuel residues, and for generally keeping the engine room in good order. Wipers usually go on to become oilers once they are familiar with the ship's machinery and the engine room's specific routines.

Now that we have explored the engine room as a system or, rather, as a collection of systems, we can begin to examine each of the individual machineries and equipment that, when combined, function as the ship's engine room.

NOTES

1. All merchant navy officers below the rank of master and chief engineer must stand watches. Deck officers stand watch on the bridge, and marine engineers stand watch in the engine room. The typical watch period is four hours and begins at 0000 to 0400 hours, then 0400 to 0800 hours, and 0800 to 1200 hours. The roster then repeats for the next 12-hour period.
2. The rank of master and captain are the same and often interchangeable. The master holds ultimate authority and responsibility for the vessel, her crew, and cargo. Although nominally part of the deck department, the master does not stand watch and is senior to the chief engineer and the chief officer.

Part I

Steering and propulsion systems

Chapter 1

Rudder and steering gear

Let us begin by asking a question – have you ever wondered why ships, unlike aeroplanes, have different types, sizes, and shapes of rudders? If so, you have stumbled upon an interesting dilemma faced by many a naval architect. The choice of type of rudder is as crucial to the effectiveness of the vessel as the location of the rudder behind the ship's propeller. The location of the rudder should be such that it is properly positioned within the region of water that is expelled by the propeller. This is called the propeller's outflow. This is important as the rudder is instrumental in producing the required turning moment for the ship. Even a slight change in the rudder type, dimensions, and rudder position, can bring about a substantial variation in the ship's response. This would, as we can easily imagine, impact on the turning ability of the ship. To broadly categorise, there are three types of rudders: spade or balanced rudder, unbalanced rudder, and semi-balanced rudder. The spade or balanced rudder (Figure 1.1) is a type of rudder plate that is fixed to the rudder stock (or axis) at the top of the rudder assembly. This means the rudder stock does not extend the full span of the rudder plate. The position of the rudder stock, in relation to the chord of the rudder (i.e., the width of the plate from the fore to the aft of the rudder) determines whether the rudder is of a balanced or semi-balanced type. With balanced types (of which spade rudders are typical) the rudder stock is positioned such that 40% of the rudder is located forward of the stock with the remaining 60% behind the stock. The reason for this design is pure physics. The centre of gravity of the rudder lies somewhere close to 40% of its chord length when looking forward to aft. If the axis of the rudder is placed near to this position, the torque required to rotate the rudder will be considerably less than if the axis had been placed closer forward. This has a direct relation to the fuel efficiency of the ship. This is because the energy requirement of the steering gear is reduced, therein lowering the vessel's fuel consumption.

Unbalanced rudders (Figure 1.2) by comparison have their stocks attached to the forwardmost point of their span. Unlike balanced rudders, the rudder stock runs the full chord length of the rudder. The reason for this difference in design is quite simple. In this case, the torque required to turn the rudder is much higher than that needed for the balanced rudder. Therefore, the topmost part of the rudder must be fixed to the rudder spindle to prevent it from suffering vertical displacement from its natural position. Due to the complexity and inefficient design of the unbalanced rudder, they are not commonly installed on modern ships. One issue that affects both balanced and unbalanced rudders is the problem of steering gear failure. With both types of rudders, should the steering gear fail, the rudder will remain still with its angle of attack stuck in that position. This of course presents all manner of hazards for the vessel. The solution to this problem was the development of the semi-balanced rudder. This is the most common type of rudder installed on ships today.

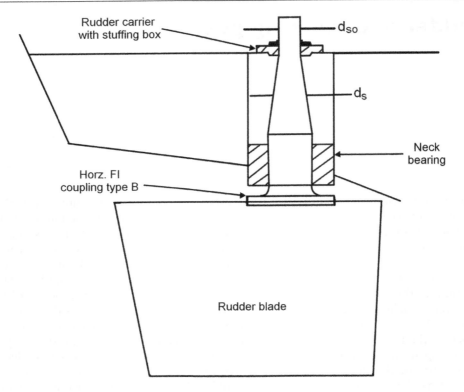

Figure 1.1 Spade or balanced rudder.

As the name implies, semi-balanced rudders are partly balanced and partly unbalanced. If we refer to Figure 1.3, we can see that a portion of the chord length from the rudder top is unbalanced, and the remaining portion of the chord length is balanced. The rationale behind this design is to help provide structural support to the rudder through vertical displacement. In addition, the balanced section of the rudder requires less torque when swinging the rudder from port to starboard. As a result, the semi-balanced rudder returns to the centreline of its own accord in the event the steering gear fails during a turn. Whilst this does not directly improve the condition of the ship – for example, by staying a steady course – the ship's navigator and helmsman can prepare a reactive response to any potential navigational hazards such as other marine traffic.

Note in Figure 1.3 the presence of the rudder horn. The rudder horn has an adverse effect on the response and torque characteristics of the rudder. Semi-balanced rudders have two designs depending on the depth of the rudder horn. A shallow horn rudder will have a horn which barely extends half the chord length of the rudder from the rudder top, whereas a deep horn will extend 50% of the chord length from the rudder top.

In addition to these three types of rudders, marine technologists and naval architects have developed various other unique and 'unconventional' rudder systems. For instance, if you have ever flown on an aeroplane, you may have noticed how, as the aircraft takes off and lands, there are small flaps that extend outwards from the wing edges. These flaps are designed to change the effective angle of attack of the entire aerofoil section of the wing. On take-off, the flaps are deployed. This helps the aeroplane to achieve lift by maximising

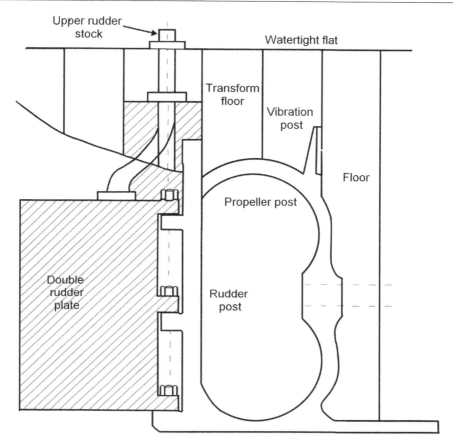

Figure 1.2 Unbalanced rudder.

the volume and flow of air which passes over the wing face. The reverse is true when the aircraft is descending to land. The same principle applies to the flap rudder (see Figure 1.4). With flap rudders, the flaps do not retract but instead alter the vessel's course by changing the angle of attack.

Another innovation in rudder technology is the *pleuger* rudder (see Figure 1.5). The pleuger rudder is a small auxiliary propeller which is accommodated within the main rudder housing and is powered by a separate motor. The propeller is mounted within the rudder structure itself and generates thrust in a direction that is orientated along the rudder. This allows ships to perform complex manoeuvres at slow speeds in confined spaces such as turning basins. Because the auxiliary propeller is operated separately from the primary propeller and rudder, the primary rudder can be used in normal operating conditions. However, when the pleuger rudder is used, the main propeller must not be used; otherwise, the pleuger may be torn off, causing catastrophic damage to the rudder assembly.

You may be wondering why propulsion has surreptitiously crept into a discussion on rudder systems ... if that is the case, then we can thank the designers of the *Voith Propulsion System*, a revolutionary concept in propulsion and steering technology. Unlike the *pleuger* rudder, which is essentially an auxiliary propeller fixed to the primary rudder,

6 Introduction to Ship Engine Room Systems

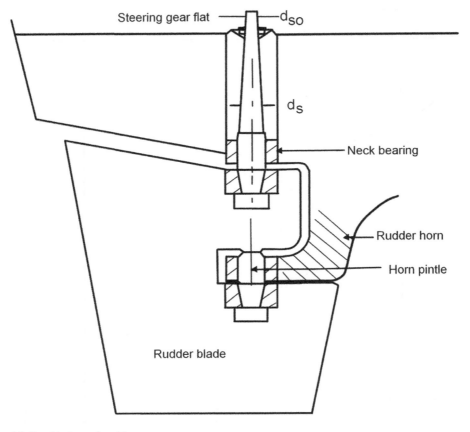

Figure 1.3 Semi-balanced rudder.

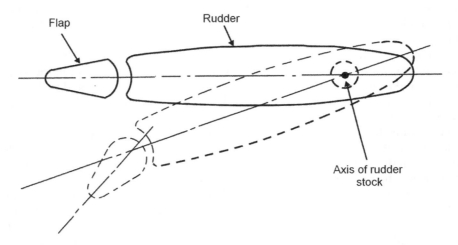

Figure 1.4 Flap rudder.

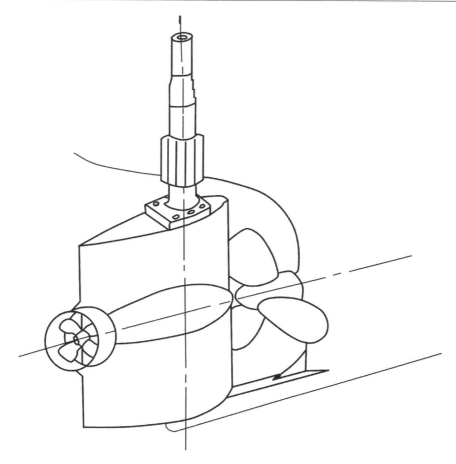

Figure 1.5 Pleuger rudder.

the *Voith* rudder combines the propeller and rudder into one singular unit. Because the propeller acts as the rudder, it does not require a rudder control surface to alter the ship's course. The *Voith* rudder consists of several hydrofoil blades mounted on a disc, which is in turn mounted to the ship's hull. The disc rotates on a horizontal plane, about a horizontal axis, thereby causing rotation of the blades.

Now that we have covered the main types of rudders, it is time to explain a little about how the rudder helps the ship to change direction. Notice this is not the same as saying the rudder *changes* the ship's direction. This is because the science behind how rudders work is slightly more complicated than might initially seem obvious.

HOW THE RUDDER HELPS THE SHIP TO TURN

Have you ever wondered why all ships have their rudders positioned behind the propellers? Or to ask that question in a slightly different way, why isn't the rudder located at the ship's bow, or even amidships? And for that matter, why is the rudder always positioned behind the propellers? These are all key questions that aspiring marine engineers have probably asked at one point or another. To answer these questions, let us imagine a ship with the rudder positioned at the bow. How effective do you think that rudder would be? Apart from looking rather silly, a bow-mounted rudder would be virtually useless! Instead, rudders are placed aft to take advantage of a phenomenon called *hydrodynamic efficiency*. To explain, let us assume a ship is making a starboard turn.[1] This means the rudder is moved to starboard. When the helmsman changes the rudder angle from zero to some angle towards starboard, at that very moment, a lift force acts upon the rudder, with the direction of the lift force initially towards port, as shown in Figure 1.6(a). This rudder force, as depicted in Figure 1.6(a), is directed along a transverse direction to the ship. In other words, this force will cause the ship to develop a sway velocity towards the port side. It is because of this that a ship will always sway to port when the rudder is turned over to starboard. It is important to clarify this port side sway is negligible in comparison to the turning moment of the ship towards starboard; even so, it does happen. In addition, the rudder force has another effect on the ship. It creates a moment of the ship's centre of gravity, as depicted in Figure 1.6(b). This happens as the centre of gravity is forward of the rudder. Given the direction of the rudder force, the moment created about the centre of gravity will be along the direction of travel.

Now, imagine the size of a rudder in comparison to the size of the ship. The rudder is incomparably smaller to the ship's hull. This raises the question of how the rudder can turn an object so many magnitudes larger than the rudder itself. The answer is as before: it does not. In fact, the rudder moment created by the rudder is so negligibly small, it could never change the ship's course. Instead, when the rudder moment acts about the ship's centre of gravity, it slightly changes the ship's orientation by giving it a drift angle (see Figure 1.7(a)). This moment is not substantial enough to turn the ship to the required heading but is strong enough to introduce a slight drift angle into the ship's movement. The ship, with this drift angle, is now moving along the initial direction. This is no longer pure surge. Rather, if we look at Figure 1.7(b), we can clearly see the ship is altering course by way of surge (longitudinal) and sway (transverse) motions. Thus, it is not the rudder that changes the course of the ship but is instead caused by the change in centre of gravity precipitated by the sway motion to port. In other words, this sway motion alters the hydrodynamic forces about the ship's hull which causes her to turn. If we look again at Figure 1.7(b), we can see how the sway velocity turns the ship towards the desired heading. With sway velocity towards the port side, the hull always sways towards port. When it does so, it exerts a force on the water particles around the port side of the hull. The water particles in turn exert an opposite force on the ship's hull caused by the inherent inertia of the water particles. The direction of this inertia force is always opposite to the sway velocity, as inertia force always opposes motion. Ergo, the hull experiences an inertia force on its hull in the starboard direction. This inertia force can be categorised in two ways: first is the part that acts on the aft portion of the ship (i.e., the inertia force at the stern) and the other which acts on the forward portion of the hull (i.e., inertia force at the bow). Following Figure 1.7(a), we can see how the inertia forces on the stern create an anticlockwise (towards port) moment of the centre of gravity. The bow inertia forces create a clockwise (towards starboard) moment of the centre of gravity. Now, the hull is designed in such a way that sway inertia forces at the bow are greater than those at the stern. Therefore, the resulting movement is towards the starboard direction as per Figure 1.7(b).

Rudder and steering gear 9

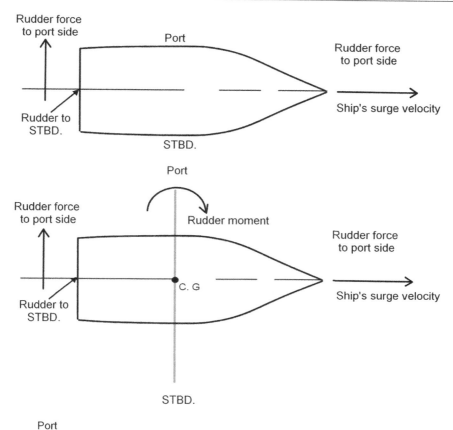

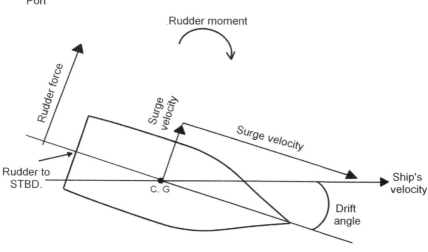

Figure 1.6 (a) Rudder force on a ship with the rudder to starboard. (b) Rudder force on a ship with the rudder to starboard. (c) Drift angle due to the rudder moment.

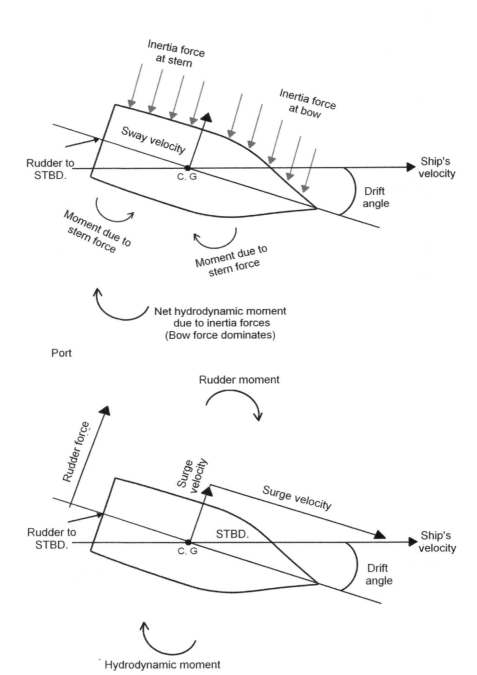

Figure 1.7 (a) Hydrodynamic moment due to sway velocity. (b) Moments acting on a ship during starboard turn.

What is important to know here is that when the hull exerts force on the water around it, during its sway velocity to port, the inertia force exerted by the water on the hull tries to achieve equilibrium. This means the magnitude of the inertia force is in the order of the ship's displacement.[2] So, when the resulting hydrodynamic moment acts on the ship, its magnitude is equal to the ship's displacement. This moment (unlike the moment caused by the rudder force alone) is sufficiently powerful enough to turn the ship. But as you can see, this hydrodynamic moment would not have come into play had the ship not attained a drift angle or a sway velocity moment, which is solely due to the action of the rudder.

In summary, the rudder does not alter the course of the ship but initiates a drift angle which results in a hydrodynamic moment. It is this moment which causes the ship to alter course. The hydrodynamic moment is the same direction of the rudder moment (as both are trying to turn the ship to starboard). The rudder angle keeps the rudder moment intact, which in turn, keeps the hydrodynamic moment intact. Once the rudder is again brought back to midships, the rudder force vanishes, which results in the diminishing of the rudder moment. It is only after the drift angle is reduced to zero, and the hydrodynamic moment becomes zero, that the ship continues in the course set.

So, what does all this have to do with the rudder being positioned aft as opposed to forward? Well, if we did not know the physics behind ship movements, then we would not be able to understand the relationship between the rudder and its location. Returning to our ship, the rudder – when turned to starboard – creates a force towards port (the 'rudder force'). Note the direction of the rudder movement that was created about the centre of gravity by the rudder force. The direction of the rudder movement was towards starboard (to create a drift angle towards starboard). Now imagine placing the rudder on the ship's bow. Given a starboard angle to the rudder, the rudder force would still be in the port direction, but what of the effect on the centre of gravity? The rudder moment would be towards port, causing a drift angle towards port. The net hydrodynamic moment would cause the ship to alter to port. If we turned the rudder to starboard, the ship would alter course to starboard. In addition, there is another reason why the rudder is never positioned forward. Most collisions at sea occur head-on between one vessel and another or between a vessel and a stationary or floating object. By positioning the rudder at the far stern, it is better protected from the hazards of collision. This means even a ship with substantial bow damage can still be manoeuvred for as long as the rudder and propulsion systems are intact.[3]

Before we move on to the steering gear, there is one further question worth asking, and that is why is the rudder always placed behind the propeller? The answer to this question lies in the function of the propeller. The propeller serves no other purpose than to increase (or decrease) the velocity of water that flows out of its slipstream. The volume of lift generated (i.e., the rudder force) is proportional to the velocity of the water falling on it. Because the rudder is placed aft of the propeller, the increased velocity of the propeller outflow results in a greater lift force. Subsequently, the rudder is always positioned after the propeller. Alternatively, were the propeller to be positioned forward of the propeller, it would generate the same turning effect with respect to direction; however, the magnitude of force would be severely diminished, as the rudder would inhibit the water flow of the propeller slipstream.

Now that we are familiar with the rudder – it helps the ship alter course – we can begin to explore the role and function of the ship's steering gear.

THE STEERING GEAR

The steering gear, when integrated with the rudder, forms what is called the '*complete turning mechanism*'. The steering gear has been an indispensable component of the ship's machinery for as long as professional mariners have sailed the world's oceans. On traditional sailing ships, the rudder was connected to the steering wheel by way of tiller ropes and pulleys. The steering wheel, located high aft for best visibility, consisted of a large pedestal, axle/spindle, wheel, and tiller ropes (Figure 1.8). Today, the ship's steering gear is almost exclusively mechanical and must meet stringent requirements set by the vessel's classification society[4] or Class. As per standard regulations, the steering gear should be capable of altering the ship's course from 35 degrees port to 35 degrees starboard and vice versa – with the vessel moving forwards at a steady head-on speed for maximum continuous rated shaft rpm and with a summer load waterline – within a maximum of 28 seconds. Moreover, with one of the power units inoperative, the rudder should be capable of turning the ship from 15 degrees port to 15 degrees starboard (and vice versa) within one minute with the vessel moving at half its rated maximum speed, or 7 Kt (8.05 mph, 12.9 km/h) (whichever is greater), with a summer load waterline. Finally, the ship's major power units and the control systems must be duplicated for redundancy so that if one fails, the other can easily substitute as a standby. If the ship is operating on emergency power, the steering gear system must be provided with an additional power unit connected to the emergency power supply from the emergency generator. This should be capable of turning the rudder 15 degrees from one side to the other (and vice versa) within one minute with the vessel moving at her maximum service speed, or 7 Kt (8.05 mph, 12.9 km/h), whichever is greater (Figure 1.9).

Figure 1.8 Ship's steering gear.

Rudder and steering gear 13

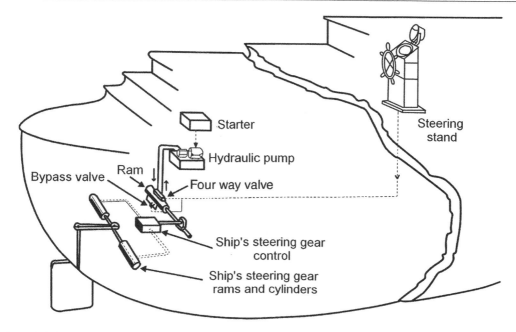

Figure 1.9 Ship's steering gear system.

As ships have grown and become faster, modern systems easing human effort have gradually developed. Today, there are two types of steering gear commonly used on modern ships: hydraulic and electrohydraulic. Yet despite marine technology advancing head-on, the working principles of the ship's steering gear have remained largely unaltered since the days of the wooden wheel described earlier.

The main control of the steering operations is given from the helm, which is located on the ship's bridge. This is not dissimilar to the cockpit of a car. The 'control force' for turning the vessel is triggered by turning the wheel at the helm, which is connected to the ship's steering gear system. The steering gear system generates a torsional force of certain scale, which is then, in turn, transmitted to the rudder stock. This turns the rudder. The intermediate steering systems on modern merchant ships can be multifarious, with each small component having its own unique function. For simplicities sake, we will refrain from going into too much detail; however, Figure 1.10 provides an illustration of the simple rudder system (Figure 1.11).

The rudder system consists of the following main components:

- Rudder actuators
- Power units
- Other auxiliary equipment needed to turn the rudder by applying torque
- Hydraulic pumps and valves

With hydraulic systems, pressure is provided through pure mechanical means. Electrohydraulic systems use hydraulic pressure developed by hydraulic pumps which are driven by electric motors. On most modern ships, advanced electrohydraulic systems are

14 Introduction to Ship Engine Room Systems

Figure 1.10 Modern-day advanced steering control at helm.

dominant. These hydraulic pumps play a crucial role in generating the required pressure to create motions in the steering gear which trigger the necessary rotary moments in the rudder system. These pumps are of two main types:

- Radial piston type (Hele-Shaw)
- Axial piston type (Swash plate)

Actuators mediate the coordination between the generated hydraulic pressure from the pumps (which are electrically driven) and the rudder stock. They do this by converting the mechanical force into a turning moment for the rudder. These actuators can be of two types:

- Piston or cylindrical arrangement
- Vane type rotor

Rudder and steering gear 15

Figure 1.11 Simple representation of a ship's rudder.

The type of actuator system is indicative of the steering system, of which there are two types:

- Ram type
- Rotary vane type

The ram-type steering gear (see Figure 1.12) is one of the most common steering gears, albeit the most expensive. The basic principle is the same as that of a hydraulically driven motor engine or lift. There are four hydraulic cylinders attached to the two arms of the actuator disc and on both sides. These cylinders are coupled directly to electrically driven hydraulic pumps. These generate hydraulic pressure through the gear pipe assembly. The hydraulic pressure field present in the pumps imparts motion to the hydraulic cylinders, which in turn corresponds with the actuator, to act upon the rudder stock. As we know, the rudder stock is an indispensable component of the steering gear arrangement and dictates the exact behaviour of the rudder response. The motion of turning the rudder is guided by the action of the hydraulic pump. A summary of the ram-type steering gear is illustrated in Figure 1.12.

In Figure 1.12, we can see the cylinders annotated as A and C are connected to the discharge side of the pump. This generates positive pressure in the piston cylinders. On the contrary, the other two cylinders – B and D – are connected to the suction side of the pump. This creates negative pressure in the cylinders. The resultant forces create a

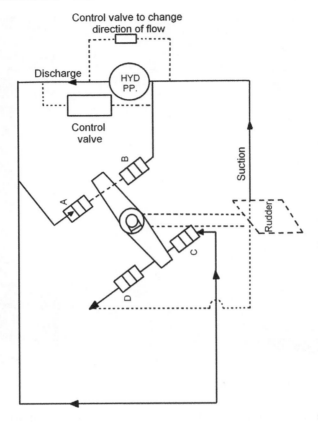

Figure 1.12 Ram-type steering gear.

clockwise moment in the rudder. Put simply, the positive and negative pressures from the pumps generate lateral forces on the rams which create a couple for turning the rudder stock. Similarly, to put it in an anticlockwise turning sense, the reverse is achieved, viz. the discharge points of the pumps are connected to cylinders B and D, while the suction sides of the pumps are attached to A and C. This reverse pressure flow from the hydraulic pumps is achieved with the help of control valves operated from the bridge. The ram-type steering gear arrangement produces high-value torque for a given applied power rating. The hydraulic oil pressure varies from 100 bar to 175 bar depending on the size of the rudder and the required torque. With the rotary vane steering gear, a fixed housing accommodates two vanes which rotate. The housing, together with the vanes, form four chambers. The principle behind its operation is like the ram type but with one small difference. When chambers A and C are pressurised, the vanes rotate in an anticlockwise direction. Cylinders A and C are connected to the discharge side of the pump whilst chambers B and D are connected to the suction side of the pump. Similarly, when clockwise rotation is required, B and D are connected to the discharge side of the pump whilst A and C are connected to the suction side of the pump. As previously mentioned, this is also operated by specialised control valves. Thus, differential pressurisation of the chambers causes rotational moments in the vane. Rotary vane type arrangements are used when the pressure

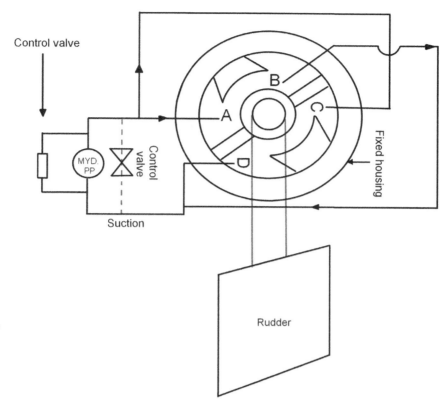

Figure 1.13 Rotary vane type steering gear.

requirement is between 60 to 100 bar of torque. This is the main advantage of the rotary vane type steering gear – it requires less hydraulic pressure and power to produce the same amount of torque as the ram type (Figure 1.13). Moreover, this arrangement has several other advantages such as lower installation costs, less weight, and a smaller sized footprint.[5]

EVALUATING THE STEERING GEAR

The safety of the ship should be the prime concern for all seafarers on board and never just the captain or the senior management team,[6] of which the chief engineer is a member. Maritime safety is also a primary concern for governing authorities such as the International Maritime Organisation (IMO) and Port State Control (PSC) as well as Class. Sadly, there have been many ship accidents in the past in which collisions, groundings, and strandings were caused by failures of the steering gear. It is for this reason that the steering gear is often subject to intense official scrutiny. Specifications and guidelines are laid down by the authorities and written into the ship's safety management system (SMS). Moreover, the guidelines for steering gear tests and drills are provided in SOLAS 1974/78, chapter V, regulation 26. In summary, regulation 26 states that within 12 hours of the vessel departing port, the following systems must be checked and relevant tests performed:

- Main steering gear and system
- Auxiliary steering gear and system
- The remote-control systems of the steering gear
- The steering position indicator on the bridge
- The emergency power supply to one of the steering units
- Rudder angle indicators showing the actual position of the rudder
- Power failure alarms for the remote steering gear control system
- Power unit failure alarms for the steering gear unit
- Automatic isolating arrangements and other automatic equipment

Further to the aforementioned, the following procedures must be carried out together with the previously listed checks and tests:

1. The full movement of the rudder as per the required capabilities of the steering gear system present onboard
2. A visual inspection of linkages and connections in the steering gear
3. The means of communication between the steering gear room and the bridge must always be operational

In addition, further other important requirements related to the steering gear include the following:

- A block diagram displaying the steering system, the changeover procedure from remote to local steering, and the steering gear power unit indicating the emergency supply unit.
- This diagram must be readily available on the bridge and in the steering gear compartment.
- All officers and crew engaged with the operation and maintenance of the steering gear system must be familiar with the changeover procedure from one mode of operation to the other.
- Emergency steering drills must be conducted regularly, and not less than once every three months.
- The date and time for the tests, checks, and drills conducted on the steering gear system must be recorded and the logs kept for the mandated period.

Despite the requirement to regularly evaluate and check the steering gear system, faults and malfunctions do occur. In this concluding section, we will look at some of the most common causes of steering gear failure.

STEERING GEAR SYSTEM FAILURE

To put this into context, we can refer to an actual incident which occurred on a vessel under pilotage.[7] In this instance, the vessel was departing a port in ballast and was downward bound in a restricted waterway. The engine control was set to bridge control, and the helmsman was using manual full follow-up (FFU) steering. At one point, a port alteration was requested. The rudder angle indicator showed 10 degrees to starboard. Several port and starboard helm inputs were attempted with the FFU, but no rudder movement could be observed on the rudder angle indicator. The pilot then ordered the engine to be stopped and that the anchor be readied. The master arrived on the bridge just as the vessel was leaving the buoyed channel. He went directly to the steering control and transferred the steering system actuator switch from the port system to the starboard system. This action restored the steering control, but it was too late. The vessel ran aground at an estimated speed of 8 Kt (9.2 mph, 14.8 km/h), causing severe damage to the ship's hull. Sadly, as this incident shows, accidents caused by steering gear failure are all too simple and all too common in the maritime industry. It simply cannot be stressed enough that good maintenance procedures and regular checks are extremely important for ensuring the smooth functioning of a ship's steering gear. Even so, there are several common issues which can occur despite the best efforts of the ship's staff. These include the following:

- *Oil leaks.* PSC authorities have strict policies regarding leaks from the ship's steering gear. Many ships have been fined and detained due to steering gear faults identified during PSC inspections. Still, oil leaks from the steering gear are one of the most common problems engineers need to tackle in the engine room. This is nothing inherently to do with the steering gear, per se, but more to do with the number of machineries dependent on hydraulic oil. Some of the principal areas to suffer leaks are the cylinder-ram seals found in the hydraulic ram-type steering gear and the chamber seals of rotary vane pumps. Irrespective of the type of steering gear, any form of leak from the steering gear system must be investigated and rectified immediately.
- *Difference in the actual rudder angle and the ordered helm angle.* Another widespread problem observed in the steering gear system is the difference in the angle given at the helm and the actual rudder angle. This occurs due to erroneous or insufficient adjustment of the control and repeat back lever. To rectify this problem, turn the buckle attached to the control rod. Repeat for the back lever to ensure the gear is precisely adjusted.
- *Unsatisfactory steering.* A ship's fuel consumption largely depends on the efficiency of its steering gear operation. If the operation of the steering gear is unsatisfactory, this will lead to a delay in the ship's progress and an increase in fuel consumption. Common causes for this type of fault include malfunctioning of the safety valves or bypass valves in the system. Any problems with the control and repeat back lever will also lead to unsatisfactory steering. To resolve this issue, the safety and bypass valve operation must be checked at regular intervals. If any problems are identified, these must be investigated and rectified immediately. To amend the control and repeat back lever, precisely adjust the turnbuckle attached to the control rod and the repeat back lever.
- *Excessive noise from the steering gear.* Excessive noise and vibrations from the steering gear indicate trapped air in the system. When air bubbles develop in the lubricating oil, the pumps and pipes can become subject to air hammering. This can lead to heavy vibrations and loud noises. Any air must be removed from the system using the vent valve provided in the cylinder and pump. This is especially important after the system is replenished with new oil. If the valve located in the oil supply tank of the steering gear is throttled or closed, it will again develop air bubbles in the system. Ensure that this valve is always left open when the system is in operation.

High oil temperature. Oil is the operating media for the steering gear system. Any abnormality in the parameters of the lubricating or hydraulic oil will lead to other operational-related issues. If there is an increase in the oil temperature, this will adversely affect the viscosity of the oil, hampering the steering operation. The most common cause for increased oil temperature is low oil level. Thus, always ensure that the low oil level alarm for the tank is working and replenish with fresh oil when required.

Rudder movement is under or beyond the expected limit. SOLAS requirements for the steering gear state that the system must be capable of putting the rudder over from 35 degrees on one side to 35 degrees on the other side of the ship at its deepest seagoing draught whilst running at maximum ahead service speed. It may sometimes happen that the maximum angle reached by the rudder is less than prescribed or the rudder is overshooting the 35-degree angle mark. One of the main reasons for this problem is the malfunctioning of the limit switch fitted on the repeat back unit or on the autopilot. Replacing the malfunctioning limit switch or adjusting the limit switch to the maximum prescribed rudder angle will usually fix this issue.

No steering from remote control. There must always be provision for local manoeuvring of the steering gear in emergency situations. This is usually needed if the remote-control operation fails. Some of the common causes for the failure of the remote control include
- Breakdown of the hydraulic pumps (in which case the other hydraulic pumps should be started),
- Malfunction of the transfer valve,
- Malfunction of the hydraulic pump bypass valve,
- Oil leaks, or
- Other unidentified problems with the remote-control system.

Rudder angle transmitter and tiller link failure. On 29 April 2011, the Panamanian registered bulk carrier *Dumun* ran aground whilst departing the port of Gladstone in Queensland, Australia. Prior to grounding, the ship's steering appeared to have stopped responding to bridge commands when the link between the tiller and rudder angle transmitter became detached. The steering gear continued to operate normally, but the transmitter lost its input signal and, as a result, the bridge-mounted rudder angle indicator stopped working. Cases such as this are rare, but they do occur. To avoid such accidents, ensure that the duty marine engineer conducts engine room rounds before every manoeuvre or departure from port and checks the links in the engine room.

In addition to those previously discussed, it is also worth noting that steering gear issues have been caused by electrical system failures, improper maintenance, collisions, and groundings. These are some of the most common issues to affect the steering gear system on board ships. When the main steering gear system fails, it is almost always necessary to engage the emergency steering system. In the final section of this chapter, we will briefly examine the role and function of the emergency steering system and how it is used in emergency situations.

EMERGENCY STEERING SYSTEM

The emergency steering system, as the name suggests, is a system which is used when the main steering system of the ship fails. As we know, ships have an electromechanical steering gear unit which steers the vessel from one port to another. Normally, the steering gear unit consists of a two- or four-ram electro-hydraulically operated unit with two or more hydraulic motors for the ram movement. Sometimes, a situation can occur in which the remote-control operation may fail to work, causing a sudden loss of steering control from the bridge. This can be caused by a sudden power failure or an electrical fault in the system or the control system, which includes faulty telemotors or servo motors used for transferring signals from the bridge to the steering unit. When steering control from the bridge is lost, manual steering from the steering gear room must be used instead. In these situations, the following procedures should be followed:

1. The procedure and diagram for operating the emergency steering should be displayed in the steering gear room and on the bridge.
2. Even in emergency situations, we cannot turn the massive rudder by hand or indeed by any other means. For this reason, a hydraulic motor is given a direct supply from the emergency generator via the emergency switchboard (in accordance with SOLAS regulations, which should also be displayed in the steering room).
3. Ensure direct communications are established between the bridge, the steering gear room, and the engine control room, usually via very high frequency (VHF) radio or the ship's telephone system.
4. Usually, a switch is provided in the power supply panel of the steering gear for the telemotor; switch off the supply from the panel.
5. Change the mode of operation by selecting the switch for the motor which is supplied with emergency power.
6. There is a safety pin at the manual operation helm. During normal operation, this pin remains in place to prevent accidental manual operation – remove that pin.
7. A helm wheel is provided which controls the flow of oil to the rams, which is connected to the rudder angle indicator. This wheel can be turned clockwise or anticlockwise port or starboard or vice versa.
8. If there is a power failure, use VHF to receive orders from the bridge for the rudder angle. As soon as orders are received, turn the wheel and check the rudder angle indicator.

Remember, routine checks should always be conducted to ensure the efficient and proper working of the main steering gear system and the manual emergency steering system. In accordance with SOLAS, these checks should be performed no less than once every three months, though good practice suggests emergency steering drills are best carried out monthly.

We have now concluded Chapter 1. In the next chapter, we will turn our attention to the second component of the steering and propulsion system: the ship's propeller.

NOTES

1. Port and starboard are nautical terms referring, respectively, to the sides of a vessel. Port and starboard unambiguously refer to the left and right side of the vessel, not the observer. That is, the port side of the vessel always refers to the same portion of the vessel's structure and does not depend on which way the observer is facing. The port side is the side of the vessel which is to the left of an observer aboard the vessel and facing the bow – that is, facing forward towards the direction the vessel is heading when underway, and starboard side is to the right of such an

observer. This convention allows orders and information to be given unambiguously without needing to know which way any crew member is facing.
2. Displacement or displacement tonnage is the weight of water that a ship pushes aside when it is floating, which in turn is the weight of a ship (and its contents). It is usually applied to naval vessels rather than commercial vessels and is measured when the ship's fuel tanks are full and all stores are on board.
3. Assuming, of course, the vessel is still seaworthy and watertight.
4. Classification societies are organisations which develop and apply technical standards for the design, construction, and survey of ships and which carry out surveys and inspections on board ships.
5. Ships generate revenue based on the amount of space available for carrying cargo. This means engine rooms are notoriously cramped. It stands to reason therefore that machinery should be as small and compact as possible.
6. The senior management team will differ across vessels and companies but typically includes the master, chief officer, and chief engineer. The bosun (most senior rating on board) may be invited to attend senior management team meetings where there is a perceived need to gauge the attitudes of the deck and engine room ratings.
7. As defined by the Pilotage Act 1987, pilotage refers to activities related to the navigation of vessels in which the pilot acts as an advisor to the master of the vessel and as an expert on the local waters and their navigation.

Chapter 2

Propeller design and function

In the previous chapter, we looked at the role and function of the rudder and steering gear. As we know, the rudder is used to help the ship change course. We also know the rudder is manipulated by the steering gear. We touched very lightly on the propeller, which propels the ship in the direction of the rudder. In this chapter, we will look at the design, role, and function of the propeller in much greater detail. Today, propellers are used as the primary form of propulsion for most commercial vessels, irrespective of the vessel's type and size. Although the concept of 'pushing' or 'propelling' a ship forward existed since Ancient Greece, the first screw propeller was not fully introduced into service until 1836. Most, if not all of us, are familiar with seeing propellers fitted to the stern of ships, but have we ever really thought about the shape and appearance of the propeller? It is fair to say most people rarely consider the physics behind the unusual nature of the propeller, which makes them so different to the normal flat-bladed fans we are accustomed to seeing in our day-to-day lives. If that person sounds like you, then you have picked up the right book, as these are concepts which we will consider in this chapter. The physics of how the marine propeller works is based on two primary functions. The first, to facilitate manoeuvrability and variation of speed, and second, to overcome fluid resistance of water. Resistance, as we are aware, is a principal phenomenon inevitable in all bodies floating in real fluids. As seawater has both viscous effects and waves, this resistance is substantial. Henceforth, we simply cannot imagine large ships and submarines operating without propellers. Indeed, propulsion is a vast subject area in the specialised field of naval architecture. But before we delve any deeper, it is necessary to first uncover some of the fundamental principles in the geometry of the marine propeller. The marine propeller is similar in form and function to the standard table fan. At the front of the table fan is a central hub or *boss*. This forms the core element of the propeller arrangement. The boss mates with a rotating plate which is linked to a motor via a rotating shaft. The rotating plate holds the fan blades. The motor transmits power via the shaft to the rotating blade. This power forces the shaft, plate, and boss to rotate causing a regular flow of cool air. The ship's propeller works in the same way, albeit in significantly larger proportions. Instead of a motor, the rotating shaft is connected to the main engine. Interestingly, detailed studies of the hydrodynamic behaviour of various propellers in different water conditions have shown that the delivered thrust and the resulting efficiency are inversely proportional to the size and diameter of the boss. Subsequently, modern designs have evolved with a smaller-sized but stronger boss. This helps maintain a safety trade-off in terms of strength versus power. Figure 2.1 provides a simple illustration of a boss-mounted propeller blade.

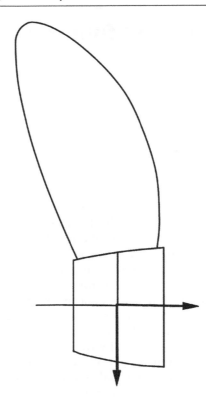

Figure 2.1 Propeller blade mounted on the boss.

PRINCIPLES OF PROPELLER DESIGN AND FUNCTION

To the uninitiated, propeller blades are often synonymous with the propeller itself, whereas in fact, the blades are mounted on the boss which creates the hydrodynamic lift required to produce thrust. Irrespective of the aesthetics of the blade design, all propeller blades share common design features. These include the *blade face* and *blade back*. The marine propeller has two hydrodynamic surfaces:

1. Face
2. Back

Put simply, the cross-section of the blade coupled with the boss when looked at from behind the ship is called the *face*. Some engineers refer to this surface as the *palm*. The opposite surface is the *back*. Figure 2.2 illustrates this better.

> *Leading edge and trailing edge.* There are two edges to a marine propeller blade. The edge which pierces the water surface first in order of succession is called the *leading edge*. Depending on the sense of rotation of the propeller (i.e., whether it rotates clockwise or anticlockwise), either of the two edges can become the leading edge. The opposite edge, which follows or 'lags behind' the leading edge is called the *trailing edge* (as shown in Figure 2.3).

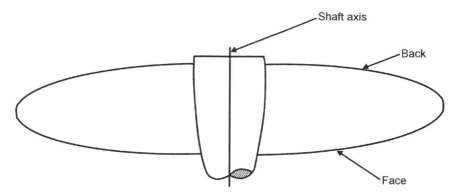

Figure 2.2 Plan of a propeller blade section showing the face and back.

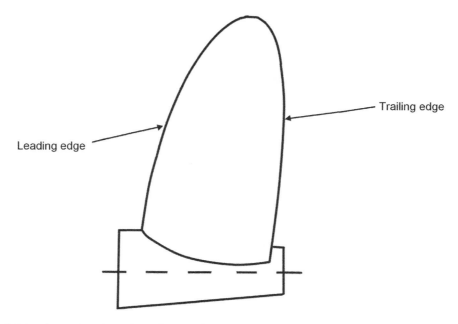

Figure 2.3 Leading and trailing edges of the blade.

Root and tip. The point of attachment for the blades with the boss is called the *root*. The *tip* is the furthest point of the propeller blade from the root and tapers like a leaf. It has the smallest section width and joins the leading and trailing edges. The root remains the same for the face and back. This is illustrated in Figure 2.4.

Every physical entity must be defined with respect to a suitable reference frame. With propellers, a uniform Cartesian coordinate system must be defined at the start of the design process. Although the choice of reference in the x, y, and z planes is arbitrary, a common 'convention' is such that the x-axis is placed along the direction of the shaft axis with the y-axis perpendicular to the shaft axis (i.e., sideways) and the z-axis situated in-plane to the ship propeller blade area. This is shown in Figure 2.5.

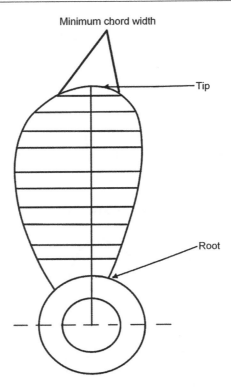

Figure 2.4 Root and tip of a propeller blade.

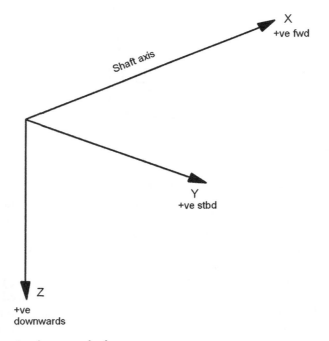

Figure 2.5 The conventional system of reference.

Pitch. Pitch is defined as the lateral distance traversed by a fixed point when a screw turns about its own axis. The same principle applies to the propeller – hence the name *screw propeller*! Pitch, in the case of a propeller, is a measurement of how much the propeller 'drives' or 'pushes forward' when it is freely turned about its own axis. Bearing this in mind, an interesting question to ask is whether the propeller *actually* drives the ship forward. As is commonly the case with marine science, the answer is an emphatic no. By way of explanation, the propeller is coupled through an intrinsic shafting mechanism to the main engine. In accordance with Newton's Third Law of Motion,[1] the turning motion of the propeller generates a reaction force in the wake region astern. It is this motion which thrusts the vessel forward. In other words, pitch can be described as 'the unit of distance moved by a point on a propeller when the propeller completes one revolution'. Deeper analysis of pitch involves more complex mathematical paradigms, which for reasons of simplicity we will not discuss here. But suffice it to say that one crucial aspect to consider is the fact that the distance calculated from normal pitch is often overestimated when compared to the actual distance travelled by a ship in one propeller revolution. The reason is obvious. There are unavoidable losses caused by resistance (viscous and wave-making) and various other factors such as losses sustained in the engine-shafting mechanism, as well as wave-induced events in the slipstream, and of course, cavitation.[2]

Rake. When the propeller is viewed from the side, we can see that the blades of the propeller are not perpendicular to the surface of the boss. Instead, it is 'tilted' at an angle either towards the fore or aft of the ship. This is called the *rake* (see Figure 2.6). The angle or the inclination of the rake in its profile is dependent on the vessel design, capacity, speed, and various other structural factors. One of the primary reasons for the rake is to allow for higher clearance between the blades and the vicinal hull

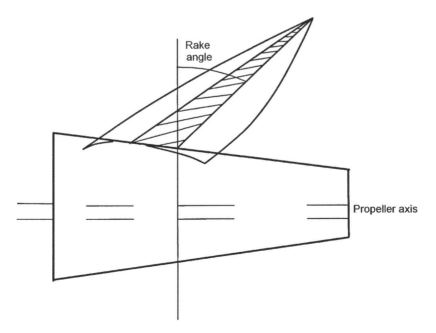

Figure 2.6 Profile of a propeller blade illustrating rake.

surface. If the blade tip is too close to the hull plating (which is very prone in case of zero or minimal rake), there is a chance of induced vibration due to the ship's propeller action. This is obviously not congenial from the perspective of propulsion or from a structural perspective. Indeed, for all vessel types, there is a maximum and minimum allowable clearance for the blade tip. As discussed earlier in the context of the reference frame, the rake is a deviation in the *x-z* longitudinal plane of the shaft. When the blade is raked forward, that is in the direction of the shaft axis towards the fore, the blade is said to have *negative rake*. Equally, when the blade is raked aft, it is said to have *positive rake*.

Skew. If we shift our reference to the *y-z* plane (which appears when we look at the propeller surface from behind) the blade is said to be 'skewed' – that is, bent or twisted sideways. The skew of the blade is shown in Figure 2.7.

There are two types of skews: *balanced* and *unbalanced*. Balanced skew occurs within lesser limits, where the generator line intersects the datum line at a minimum of two points. These types of propellers are known as '*moderately skewed.*' Alternatively, an unbalanced skew has a higher degree of deviation where the generator line intersects the reference line at not more than one point (creating a higher range of skew). These are said to be '*heavily skewed.*' Figure 2.8 gives an illustration of the two types of skews:

The flow of water in the field of wake is unsteady and unpredictable. This is further exemplified by the propeller's actions. Following years of experimentation, analysis, and sea trials, it is widely recognised by naval architects that by skewing a ship's propeller, it is possible to mitigate or at least minimise the extent of unsteady hydrodynamic loading within the field of wake. This indirectly reduces the hull resistance caused by the viscous 'drag' effect of the water. Thus, we can conclude by saying that in the ship design and building process, the design and type of propeller are critical to enhancing or impeding the ship's efficiency, and therefore the ship's profitability. Now that we have covered the basic working principles of the propeller, and the indirect role of the propeller in driving the ship forward, we can begin to examine

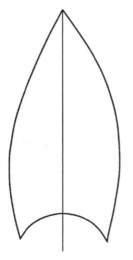
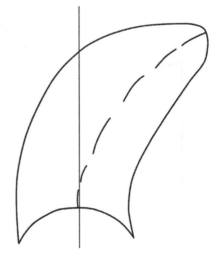

Figure 2.7 (a) No skew. (b) Skew.

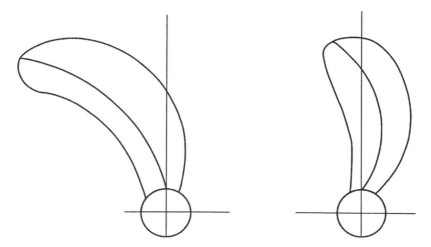

Figure 2.8 (a) Balanced. (b) Heavily skewed.

the various types of propellers and their construction. This is important for marine engineers to understand, as they are fully involved in supervising the overhauling and maintenance of the propeller whenever the ship is in dry dock.

As we know, the propeller is a rotating fan-like structure that is used to help propel the ship by using the power generated and transmitted by the vessel's main engine. This power is transmitted from the engine to the boss via the propeller shaft. When the boss is charged, it converts this energy from a rotational motion to generate thrust which imparts momentum to the water. This results in a force that acts on the ship, pushing it forward. This motion of propulsion works is based on Newton's Third Law of Motion and Bernoulli's principle.[3] In essence, a pressure differential is created on the forward and aft side of the blade, forcing the water to accelerate backwards. The thrust from the propeller is transferred through a transmission system which consists of a rotational motion generated by the main engine crankshaft, intermediate shaft and its bearings, stern tube shaft and its bearing, and, finally, by the propeller itself. A ship may be fitted with one, two, or even three propellers depending on the speed and manoeuvring requirements of the vessel. Typically, marine propellers are made from corrosion-resistant materials. This is necessary, as seawater is a corrosion accelerator. Most propellers are manufactured from metal alloys such as aluminium and stainless steel. Other popular alloys include nickel, aluminium, and bronze, which are 10%~15% lighter than other materials and have higher strength profiles. The construction of the propeller involves attaching the blades to the boss by either welding or forging the propeller and blades as one piece. Forged blades are more dependable and have greater strength than welded blades but are more expensive.

CLASSIFICATIONS OF PROPELLER

Propellers are classified according to several factors, with the most common being the *number of blades attached*. Most large container ships, tankers, and bulk carriers (i.e., ships over a certain size and deadweight) are usually fitted with between four and five blades, with five blades being the most common. Despite propeller efficiency being dependent on a lower number of blades, a trade-off is often required by offsetting efficiency against thrust. The latter is particularly important for large oceangoing vessels that may be fully laden during heavy seas. *Three-bladed propellers*. All three-bladed propellers share the following characteristics: the manufacturing cost is lower when compared to four-, five-, and six-bladed propellers; they are ordinarily made of aluminium alloy; they provide efficient high-speed performance; although acceleration is comparatively better, low-speed handling is markedly worse than with all other propeller types. *Four-bladed propellers*. Four-blade propellers are normally manufactured from stainless steel alloys. This provides better strength and durability, improved low-speed handling and performance, and better holding power in heavy seas. Moreover, a four-bladed propeller provides the best fuel economy when compared to all other propeller types. *Five-bladed propellers*. The manufacturing cost of five-bladed propellers is the highest of all types. That said, the vibration qualities are the lowest, and they provide markedly improved holding power when compared to four-blade propellers, especially in heavy seas. *Six-bladed propellers*. Perversely, six-bladed propellers are not as expensive as five-bladed propellers, though they are more so than four or three-bladed propellers. Moreover, the vibrations are lower than all other propeller types, and the holding power in heavy seas is best. With six-bladed propellers, the induced pressure field over the propeller decreases, which is why they are fitted to large container ships which require high-speed thrust.

The second way for classifying propellers is by the *pitch of the blade*. We already covered pitch earlier, but to remind ourselves, propeller pitch is defined as the displacement that a propeller makes for every full 360-degree revolution. The first category of propeller based on pitch is the *fixed pitch propeller* or *FPP*. Here, the blades are permanently attached to the boss. The position of the blades, and hence the position of the pitch, is permanently fixed and cannot be changed during operation. They are typically manufactured from copper alloy. FPPs are robust and dependable, as the system does not incorporate any mechanical or hydraulic connections. The manufacturing, installation, and operational costs for fixed pitch propellers are significantly lower than for *controllable pitch propellers (CPP)*, though manoeuvrability is not as good. FFPs are usually installed on vessels which do not require sensitive manoeuvrability (such as tankers and bulk carriers). The second category is the CPP. With a CPP, it is possible to alter the pitch by rotating the blade about its vertical axis by means of a mechanical and hydraulic arrangement. This helps in driving the propulsion machinery at constant load with no reversing mechanism required. This is because the pitch can be altered to match the required operating conditions. This improves manoeuvrability and increases engine efficiency. The main disadvantage of CPPs is the potential for oil pollution caused by leaking hydraulic oil in the boss. Moreover, the CPP is far more complex and expensive to install, operate, and maintain. In the worst cases, the pitch can get stuck in one position, making it difficult to manoeuvre the vessel. It is also worth noting that the propeller efficiency for CPPs is slightly lower than the same size FPP due to the larger boss needed to accommodate the blade pitch mechanism and piping.

Propeller dimensions

As a rule of thumb, a larger diameter propeller will be more efficient. But the real dimension of the propeller will depend on the type and design of the ship onto which the propeller is installed. The main factors which determine propeller dimensions are the aft body design and construction of the ship, the clearance required between the tip of the propeller blade and the hull of the ship, the general ballast conditions of the ship (for tankers and bulker carriers, the propeller size will be smaller compared to container ships), and the designed draught of the ship. As such, propeller dimensions can be calculated using the following formula:

For container ships:

$d/D = 0.74$

For tankers and bulk carriers:

$d/D = 0.65$

Where:
   d equals the diameter of the propeller
   D equals the design draught

PROPULSION SYSTEM ASSEMBLY

Propeller shaft

The ship's engine is connected to the propeller via a series of shafts which are all interjoined. These shafts are commonly referred to as the thrust shaft, the intermediate shaft, and the tail shaft. *Thrust shaft*. The crankshaft of the engine is first connected to the thrust shaft which passes through the thrust bearing, whose main function is to transfer the thrust to the ship's structure. The casing of the thrust bearing is similar in construction to that of the main engine bedplate, and the bearing is lubricated by the main engine lube oil system. The material of the thrust shaft is usually solid forged ingot steel. *Intermediate shaft*. The thrust shaft is then connected to a long intermediate shaft which comes in parts and is assembled using solid forged couplings. The length and number of intermediate shafts which are joined together depend on the location of the main engine. Larger ships will have more distance between the main engine and the propeller, meaning they will have more intermediate shafts. The material of the intermediate shaft is usually solid forged ingot steel. *Tail shaft*. The tail shaft, as the name suggests, forms the end part of the shaft arrangement and carries the propeller. The tail shaft itself is carried in a lubricated stern tube bearing. The tube is sealed as it connects and protrudes out of the engine room and into the open water. Lubrication is provided either through lube oil or water. The tail shaft transmits the engine power and motion drive to the propeller. The material of the tail shaft is high-strength duplex stainless steel alloy.

Each shaft is joined using a *coupled bearing*. The coupling is achieved by virtue of joints that are usually rigid and do not flex. The coupling units are bolted to each other using high-strength fasteners that can withstand large vibrational stresses. *Shaft bearings* are components used to support and bear the load of the shafts. They run the length of the shaft and ensure smooth rotation. These bearings are constructed according to their specific location. The *thrust blocks* are the last part of the propeller shaft system. These blocks support the propeller shafts at regular intervals. The blocks transfer excess power from the shafts into the ship's hull. This is necessary, as the shafts rotate at extremely high speeds, creating large vibrations. If these vibrations are not dissipated, jarring shocks may compromise the structural integrity of the vessel. Thus, by using specialised bearings, these shocks can be dispersed safely throughout the ship's hull. To anchor these thrust blocks to the bed of the ship, a reinforced frame is needed. This frame consists of a primary thrust block placed aft of the engine crankshaft, which disperses most of the shock into the hull girders and hull structure. In addition to the main components discussed earlier, there are a variety of smaller parts, such as sealants and bearings.

The design and construction of the thrust shaft are important, as they ensure structural strength. With shaft speeds reaching anywhere between 300 to 1,200 rpm, care must be taken to control material fatigue and to reduce damage caused to the components of the ship. Furthermore, the construction of the shaft bearings is critical, as these hold the entire weight of the propeller shafts. There are two main types of bearings used in the shaft assembly: the *full case bearing*, located at the stern, and the *half case bearing*, located in all other positions. The full case bearing provides a complete bearing for the weight of the shaft and forms an integral part of the shaft assembly. The reason it is located at the stern is to account for both catenary weight forces and to counteract any buckling or reverse thrust forces which may occur at the stern due to the motion of the propellers. This bearing is also known as the aftmost tunnel bearing, as it encases the shaft like a tunnel. The other shafts only account for weight, and hence do not require an upward casing unit. These bearings must be constructed from high-strength metals that do not easily buckle or deform under high stress. That said, low levels of tolerances are expected during the manufacturing process. Special bearing pads are fitted into slots on the connecting inner face of the bearings such that they allow for smooth rotation. To lubricate the shaft bearing, an oil dip arrangement is provided. By coating the rotating surface with oil from an oil thrower ring at regular intervals, a thick coat of lube oil is always maintained. The coolant used to prevent overheating and damage is typically water circulated about the shaft bearing. This is stored in specialised tubes that run the length of the bearing and shaft. Tanks stored above the engine platform house coolant that is circulated around the propulsion machinery and systems.

With the intermediate shaft arrangement, the thrust blocks are used primarily to dampen and absorb forces from the rotating propeller shafts. These forces are redirected into specialised frames that make up the bed of the engine compartment. The energy in these frames is further distributed into the hull through hull girders. The hull girders serve as the framework upon which the hull of the ship is built. The thrust blocks must be rigidly mounted in place to prevent any form of vibration during the passage. Also, the primary thrust block can either be an independent unit that is built separately or may be integrated into the marine engine itself. Integrating the block into the engine helps reduce space requirements and maintenance costs when underway. However, maintenance, while berthed, can be more problematic, as it requires opening the engine block casing. The casing that makes up the thrust blocks is built in two parts: an upper half that is detachable and a lower half that supports the shaft. The shaft is laid onto the lower block, and the upper half is then bolted into place using specialised shock-absorbing fasteners. To lubricate the rotating shaft, the rotating surface is regularly coated with lube oil. This is achieved in a comparable manner to the shaft bearings. An oil thrower and deflector are placed to maintain a constant supply of oil from a storage unit located on the lower half of the thrust block. The operating temperature is controlled using cooling coils that circulate coolant throughout the block. This also draws coolant from the central propulsion cooling system.

To absorb vibrations and shocks, bearing pads are attached to the blocks. These can be of two types: tilt pads, or pivotal pads, both of which are held in specialised holders built into the thrust block. The thrust pads transfer energy to the lower half of the casing, which is constructed to withstand larger shock impacts. A thrust collar is also used to absorb thrust from the propeller shaft. The thrust blocks incorporate integral flanges that help bolt the block to other surfaces. For instance, the block can be connected to the gearbox or engine using this flange. It can also be used to connect the engine thrust shaft to the intermediate shafts. Where the thrust block is built into the engine block, it is made of the same casing material as the engine base plates. In addition, they use the same lube oil and coolant as the engine. The integrated block is like the normal thrust block regarding most other features. It is interesting to note that the thrust block is integrated into the engine on most ships, except for smaller vessels, which have considerable space constraints. The shafts themselves must be built from robust materials with high yield strength, with a low probability of buckling. Each shaft starting from the thrust shaft must be built into small and manageable components that can be disassembled whenever the need arises. In addition, seals and stuffing boxes are also built from appropriate materials that provide an effective seal for the inner working machinery from external water. High-grade materials must be used when manufacturing the propeller shafts, as these components are extremely sensitive and need to manage large stress forces.

Marine gearbox

The marine gearbox is an integral component that is located between, and is attached to, the tail shaft and the intermediary shaft. It is used to manipulate the torque transferred from the engine crankshaft to the propellers located at the stern. It is mostly found on large vessels with high-speed engines. The marine gearbox works in the same manner as a standard car gearbox. It uses a system of robust gear arrangements which includes a clutch disc and pads to control torque. Constant lubrication is necessary to prevent friction-related damage.

Stern tube and propeller shafts

The stern tube arrangement refers to the way the tail shaft is borne by the stern tube, which is located aft of the vessel. The stern tube is a hollow, horizontal metal tube that serves as the primary connection between the propellers and the rest of the vessel. Attached to the stern frame, the stern tube acts as a plug at the rear of the vessel. The stern frame is the primary structural member that supports the stern overhang that lies above the propellers and rudder. The stern tube houses the tail shaft of the marine drive shaft system and serves two main purposes: to withstand load and to seal the vessel at the aft portion. Since the stern tube serves as the primary link between the vessel and the propeller, it must be able to withstand tremendous forces exerted by the suspended propellers. In addition, it should provide sufficient room for the propeller boss to move without creating friction. To manage the load, white metal[4] is commonly used, as it can withstand the required loads. Lubrication is provided within the stern tube to ensure smooth function of the marine propulsion system. Along with supporting the structural weight and forces of the propeller, the stern tube also needs to be able to effectively seal the vessel against external seawater. This means preventing water from entering through the aft section. It achieves this by using a combination of seals along its entire length. The stern tube has two main seals located at the aft and fore regions. This serves as dual protection against leaks that may occur over extended periods. These seals can be of three main types: stuffing boxes, lip seals, and radial face seals. *Stuffing boxes* are made from a variety of packing materials that are used to plug the stern tube. *Lip seals* are a type of gland seal that are used to prevent lubricants from seeping out into the water. They also serve the dual purpose of preventing water from entering the stern tube. *Radial face seals* extend in a radial manner out from any points of ingress and use a spring system to seal the entire structure. These are composed of two components that join to completely seal the rear portion.

In summary, the stern tube plays a key role in marine propulsion, as it absorbs and dampens a considerable amount of power from the propeller. As we know, the propeller is supplied with engine power to rotate and propel the ship in the desired course. If, however, the amount of power provided to the propeller is not generating sufficient revolutions, for example, because of a misaligned shaft, the propeller is said to be in a *heavy running state*. This may be a consequence of damage to the propeller blades; an increase in hull resistance due to hull fouling (resulting in a change in the field of wake); during passages through rough or heavy seas, when the ship is sailing against the current; when the ship is in a light ballast condition; when sailing in shallow waters; or where the ship is designed with a flat stern.

Propeller boss

The propeller boss or propeller hub is required irrespective of the way the propeller shaft exits the hull of the ship. There are three main types of bosses:

1. Shaft bossing
2. 'P' bracket holder
3. 'A' bracket holder

Shaft bossing refers to an arrangement in which the boss is placed right at the mouth of the stern tube such that there is almost no portion of the marine shaft located externally. Alternatively, the *'P' and 'A' bracket holders* are designed as overhang appendages that are located aft of the stern tube. They are more common on cruiser-type sterns as compared to transom sterns.[5] The marine shaft passes through the stern tube and then through the bracket supported by either a 'P' or 'A' type holder. The shaft terminates astern of the bracket at the boss. Choosing between the different arrangements depends on the vessel type and any restrictions on shaft exposure. With shaft bossing, the boss is partially exposed to external fluids (i.e., seawater). Thus, it must be internally waterproofed, with special gland systems used to prevent the leak of fluids across the boss. In addition, the boss must be well lubricated to reduce friction within the stern tube. For 'P' and 'A' bracket arrangements, the entire boss is exposed to seawater. Due to the extended length of the propeller shafts, vibrational and catenary forces act on the boss. Thus, the boss must be constructed in such a way as to withstand large vibrational shocks.

Depending on the boss and blade configuration, there can be two types of propellers. These are the solid propeller and the built-up propeller. If the propeller blades are directly integrated with the boss, the design is referred to as a *solid propeller*. If the blades are bolted into place on the boss, the design is a *built-up propeller*. Each comes with its own pros and cons depending on the type and classification of the vessel. For instance, solid propellers take less time to manufacture, as the blades and boss are cast in one single operation. Integrating the blades onto the boss is achieved either by casting them together or by welding the blades separately. Fuse welding the blades is not preferred as the joints form the weakest points in the structure, as they receive the largest reaction forces. On the other hand, casting the entire propeller unit requires considerable expertise and is costly. Choosing the best option depends on the type of use and force limits that the propeller is expected to encounter. Although solid propellers take, comparatively, less time to design and build, the casting must be successful in the first operation; otherwise, any cracks or fissures will render the entire cast useless. This makes the solid propeller type particularly expensive. With built-up propeller units, the blades are separate from the propeller boss and must be bolted in place. Specialised fasteners are used to secure the joint and are made waterproof to prevent the accumulation of fluid within the boss interior. The benefit of using built-up propellers is that during lay-up and dry dock, the entire propeller assembly need not be completely disassembled, as only the required regions need be removed. In other words, if one blade needs to undergo maintenance, it is only that blade which is removed, rather than the entire assembly. This saves considerable time, effort, and money. With solid propellers, the opposite is true. The entire propeller assembly needs to be removed, which is time-consuming and extremely expensive.

A further advantage of using built-up propeller hubs is that the angle of pitch of the blades can be changed to cover a wide variety of thrusts. Such units are known as CPPs and are covered in detail in the last section of this chapter. The propeller boss plays a key role in the CPP unit, as it accommodates the propeller's essential machinery. The preferred materials for casting propeller bosses are copper and bronze alloys for large ships. Smaller vessels tend to use aluminium, bronze, and nickel alloys. Bronze and copper are chosen for their high tolerance to rusting and corrosion, as well as their strength and durability. Aluminium is chosen as it is extremely lightweight while also possessing high structural strength. In fact, aluminium has one of the highest strength-to-weight ratios amongst all metals, a property that is preferred by ship designers. The glands and other sealing materials that prevent leakage form part of the stern tube but can be integrated into the propeller boss to increase efficiency. These sealants can vary depending on ship configuration and structure. Sealants such as packing boxes and lip seals are used at the junction where the boss connects to the propeller shaft. The boss is an important rotational component that requires regular servicing to keep the vessel sailing smoothly. As we know, the propeller assembly, including the boss, is acted upon by various forces, including vibration, submerged water pressure, and centrifugal forces. To access the propeller boss, the vessel needs to be taken into dry dock such that the keel portion is exposed. Then, heavy-duty cranes must be employed to hold the propeller unit in place whilst it is disconnected and removed from the propeller shaft. In the case of solid propellers, special lifts are attached to the boss which gradually shift the entire unit off the vessel. For built-up propellers, if repair is only needed for one propeller blade, the boss and the remaining blades are left in place whilst a heavy-duty crane supports the affected blade. Once the bolts are disconnected, the blade can be safely removed. If, however, the entire unit needs to be removed, the procedure is the same as the solid propeller unit. Common repair operations required for the propeller boss include regrinding and resmoothing the surface of the boss. In the event of any major defects, the area must be again recast or filled in with alloy. Checks are also conducted on the welds to assess their integrity. Depending on the type of maintenance and the extent of damage sustained, the repair period may take anywhere from a few hours to several weeks. Checks on the internal machinery for CPP units and internal sealing glands of the hub are performed in addition to working on the exterior structure of the hub.

BOSS CONTROL ON CPP

The CPP is a design of propeller that involves being able to control the angle of pitch of the propeller blades. As the blades are housed on the propeller boss, the boss plays a key role in manipulating the angle of pitch. For this reason, CPPs are also known as *variable pitch propellers* (VPP). Pitch refers to the angle change that takes place when a propeller blade rotates about its long axis. In the case of ship propellers, the long axis extends radially outwards from the centre of the boss and along the longest section of the blade. The shape of the blade is such that there are two distinct edges: the leading edge, which meets incoming fluid particles, and the trailing edge, which redirects the outgoing fluid mass for increased thrust and acceleration. By rotating the blade, the angle at which fluid exits the propeller is altered which further alters the speed and control of the vessel. The CPP machinery is housed within the propeller boss. This machinery includes motors and sensors to report information to the bridge. The blades are in the form of a built-up propeller unit but are mounted on specialised bases built into the boss. Along with fasteners, precise motors within the boss gradually rotate the blade about its long axis to achieve the required pitch change. Pitch changes may also be automatically driven by an autonomous system that aids in navigation, steering, and other operations. The advantage of using controllable pitch propeller designs is that the efficiency of the engine is drastically improved. The operating rpm can be optimised to suit the need of the vessel while also providing improved acceleration and deceleration. In addition, for vessels in which negative pitch can be achieved by rotating the blade in the opposite direction, the vessel can even move astern (i.e., in reverse) without needing specialised engines. Thus, it has several advantages which make it an attractive choice for large ships. However, CPP can be expensive due to the advanced technology and skill needed to design and manufacture the propeller assembly.

PROPELLER MAINTENANCE AND OVERHAULING

Boring and sighting

The main engine of a ship is coupled to the propeller by means of a shaft. The translational motion of the pistons induces a rotatory motion on the crankshaft, which is in turn coupled to the propulsion shaft. The shaft then passes through the stern tube. At the outer region of the stern tube sits the propeller boss and blades. To work efficiently and effectively, the centreline of the crankshaft must be along the centreline of the propulsion shaft and the propeller. If that fails, the propeller will quiver about its position during running conditions. Even just a few millimetres of quiver can result in the perpetuation of high stresses in the shafting arrangement. If left unchecked, these can lead to structural failure. Bearing this in mind, let us suppose during the construction of the ship, the shaft was not positioned exactly along the crankshaft centreline. Given the fact that the shafts are long enough, up to more than 7 to 10 m (22.9 ft–32.8 ft) on an average ship, the offset of the shaft centreline at the aft end would end in the order of centimetres. That is not necessarily a design failure, but a failure in the production method. How then do shipbuilders ensure the alignment of the shafts exactly as per the architect's design? To ensure the shaft is correctly aligned, the builders follow a method called *boring and sighting of the stern tube*. The stern tube consists of two bearings. One bearing at its forward end (called the *forward*

Propeller design and function   37

bush bearing) and the other at its aft end (called the *aft bush bearing*). It is through the aperture of these bearings that the propulsion shaft passes. The clearances between the bearings and the shaft are very minute, and hence, the shaft centreline must be correctly established in line with the centres of each bearing. By maintaining this, the builder can ensure that the shaft centreline matches the centreline of the bearings and the crankshaft. Again, the bearings are fitted within bosses (discussed in detail later). The stern frame of the ship is the aftmost structure of the hull and is forged separately and then attached to the remaining hull structure. The stern frame also houses the stern tube. The stern tube, in turn, houses the aft bearing. When the shipyard orders the aft bush bearing from the manufacturer, they will always include a machining allowance based on the internal diameter. What this means in practicality is that if the required internal diameter of the bearing is 0.5 m (50 cm or 1.64 ft, 19.68 in), the manufacturer will provide an internal diameter of 0.49 m (49 cm or 1.60 ft, 5.88 in). When the shaft is passed through the bearing, the shipbuilder will machine the internal diameter to 0.5 m (50 cm or 1.64 ft, 19.68 in) to meet the design requirement. The next question to ask ourselves is, How do these bearings fit within the stern tube? The stern bearings are fitted within hollow steel cylinders within the stern tube, called bosses. The shaft is accommodated within the bearings, which are housed within bosses, which again, are housed within the stern tube. Therefore, the aft boss houses the aft bearing, and the forward boss houses the forward bearing. To be able to match the centreline of the bearing with the bosses, and that of the bearings, the bosses are ordered with a machining allowance for their internal diameter (for the same reason as the bearings have machining allowances in their internal diameter). The stern frame is welded to the hull structure, and the stern bosses are welded to the stern tube. Now arises a problem. Because of multiple welds on the hull structure, and because of the cutting allowances for each steel plate on the hull, the geometric centreline of the aft and forward bosses will not match the required centreline as specified in the design drawing. A telescope is placed at the required height which matches the height of the design centreline. Multiple targets are placed at the aft and forward ends of the aft boss, forward and aft end of the forward boss, and along the centreline of the engine output flange. The arrangement is then viewed through the telescope, and the positions of the targets are aligned accordingly until all the centrelines of all the targets appear to be in one line. The centres of the forward and aft boss are then marked. These centres should now match the centrelines of the forward and aft bush bearings, respectively. In accordance with the obtained centres of the bosses, the internal diameter of the bush bearings is machined to the required internal diameter to be able to house the propulsion shaft. Again, therefore the shipyard will always order the bearings with a machining allowance on the internal diameter. Care must be taken to ensure the correct internal diameter of the bush bearings. If the internal diameter is too large, the shaft will quiver within it, and the centrelines will not match. If the internal diameter is too small, it will not be able to accommodate the shaft within the bearing. In fact, the forward and aft bush bearings are ordered with 5 mm (0.064 in) of machining allowance on their outside diameters. The outer diameter of these bearings is machined so that there is a difference of about 0.01 to 0.02 mm (0.0003 in–0.0007 in) between the internal diameter of the bosses and external diameter of the bearings. This allows the bearings to be pressed into the bosses of the stern tube. Once the centreline is achieved, the propulsion shaft is fed into the bearings for installation. Even though the shafting system is aligned so precisely during the building process in the shipyard, the shaft may still deflect from its original alignment due to the bending of the hull girder. Different bending scenarios may occur, depending on the loading conditions and the sea states the ship is sailing in. It is important for designers, therefore, to consider the effect of hogging and sagging of the hull girder on the change in alignment in the shaft system. To put this into context, we need to understand the underlying principles of Euler's beam-bending theorem. During the design stage of the shaft for a ship, designers estimate the torsional, bending, and shearing loads on the shaft. This indicates the critical points for bending. Accordingly, the position of the bush bearings (aft and forward) is decided so to ensure deflection in the shaft is as low as possible in the worst loading conditions. Classification societies, being responsible for the development of structural safety rules for ships on an initiative-taking basis, have been involved in developing rules considering and countering this effect. It is also especially important, and necessary, to conduct regular checks for bearing clearances between the bush bearings and the propulsion shaft. Due to prolonged use in various loading conditions, the inner linings of the bearings tend to wear out, thereby increasing the clearances between the shaft and bearing metal. If left unchecked, this may also lead to the shaft quivering. During tests for checking the shaft alignment and deflection, the observations should be taken in a light ship condition (in which case the shaft deflection will be minimum and will exhibit the inherent deflection in the shaft) and in a fully laden draft condition (wherein the deflection will be at its maximum owing to the additional deflection due to the bending of the hull girder).

Renewing the propeller

There are assorted reasons why a ship's propeller should be renewed, including damage to the propeller, where the propeller is dynamically unbalanced, and when renewing the existing propeller with a new propeller that is of a better design, material, has a larger blade area. When renewing the propeller, and before decoupling the intermediate shaft and the propeller shaft (tail shaft), it is important to conduct thorough inspections of the entire propulsion system. This means performing a shaft jack-up test prior to removing the intermediate shaft. *Jack-up test.* When performing a jack-up test, the shaft is jacked up using a hydraulic jack. This is done to evaluate the bearing reaction. For the most accurate results, it is advisable to position the hydraulic jack as close to the bearing as possible. To analyse the results of the test, a comparison is made against the American Bureau of Shipping (ABS) Reaction Coefficient Matrix, wherein the expected gradient of the average line of the curve is tabulated. Once the jack-up test is complete, the intermediate bearing cap and the intermediate shaft are removed prior to which the tail shaft is secured. It is especially important to protect the bearing area, which can be achieved by welding pad eyes in place. Always ensure redundant pad eyes are used and the lifting arrangements are sufficient to accept the load of the intermediate shaft. Once the intermediate shaft is removed and stowed in a secure location, inspect the condition of the plumber block bearings. Look for any white metal debris or contact marks. Inspections of the tail shaft can only be done when the ship is in dry dock. For this, the first step is to cut the rope guard and stow it aside. The propeller cone is then removed. Drain the stern tube of all oil and then slacken the aft stern tube seal. Remove the propeller nut (ensure the reference point is marked on the tail shaft and the propeller is secured prior to the tail shaft removal). Remove the forward stern tube seal and then pull the tail shaft and shift to the lifting zone; in some situations, the tail shaft may be removed through the skylight but in most cases requires cutting through the ship's shell for access. The tail shaft is carefully pulled out through the stern tube and removed through the access. It is strongly advised to perform non-destructive testing (NDT) on the tail shaft taper area. An evaluation of the propeller and tail shaft bedding will reveal how good the contact is. This is done by applying Prussian blue to the tail shaft tapered area. The tail shaft is then lifted vertically and matted to the propeller boss, which is positioned horizontally. The contact area in the boss is visually examined to confirm the contact meets class requirements (typically 70%–80% of contact). The Prussian blue mark zone on the boss is polished, and the process is repeated until a 70%–80% fit is achieved.

On completion of the propeller bedding test, the tail shaft and other accessories that were removed are reinstalled. The propeller is then mounted. This is a significant stage and careful attention must be paid throughout the process. In this section, we will discuss the oil injection method for a keyless propeller mounting. First, it is necessary to clean the propeller boss and the propeller shaft. Record the temperature of the propeller boss and propeller shaft. Slide the propeller onto the propeller shaft and align to the match marks. Screw in the pilgrim nut. For a *dry fit*, actuate the high-pressure pump connecting the pilgrim nut and allow the propeller to slide in a certain distance, and set the dial indicator to zero. Most manufacturers recommend an initial load. For a wet fit, actuate the high-pressure pump connecting the propeller boss expansion oil port and simultaneously actuate the pilgrim nut pressure pump. Raise the pressure gradually until the predetermined push-up length is achieved. Once the required push-up length is achieved, gradually release the pressure of the boss expansion port and then release the pilgrim nut pressure. Remove the connections and plug both the propeller boss port and the pilgrim nut port. Remember to perform a shaft jack-up test once the installation is complete. Verify the coupling alignment using the sag and gap method.

SIDE THRUSTERS

Side thrusters are a type of propeller-shaped system fitted either on the bow (bow thruster) or at the stern (stern thruster) of the ship (see Figure 2.9). They are smaller in size when

Figure 2.9 Example of a typical open-side thruster.

compared to the ship's propeller and help with the ship's manoeuvrability at lower speeds. Most ships of a certain size or type typically have bow thrusters, which are used for manoeuvring the ship in coastal waters, channels, or when entering or leaving a port and when experiencing strong currents or adverse winds. Stern thrusters are less common but where fitted are found on larger vessels such as container ships and bulk carriers. The thrusters come into their own when assisting tugboats to bring the ship alongside its berth. Moreover, the presence of bow thrusters eradicates the need for two tugs when entering and leaving port. This saves considerable time and money. Unless the port authority stipulates otherwise, the thrusters may negate the need for tugboats entirely. The requirement for the number of thrusters to be installed depends on the length and the cargo capacity of the ship. The route of the vessel also plays a principal factor, as many countries have local regulations regarding the compulsory use of tugboats when entering or leaving their port limits.

For the installation of the side thrusters, the following points are important to note. The thruster compartment, also known as thruster room, should be easily accessible from the open deck by the ship's crew. As most seagoing vessels use an electric motor to power the thruster, which is a heat-generating machine, it must be positioned in a dry and well-ventilated space. The thruster room should be fitted with a high-level bilge alarm with the indication provided both in the engine control room and on the bridge. For safety, the thruster room should be well lit and provided with at least one light supplied from an emergency power source. The thruster room should never be used to store flammable products, and flammable products should never be brought into proximity of the electric motor. The installation of the tunnel or conduit containing the propeller must be positioned perpendicular to the axis of the ship in all directions. The propeller should not protrude from the conduit. Grid bars may or may not be fitted at both ends of the tunnel (considering how much debris the ship bottom will experience when at sea). The number of bars should be kept to a minimum, as these tend to reduce the thrust force impacting on the overall performance of the thrusters. Any sharp edges on the grid bars should be avoided. A trapezoidal shape with no sharpness is considered a viable choice of design for grid bars installed perpendicularly to the direction of the bow wave. The design and position of the thruster tunnel should not interfere with the water flow under the hull or should not add to the hull's resistance.

DESIGN AND FUNCTION OF THE THRUSTERS

The bow and stern thrusters are in through-and-through tunnels which open at both sides of the ship. There are two such tunnels: one forward and one aft. The thruster takes suction from one side and pushes it out through the other side of the vessel. This force pushes the ship in the opposite direction of the inflow. This can be operated in both directions, i.e., from port to starboard and starboard to port. The bow thrusters are placed below the waterline of the ship. For this reason, the bow thruster room should be regularly checked for water accumulation. The bow and the stern thrusters can be electrically driven, hydraulic driven, or diesel driven, however, the most common type is electric driven, as hydraulic- and diesel-driven thrusters may leak. Also, with diesel-driven thrusters, the amount of maintenance required is higher. This detracts the engineers from their core duties in the engine room. The thruster consists of an electric motor which is mounted directly over the thruster using a worm gear arrangement. The motor runs at a constant speed, and whenever there is a change required in the thrust or direction, the controllable pitch blades are adjusted. These blades are moved, and the pitch is changed with the help of the hydraulic oil which moves the hub on which the blades are mounted. As the thruster is of controllable pitch type, it can be run continuously, and when no thrust is required, the pitch can be made to zero. The thruster is controlled from the bridge, and the directions are given remotely. In the event of remote failure, a manual method for changing the pitch is provided in the thruster room and can be operated locally. Usually, the hydraulic valve block which controls the pitch of the blades is operated in the thruster room for changing the blade angle in an emergency. When the bow thruster is operated alone, and the signal is given to operate the pitch at port side, the thrust will result in turning the ship towards the starboard side from the forward part. Similarly, when the bow thruster is operated alone, and the signal is given to run the pitch at starboard side, the thrust will result in turning the ship towards the port side from the forward part. When the stern thruster and bow thruster are operated together on the same side, the ship will move laterally towards the opposite side (Figure 2.10).

The thruster assembly consists of the following components:

1. Electric motor with safety relays
2. Flexible coupling between motor and thruster
3. Mounting and casing for the electric motor
4. Connecting flange and shaft
5. Motor casing seal

Figure 2.10 Example of a typical grated side thruster.

6. The tailpiece with shaft seal
7. Bearings
8. Propeller shaft
9. Zinc anodes
10. Grid with bars at both ends of the tunnel

SIDE THRUSTER MAINTENANCE

The side thrusters are an integral and important part of the ship's machinery, as they aid the vessel during difficult manoeuvres and in heavy sea conditions. It is the responsibility of the marine engineers to ensure the thrusters are kept in good working condition. Although there is little the engineers can do about the outer regions of the thrusters in which the ship is submerged (i.e., not in dry dock), there are several key maintenance tasks that need to be performed regularly as part of the ship's preventative maintenance schedule. These include the following: (1) the insulation needs to be checked regularly and should always be kept dry. This is because the thrusters are used infrequently, which increases the opportunity for damage through moisture. Moreover, because of the frequent idle state of the thrusters, there can be a reduction in insulation resistance, especially so in colder regions. This includes the space heater to ensure the insulation is kept dry. (2) The bearings of the motor and the links must be greased every month; (3) the condition of the hydraulic oil should be checked for the presence of water in the oil and samples sent ashore monthly for laboratory analysis; (4) the thickness of the contactors should be checked periodically; (5) the thruster room should be inspected periodically for the presence of water, which may indicate a leaking seal; (6) the flexible coupling between the motor and thruster should be checked together with the cable connections for cleanliness and tightness, and the motor grid should be vacuumed and blown clean to rid any carbon grit deposits. The major overhauling and maintenance of the bow and stern thrusters are done during dry dock when the ship's hull is out of the water, and the thruster blades and tunnel can be easily accessed. During this period, it is typical to (1) replace the 'O' rings and the sealing rings; (2) remove the pinion shaft; and (3) inspect, maintain, and if needed, replace the gear set; replace the bearings; perform repairs, cleaning, and replacement of the thruster blades; inspect the boss and carry out any repairs if needed; and inspect and overhaul the oil distribution box (which is used for the operating propeller blades).

In this chapter, we have covered a lot of ground pertaining to the design, construction, function, and maintenance of the ship's propeller and its associated fixtures. In Chapter 3, we will begin to look at the ship's propulsion plant – the marine main engine. As we have discussed previously, the main engine provides the power which turns the shafts which rotates the propeller. Although the engine produces the torque and thrust which powers the propellers, without the propeller, the ship would be unable to move forward. Such is the critical integration of so many of the ship's machinery.

NOTES

1. Newton's Third Law states that when two bodies interact, they apply forces to one another that are equal in magnitude and opposite in direction. The third law is also known as the law of action and reaction.

2. Cavitation is a phenomenon in which the static pressure of a liquid reduces to below the liquid's vapour pressure, leading to the formation of small vapour-filled cavities in the liquid.
3. In fluid dynamics, Bernoulli's principle states that an increase in the speed of a fluid occurs simultaneously with a decrease in static pressure or a decrease in the fluid's potential energy. The principle is named after the Swiss mathematician and physicist Daniel Bernoulli, who published it in his book *Hydrodynamica* in 1738.
4. White metals are a series of often decorative bright metal alloys which may be used as a base for plated silverware, ornaments, or novelties, as well as any of several lead-based or tin-based alloys for things like bearings, jewellery, miniature figures, fusible plugs, and some military medals.
5. A transom stern is a stern shape characterised by a flat shape extending to the waterline. The transom stern offers greater deck area aft, is a simpler construction, and can also provide improved flow around the stern. By comparison, the cruiser stern was initially designed only to lower the steering gear below the armour deck. A cruiser stern is characterised by an upward curved profile from the aft perpendicular to the main or poop deck.

Chapter 3
Introduction to the main engine

Marine propulsion is the mechanism or system used to generate the thrust needed to move a ship or craft through water. Whilst paddles and sails are still commonly used on many smaller boats and pleasure craft, most modern ships over a certain size are propelled by mechanical systems consisting of an electric motor or engine which turns a propeller, or less frequently, as is the case with pump-jets, an impeller. Manpower, in the form of paddles, and wind power, in the form of sails, were the first forms of marine propulsion. Rowed galleys, with some equipped with sails, were the first major type of seagoing vessel. As technology evolved with advancements in science, manual paddling was replaced by a combination of paddles and sails, and eventually sails only. The first advanced mechanical means of propulsion came with the marine steam engine, introduced in the early nineteenth century. The marine steam engine remained dominant until it too was replaced in the mid-twentieth century with the two-stroke and four-stroke diesel engines on larger commercial vessels, outboard motors on smaller craft and boats, and gas turbine engines on naval ships. Marine nuclear reactors, which first appeared in the 1950s, have been shunned by commercial shipbuilders; however, they are extensively used in naval ballistic submarines, large surface warships, and icebreakers. Most naval submarines use electric batteries or a combination of electric-diesel engines for propulsion and power. Over recent years, liquefied natural gas (LNG) fuelled engines have become increasingly popular, having gained recognition for their low emissions and operational cost advantages. Until the application of the coal-fired steam engine in the early nineteenth century, oars and wind were the principal means of watercraft propulsion. Whereas merchant ships relied on sails, military vessels employed the hard graft of oarsmen (who were usually slaves) to ram one ship into another. The Greek navies that fought in the Peloponnesian War used triremes, as did the Romans at the Battle of Actium. The development of naval gunnery from the sixteenth century onwards meant that naval vessels could attack each other from the broadside rather than head-on, as was previously the case. This change in naval tactics meant greater emphasis was placed on broadside weight rather than manoeuvrability, leading to the dominance of wind-powered sail ships over the next three centuries. Human propulsion has not been entirely eradicated, however, as many smaller boats still rely on manpower as an auxiliary propulsion system, typically involving the push pole, rowing oars, and even pedals. Sail propulsion of a large natural or artificial cloth sail hoisted on an erect mast. This mast is then supported by stays, and the direction of the sail is controlled by lines made of rope. Sails were the dominant form of commercial propulsion until the late nineteenth century and continued to be extensively used worldwide into the twentieth century in regions where wind was assured and coal supplies were questionable (for example, on the South American nitrate trade). Today, sails remain a popular form of propulsion

Figure 3.1 Engine control room.

for leisure and sporting craft, although innovations in technology are reintroducing sail power through applications including turbosails, rotorsails, wingsails, and windmills. In the second half of the twentieth century, rising fuel costs led to the demise of the steam turbine. Since the early 1960s, all new ships have been designed with diesel engine propulsion systems. These are built as either two or four stroke. The last major passenger ship to be launched with steam turbines was the *Fairsky*, in 1984 (and scrapped as *Atlantic Star* on 14 April 2013). Similarly, many ships which were designed and constructed as steam turbine powered were later re-engineered to improve fuel efficiency. The most famous example of this is Cunard's *Queen Elizabeth II*, which had her steam turbines replaced with a diesel-electric propulsion plant in 1986. Today, most new-build ships with steam turbines are specialist vessels such as nuclear-powered vessels, and certain types of merchant vessels (notably LNG and coal carriers) where the cargo can be used as bunker fuel (Figure 3.1).

TYPES OF MAIN ENGINES

Steam engines

There are two main types of steam-powered engines: reciprocating (where the steam drives pistons connected to a crankshaft) and turbine (with steam-driving blades attached radially to a spinning shaft). The shaft power from each can either go directly to the propeller, pump jet, or other mechanism or go through some form of transmission be it mechanical, electrical, or hydraulic. In the 1800s, steam was the main source of power for marine propulsion. In 1869, there was a large influx of steamships as the steam engine underwent a period of considerable technological advancement.

Reciprocating steam engine

The development of piston-engine steamships was a complex process. Early steamships were fuelled by wood and later vessels by either coal or fuel oil. Early ships used stern or side paddle wheels, which eventually gave way to screw-type propellers. The first commercially successful steamboat is attributed to Robert Fulton's North River Steamboat (often referred to as the *Clermont*), which was launched in the US in 1807. The *Clermont* was soon followed in Europe by the 14 m (45 ft) *PS Comet* in 1812. Following the launch of the first steamboats, steam propulsion technology advanced quickly over the course of the nineteenth century. Notable development included the steam surface condenser, which eliminated the use of seawater in the ship's boilers. This, together with improvements in boiler technology, permitted the use of higher pressured steam, which in turn enabled higher efficiency multiple expansion (compound) engines. As propulsion technology continued to develop, so did ship designs. Within a matter of decades, the paddle wheel was to give way to the more efficient screw propeller. Multiple expansion steam engines became widespread in the latter part of the nineteenth century. These engines worked by exhausting steam from a high-pressure cylinder to a lower-pressure cylinder, providing a larger increase in efficiency.

Steam turbine engines

Steam turbine engines were fuelled by coal or, later, by fuel oil or nuclear power. The marine steam turbine developed by the Anglo-Irish engineer Sir Charles Algemon Parsons raised the power-to-weight ratio. Parson demonstrated his modern design by installing it on the first steam turbine–powered ship, the 30 m (100 ft) *Turbinia*, at the Spithead Naval Review in 1897. This facilitated a generation of high-speed liners in the first half of the twentieth century and rendered the reciprocating steam engine obsolete, first in warships, and then later in the merchant fleet. In the early twentieth century, heavy fuel oil came into more widespread use, gradually replacing coal as the main fuel of choice for steamships. The main advantages of heavy fuel oil over coal were convenience, reduced operating costs through the elimination of trimmers and stokers, and reduced space needed for fuel bunkers.

Nuclear powered

In nuclear-powered steamships, a nuclear reactor heats water to create steam, which in turn drives the turbine. When first developed, exceptionally low prices of diesel oil limited the commercial attractiveness of nuclear-powered propulsion. The advantages of its fuel-price security, greater safety, and low emissions were unable to overcome the significantly higher initial costs involved in designing, constructing, and maintaining a nuclear power plant. Despite rising fuel costs, and a global mission to drive down emissions in the maritime industry, in 2022, nuclear-powered engines remain exceptionally rare. The main operators of nuclear-powered naval vessels are the US, UK, and Russia. The US operates a fleet of nuclear-powered aircraft carriers, with the space formerly used for stowing the ship's bunkerage now used for holding aviation bunker fuel. The US, Russia, UK, France, and several other nations operate nuclear-powered ballistic submarines. Only Russia continues to operate nuclear-powered surface combatant vessels, the *Kirov* class main battler cruiser. In terms of non-military usage, nuclear-powered propulsion has been less readily accepted.

As of 2022, the largest non-military nuclear-powered vessel in operation is the Russian *Arkitka*-class icebreaker, boasting an immense 75,000 shaft horsepower (55,930 kW). In the 1950s, an attempt was made to harness nuclear propulsion for civilian shipping. The first nuclear-powered civilian ship to launch was the Soviet icebreaker *Lenin*, on 5 December 1957. She was soon followed by the *NS Savannah*, a US-flagged part cargo-part passenger ship. Launched on 21 July 1959, the *NS Savannah* was the first nuclear-powered merchant ship. Built at a cost of $46.9 million (including a $28.3 million nuclear reactor and fuel core) and partly funded by several US government agencies, the ship proved commercially unsuccessful. In service between 1962 and 1972, the ship was deactivated in 1971, and following several moves, it has been moored at Pier 13 of the Canton Marine Terminal in Baltimore, Maryland, since 2008. Over the last few years, there has been a renewed interest in commercial nuclear shipping. With fuel prices reaching similar levels as the 1970s oil crisis, shipbuilders and ship operators are increasingly looking at new and innovative ways of reducing operating costs. Nuclear-powered cargo ships are one solution amongst many that naval architects and marine engineering technologists are currently exploring.

Diesel engines

Most modern ships use a reciprocating diesel engine as their prime mover. This is due to their operating simplicity, robustness, and fuel economy when compared to other types of prime mover mechanisms. The rotating crankshaft can be directly coupled to the propeller with slow-speed engines, via a reduction gearbox for medium and high-speed engines, or indeed via an alternator and electric motor for diesel-electric engines. On intelligent diesel engines, the rotation of the crankshaft is connected to the camshaft or a hydraulic pump. The reciprocating diesel engine first came into use in 1903 when the diesel-electric river tanker *Vandal* was put into service by the Russian company, Branobel. Although it was quickly realised that diesel engines offered superior efficiency compared to steam turbines, the power-to-space ratio was a limiting factor and remained so until the development of the turbocharger. Today, diesel engines can be broadly classified according to their operating cycle (i.e., two-stroke or four-stroke); their construction (crosshead, trunk, or opposed piston); and their speed (slow speed, medium speed, and high speed). Most larger merchant vessels use either slow-speed, two-stroke crosshead engines, or medium-speed, four-stroke trunk engines. Smaller vessels and craft tend to use high-speed diesel engines. The size of the different engines is a crucial factor in selecting the type of engine to be installed in a new-build ship. Slow two-stroke engines are much taller, but the footprint required is much smaller than that needed for an equivalently rated four-stroke, medium-speed diesel engine. As space above the waterline is of a premium in passenger ships and ferries (especially those with car decks), these ships tend to use multiple medium-speed engines, resulting in a longer, but lower, engine room compared to those needed for a two-stroke diesel engine. Multiple engine installations also provide redundancy in the event of mechanical failure and provide greater efficiency over a wider range of operational conditions. As modern ships' propellers operate at their most efficient at the operating speed of most slow-speed diesel engines, ships with these engines do not require gearboxes. Typically, such propulsion systems consist of either one or two propeller shafts, each with its own direct-drive engine. Ships propelled by medium or high-speed diesel engines may have one or two (or sometimes more) propellers, commonly with one or more engines driving each propeller shaft through a gearbox. Where more than one engine is geared to a single shaft, each engine will drive through a clutch, allowing engines not being used to be disconnected from the gearbox whilst the others remain running. This arrangement allows maintenance to be conducted whilst underway, even when the ship is far from port.

Gas turbines

Most warships have, since the 1960s, used gas turbines for their propulsion. Since the early 2000s, warships have increasingly used steam turbines to improve the efficiency of their gas turbines in a combined cycle, wherein waste heat from the gas turbine exhaust is used to boil water to create steam for driving a steam turbine. In this combined cycle, the thermal efficiency of the engine can be as good as, or even slightly greater than, a diesel engine alone. It should be noted that the grade of fuel needed for this type of gas turbine is much more costly than that needed for standard diesel engines. This means there is a pay-off between increased thermal efficiency and increased running costs. Subsequently, gas turbines are commonly used in combination with other forms of engines. For example, the cruise liner *RMS Queen Mary II* had gas turbines installed in addition to her standard diesel engines. For passenger ships, such as the *RMS Queen Mary II*, having a dual propulsion system provides various key benefits. Primarily, gas turbines produce markedly fewer emissions than diesel engines. This allows ships to operate in environmentally sensitive areas such as the Polar regions, the Caribbean, and the Baltic Sea, as well as in port. Due in part to their poor thermal efficiency, it is common for ships fitted with gas turbine and diesel engines to reserve gas turbine operation for high-speed cruising. Some private yachts, such as the Aga Khan's *Alamshar*, also have gas turbine propulsion engines (Pratt and Whitney ST40M), which enables a top speed of up to 70 Kt, which is unique for a 50 m (164 ft) yacht.

LNG engines

Shipping companies are required to comply with the IMO's and *International Convention for the Prevention of Pollution from Ships's* (MARPOL) emissions rules. One method of conforming to these rules is using dual fuelled engines. In this instance, the vessel is powered by either marine-grade diesel, heavy fuel oil, or LNG. Modern LNG engines have multiple fuel options which allow vessels to transit without relying on one specific type of fuel. Studies have demonstrated that LNG is the most efficient of the fuels available, although widespread production and use are hampered by a worldwide scarcity of LNG fuelling stations. Vessels providing services to the LNG industry have been retrofitted with dual-fuel engines, which have proven extremely effective. Benefits of dual-fuel engines include fuel and operational flexibility, increased efficiency, reduced emissions, and operational cost advantages. As the global shipping industry moves closer towards its goal of reducing and eliminating carbon emissions, LNG engines will become increasingly popular, as they provide an environmentally friendly alternative to providing power to ships. In 2010, STX Finland and Viking Line made history by signing an agreement for the construction of what would become the world's first (and at that time largest) environmentally friendly cruise ferry.

Other types of engines

We have now summarised the main types of marine engines used by most vessels. In addition to these, there are two additional types of engines which are worth discussing, albeit very briefly. The first type is the electric engine, which is employed on smaller vessels and craft, and the Swedish-designed Stirling engine.

Electric only

Battery-electric propulsion first appeared in the latter half of the nineteenth century, powering small lake boats. These relied on lead-acid batteries for the electric current to power their propellers. ELCO, the Electric Launch Company, evolved into the market leader later expanding into other forms of vessels, including the World War II era motor torpedo or PT boat. In the early twentieth century, electric propulsion was adopted for use in submarines. As their underwater propulsion was driven exclusively by heavy batteries, submarine progress was slow and of limited range. This led to the development of rechargeable batteries. Submarines were quickly fitted with a combined diesel-electric system, whereby the submarine would be powered by diesel when on the surface and electric power when submerged. When the diesel engines were running, this would recharge the boat's batteries. The first diesel-electric submarines were developed by the US Navy and were later adopted by Britain's Royal Navy. To expand the range and duration of the submarine during World War II, the German Kriegsmarine developed a snorkel system, which allowed the diesel-electric system to function even after the submarine had fully submerged. In 1952, the US Navy launched the *USS Nautilus*, the world's first nuclear-powered submarine, which eliminated the restrictions of both diesel fuel supplies and limited battery durations. On 12 November 2017, Guangzhou Shipyard International of China launched what was then the world's first all-electric, battery-powered inland coal carrier. With a deadweight of 2,000 tonnes, the ship can carry bulk cargo for up to 40 nautical miles (46 mi, 74 km) per charge. The ship carries lithium-ion batteries rated at 2,400 KWh, the same as 30 Tesla Model S electric cars.

Diesel-electric

The diesel-electric transmission of power from the engine to the propeller affords flexibility in the distribution of machinery within the vessel at a higher initial cost than with direct-drive propulsion. Subsequently, it is a preferred solution for vessels that employ pod-mounted propellers for precision positioning or reduced general vibrations via highly flexible couplings. Moreover, diesel-electric power provides the flexibility to assign power outputs to onboard applications other than those required for propulsion. On vessels with limited space, such as submarines, this provides an ideal solution.

Turbo-electric

Turbo-electric transmission systems use electric generators to convert the mechanical energy of a turbine (steam or gas) into electric energy. Electric motors then convert the energy back into mechanical energy to power the drive shafts. An advantage of turbo-electric transmission is that it allows the adaptation of high-speed turbines to slow-turning propellers or wheels without a heavy and complex gearbox. Furthermore, like the diesel-electric engine, turbo-electric engines have the advantage of being able to provide electricity for the ship's other electrical systems, such as lighting, computers, radar, and communications equipment.

Stirling engine

In the late 1980s, the Swedish engineering company Kockums developed several successful Stirling engine-powered submarines. The system works by storing compressed oxygen, which allows a more efficient and cleaner external fuel combustion when the boat is submerged. This provides the heat for the Stirling engine's operation. Currently, the Stirling engine is used on the *Gotland* and *Södermanland* class submarines of the Royal Swedish Navy and the *Sōryū*-class submarines of the Japanese Defence Forces.

A ship with an engine, but without a propeller or other means of propulsion, is not going to get far. Whilst the engine provides the means of power, it is the propeller that moves the vessel forward or astern. The propeller is a distinct piece of engine room machinery, so we will not cover it in this chapter; however, it is worth bearing in mind that the engine and propeller must work together for the ship to function. In the next part of this chapter, we will begin to look at some of the main types of marine diesel engines available in today's market. Of all the marine engine manufacturers, two stand out as market leaders: these are the Finnish company Wärtsilä Sulzer (formerly Sulzer) and the German company MAN B&W. As previously mentioned, marine prime movers come as either two-stroke or four-stroke engines. The main reason the two-stroke engine has emerged as the dominant choice for large merchant vessels lies in the greater fuel efficiency that two-stroke engines provide. Given that fuel prices are extremely volatile, shipping companies want to be sure they are maximising the power they can get from the onboard machinery whilst paying out as little as possible for fuel. Most shipping companies operate on very tight margins, which means even a slight increase in bunker fuel costs can eradicate any profit. Shipping is a famously expensive business with high crew costs, vessel maintenance, and upkeep, ever-increasing regulatory compliance demands and costs, and so forth. Added together, it is easy to understand why ship operators are so keen to reduce their fuel costs.

Two-stroke engines can burn low-grade heavy fuel oil, which can reduce the ship's operating costs. Moreover, the thermal efficiency of two-stroke engines is considered superior to that provided by four-stroke engines. This means two-stroke engines gain more thermal output for the same volume of fuel compared to four-stroke engines. Most two-stroke engines built today are comparable to four-stroke engines in size but can produce more power. In other words, they have a higher power-to-weight ratio when compared to standard four-stroke engines. Because two-stroke engines can generate more power, they enable ships to carry more cargo. Increased cargo-carrying capacity equals greater profitability, especially when each of the aforementioned factors is taken into consideration. Compared to four-stroke engines, two-stroke engines require less maintenance and upkeep. This equates to improved ship efficiency. Direction control is a major benefit which two-stroke engines can provide. Direct starting and reversing are easier, and as there are no reduction gears or speed reduction arrangements needed, two-stroke engines provide significantly enhanced ship handling. Despite these being considerable benefits, two-stroke engines are not without their flaws. For example, the ease-of-manoeuvring a two-stroke engine is less when compared to a four-stroke engine. The cost of installing two-stroke propulsion plants is higher than the running and maintenance costs of a four-stroke engine. The money saved on high-grade fuel often more than compensates for the initial high installation costs and of course the savings gained from through-life operation and maintenance.

HOW MAIN ENGINES ARE INSTALLED INSIDE THE SHIP'S ENGINE ROOM

The modern merchant ship's main engine is a massive metal structure with an average height of between three- and four-storey buildings or approximately 14 m (45 ft). These huge machines can weigh as much as 500 African elephants or 2,500 tonnes. Because of their size and weight, installing a main engine into a ship's engine room is a complex process which requires several sections divided into the various parts of the engine. The engine block is installed in parts during the ship building process. The enormous structure of the engine consists of several moving parts (both rotating and reciprocal) which transmits the mechanical power generated by the engine to the ship's propeller. It is this conversion of mechanical to rotary power that propels the ship through the water. As all the components of the main engine act under different forces, the engine must be firmly secured to the ship to avoid any damage caused by excessive vibration. Subsequently, the engine is attached to the ship's hull by way of holding down bolts and chocks. Given the weight of the engine block's hulk, and the associated mechanical fittings, the deck onto which the engine is placed must be strengthened using reinforced deck plates and additional bars and girders. The bedplate, which is the base of the engine, is attached by means of a holding-down bolt and chock arrangement. There are two main chock materials that are used for this purpose: cast steel chocks and epoxy resin chocks. Cast steel chocks require expert installation and are expensive; therefore, modern ships tend to use epoxy resin-based chocks, as these do not require any extraordinary measures and are more cost-effective. When installing the marine engine, first the whole engine (the crankshaft, intermediate shaft, and propeller shaft, along with the propeller) must be aligned in a straight line. This is achieved by following a specific procedure: (1) the area where the chocks and holding down bolts are to be fitted is cleared; (2) the chocks are prepared well before time by mixing a hardener and resin as required by the weight or volume ratio of the engine and its fixtures; (3) the holes for the bolts are pre-drilled, with the bolts made available but not inserted; (4) a foam dam is prepared for the installation of the chock; important – it is critical to ensure there is no hot work being performed within proximity of the operating location; (5) the temperature of the liquid resin when pouring must be higher than 25°C (77°F); if the temperature is lower than 25°C (77°F) the solution must be kept heated during the pouring process; (7) a holding bolt is fitted into each drill hole and sprayed with a releasing chemical agent; (8) the resin mixture is then poured into the bolt hole and each inserted with a bolt before the resin dries; (9) the holding down bolt is tightened using a hydraulic jack; (10) side chocks are then fitted in line with the main bearing girders; (11) end chocks are fitted at the aft and fore end to resist axial thrust emanating from the propeller; (12) the curing time of the epoxy resin depends on the steel temperature, which can range from a zero cure to 48 hours. Like all things, there are advantages and disadvantages to this process. First, the advantages include almost 100% contact, even on rough surfaces. Second, it provides a cheaper installation process, as there is no requirement for specialist equipment. Third, the binding is chemical resistant and non-corrosive. The disadvantages are if the engine is misaligned or the chocks are incorrectly fitted, the service life of the engine will be reduced accordingly. Overtightening or stressing of the holding bolts may lead to chock damage. The maximum temperature the epoxy resin can endure is 80°C (176°F), which means any engine room fires may result in the weakening of the engine placement. The holding-down bolts and chocks are exceedingly small components and must be inspected and checked regularly for tightness. Last of all, any loose bolts can lead to heavy engine vibration, misalignment, bearing damage, and even, in the worst situations, crankcase explosions.

Whenever there is a change or renewal in the major combustion parts of the engine, i.e., the piston or liner, or in the event the engine has undergone a complete de-carbonisation (d'carb), it must be returned to operation following a step running programme referred to as 'breaking in' and 'running in'. As the newly fitted liner, pistons, or piston rings are machine prepared ashore, they will have surface asperities and no bedding between the moving surfaces, i.e., between the liner and piston rings. Under these conditions, if a proper step running procedure is not followed, then a heavy blow past of combustion gases may result. This blow past can be dangerous, as it can lead to scavenge fires. Therefore, the step running programme is needed for newly fitted pistons, piston rings, and liners. For a complete d'carb engine, it is important to monitor the various parameters of the engine under increasing load, which can be achieved by breaking in and running in the engine. *Breaking in.* The process of breaking in involves a brief period of running the marine engines under zero load so that the piston rings are allowed to seat and lubricate properly. The breaking-in time will differ from engine to engine and is provided in the manufacturer's engine operations manual. The average breaking-in time for a two-stroke engine is around 48 hours. Breaking in is conducted to achieve the maximum wear rate, which causes asperities to break down faster. For this reason, heavy fuel oil and low total base number (TBN) oil must be used. Where low-sulphur fuel or marine diesel oil is used, the breaking-in period will increase accordingly. A low jacket water temperature must be maintained to increase the rate of wear. *Running in.* Running in follows the breaking in and is a long-run programme with step-by-step increases in engine load and speed. Just like breaking in, the running in schedules are also provided in the engine manual and differ from type to type. In two-stroke engines, cylinder lubrication is kept high in terms of oil quantity to ensure the proper lubrication of the piston rings and liner. For four-stroke engines with common sump lubrication, low TBN lube oil is used initially. Once a 30% load has been achieved, this is replaced with the manufacturer's recommended lube oil. If the proper breaking-in and running-in period are not followed, the engine may suffer from heavy scuffling, leading to increased liner wear and combustion gas blow past, which in turn can lead to scavenge fires.

In this chapter, we have begun to build an understanding of the diverse types of marine engines and how they differ in terms of design and function. We have also discussed the basics of main engine installation and post-installation breaking in and running in. In the next chapter, we will turn our attention to the components that make up the main engine.

Chapter 4

Key components of the main engine

In the previous chapter, we looked at some of the main types of marine propulsion systems and main engines. As we saw, there are many variants available depending on the size and type of vessel and area of operation. In this chapter, we will examine the key components that make up the main engine as found on most modern merchant ships. As we already know from Chapter 2, the rotational power of a ship's propeller is determined by the power produced by the marine engine, which rotates the crankshaft. The crankshaft of the main engine is supported and joined to a connecting rod via main bearings. The main function of the main bearings is to transmit the load without inducing any metal-to-metal contact. This is achieved by using special materials during the manufacturing process of the main bearings which float the journal pin of the rotating crankshaft in a layer of lubricating oil. Given the main bearings play such a vital role in the operation of the main engines, it seems logical to begin our discussion of the main components with a detailed summary of their role and function.

MAIN BEARINGS

The ship's engine comprises many heavy rotational parts which exert different forces on various sections of the engine crankshaft. The *main bearings* are one of the significant load-bearing parts of the crankshaft system. The bearings in a marine engine are subject to multiple pressure and contact forces, which include gas pressure generated inside the liner, dynamic inertial forces caused by various reciprocating and rotating motions of the engine parts, centrifugal forces caused by the different reciprocating and rotating motions of the engine parts, and friction between the crankshaft and bearing due to engine vibration. The main bearing is thus designed to manage these various forces whilst supporting the crankshaft when rotating at high speed. The material used in the manufacture of main bearings is critical for ensuring the bearing can both support the crankshaft journal and adjust to minor surface irregularities. Importantly, the main bearing cannot achieve this alone. It needs a compatible lubricating oil to bear the load and allow the smooth rotation of the crankshaft journal. The lubricating oil enables the bearing to withstand abrasive particles which cause friction between the journal and the bearing. When selecting the main bearing for a marine engine, there are several core qualities and attributes that ought to be considered. First, the material must be anti-corrosive in nature to avoid corroding the bearing material as well as its associated parts, such as the journal and bearing keeper. It should be friction resistant so that there is minimal energy loss between the bearing and the journal. It should have an excellent load-bearing capacity to compensate for the dynamic loads acting upon the bearing, and it should have good running-in and grinding-in qualities. Moreover, the bearing must be able to support the lubricating oil film to allow the smooth rotation of the journal. This means the bearing material should be such that it does not react with the lubricating oil. The bearing should have appropriate embeddability properties so that small particles embed in the bearing surface without harming the journal pin. Finally, the bearing material should have excellent compressive and tensile strength which, when combined with a thermal resistant property, avoids the bearing from sustaining heat damage when running.

Types of main bearings

Within the maritime industry, there are three types of main bearings (see Figure 4.1) used for two-stroke and four-stroke propulsion engines. These are the (1) tri-metal bearing; (2) tri-metal bearing with a cosmetic tin finish, consisting of a copper alloy with an overlay and nickel barrier; and (3) the bi-metal bearing, consisting of an aluminium bearing alloy and an aluminium bonding layer. The tri-metal with a cosmetic tin finish consists of four layers. The first is the flash layer, which is the uppermost layer of tin and lead, and has a thickness of around 0.035 mm. The flash layer is used to protect the bearing from corrosion and dust. The second layer is the nickel barrier. This layer has a thickness of approximately 0.02 mm, and its main function is to prevent corrosion and avoid the diffusion of the tin with the bearing metal. The third layer consists of a lead bronze alloy. This alloy has excellent anti-seizing properties and forms the main constituent of the bearing component. The fourth layer is the steel back. The steel back is used to provide support and the shape over which the other three layers are bonded. The tri-metal bearing consists of three

Figure 4.1 Main bearing.

metal layers: the flash layer, the overlay, and the interlay. There is also a lining and steel back. The flash layer is the topmost layer with a thickness of one micron of tin and lead. This prevents corrosion. The overlay is the second layer and is made from white metal (tin, antimony, and copper) and forms the main component of the bearing. The thickness of the overlay is usually around twenty microns. The third layer is the five-micron thick interlay which is used as an anti-corrosive layer for the overlay. The lining layer between the interlay and the steel back has a thickness of about 1 mm and consists of a bronze and lead alloy. The steel back is used for shape and support.

The bi-metal bearing consists of an initial layer of aluminium tin (around 0.5 mm to 1.3 mm thick) which forms the main component of the bearing. A bonding layer of 0.1 mm thick aluminium is adhered to the aluminium tin outer layer and the inner steel back. Like the tri-metal bearing, the steel back provides the support and shape over which the other layers are bonded. Bi-metal bearings are mostly used in four-stroke engines and rarely with two-stroke engines.

Common bearing defects

Due to their continuous use and the pressures and forces exerted on them, the main bearings often suffer defects and other related issues which require constant monitoring and remediation. Some of the most common types of defects the marine engineers need to look out for are corrosion, abrasion, erosion, fatigue, wiping, spark erosion, and crankshaft misalignment. *Corrosion.* If the oil in which the bearing is placed is acidic, it may lead to corrosion. The surface of the bearing will become discoloured and rough. *Abrasion.* If the oil is not filtered and treated correctly, and contains minute particles, which are common in engines burning heavy fuel oil, this may cause fine scratches to appear on the bearing surface. *Erosion.* When the oil supply pressure is not appropriate, or there is rapid and unusual journal movement, this will lead to the stripping of the overlay layer of the bearing. This is more common in medium-speed engines. *Fatigue.* When the engine load over the bearing is too high, this can lead to the removal of the bearing lining and the development of cracks and fissures. *Wiping.* This is the process whereby the overlay layer is removed due to elevated temperatures. When the bearing is new, wiping is required to remove the initial layer, which helps in the realignment of the bearing to the journal. However, too much metal wiping can lead to an increase in clearances, affecting the performance of the bearing. *Spark erosion.* When the propeller is at rest, the stern tube, propeller shaft and bearings are in contact with each other. Similarly, the main engine bearing, and the journal, are in contact with each other, maintaining continuity of the circuit. When the ship is running, due to the rotation of the propeller and the lubricating oil film the shaft becomes partially electrically insulated. This may also occur on the tail shaft by using a non-metallic bearing which acts as a form of insulation. The propeller is a large area of exposed metal which attracts protective cathodic currents. This produces an arc when discharging from the lubricating film. This results in spark erosion of the bearings, which can be worsened if the lube oil is contaminated by untreated seawater. *Crankshaft misalignment.* The crankshaft is a massive component when fully assembled. Initially, the complete crankshaft is aligned in a straight line (with the connection drawn from the centre of the crankshaft making a straight line) before setting it on the top of the main bearings. In time, this alignment may deviate and misalign, which can lead to damage of the main bearings (see Figure 4.2).

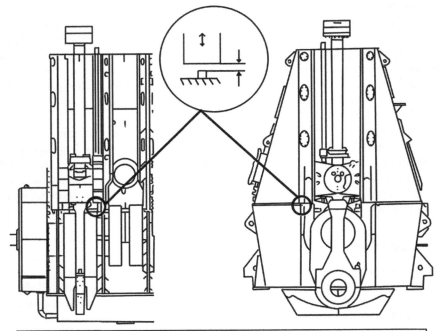

Sensors are placed on brackets mounted on the starboard side structure below both guide shoes at the bottom dead centre. The system monitors the variation in the distance betwween the guide shoe and the sensor in case of wear in main, crankpin, and crosshead bearings.

Figure 4.2 Bearing wear monitoring.

ENGINE BRACE AND STRUT

Two-stroke marine engines are the most powerful engine types used on ships, with some capable of producing the equivalent power of 108,920 horses (108,920 bhp). However, when running, they produce a vast amount of vibration. Engine vibrations can be extremely harmful to the engine and must be contained to avoid causing damage to the machinery, the ship's hull, and the strengthening members. To reduce engine vibrations, different methods and systems may be used, including struts or bracings, de-tuners, thrust pads, and chocks. Vibration is naturally present in every mechanical machine and is caused by the internal moving parts of the machine. This is more so during the fuel combustion process, which itself creates immense vibrations and reverberations. If the vibration level or amplitude increases more than the allowable limit, faults may begin to manifest including the following:

- Cracks in the attached piping
- Reduced turbocharger efficiency
- Fretting in the engine structure joints (especially between the 'A frame' and entablature)
- Loosening of the engine chocks and holding down bolts
- Damage in the intermediate shaft, bearing, or bearing support structures

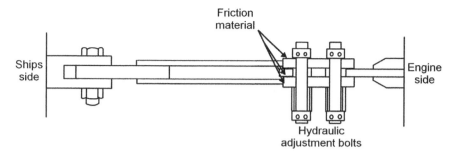

Figure 4.3 Friction-type top bracing (plan view).

- Damage to the thrust bearings
- Damage to the main bearings

To help minimise the effects of engine vibrations, engine struts or braces may be used. These are stud-like structures which are usually incorporated together with hydraulic or mechanical systems to transmit engine vibrations directly to the ship's hull. One end is fitted at the upper part of the main engine, and the other end is attached to a very stiff location in the ship's hull. This strut or brace acts as a de-tuner which increases the natural frequency of the system so that resonance occurs above the engine's rotational speed. The friction-type strut is one of the most common types used for two-stroke slow-speed marine engines. The working principle of this type of strut depends on the friction between the pads that brace the engine at the top so that the resonances with critical orders are above the speed range of the engine (see Figure 4.3). There is a frictional grip, which is responsible for the main functioning of the strut, which is hydraulically tightened. The tension on the hydraulic bolts must be regularly checked together with inspections of the strut structure to identify any signs of cracking, especially around the welds.

COMMON RAIL SYSTEM

The common rail system, as the name suggests, is a system which is common to every cylinder or unit of the marine engine. Early types of marine engines had a fuel system, wherein each unit had its own jerk pump, with the oil pressure supplied through the jerk pumps. However, with the common rail system, all the cylinders or units are connected to the rail, and the fuel pressure is shared between each cylinder. The supplied fuel pressure is thus provided through the rail. A similar type of common rail system is also fitted to the servo oil system for opening the exhaust valves. Although developed before the jerk pump, the common rail system was not initially commercially successful. It took several years of further development before the common rail system was adopted by ship designers. Common rail engines are often referred to as smokeless engines, as the fuel pressure required for combustion is the same for all engine loads and rpm. The common rail is employed for the following systems: (1) for heated fuel oil at a pressure of 1,000 bar, (2) for servo oil for opening and closing the exhaust valves at a pressure of 200 bar, (3) for the control oil used for opening and closing the valve blocks at a pressure of 200 bar, and

(4) for compressed air for starting the main engine. The common rail system consists of several main components, including the high-pressure pump which may be cam driven, electrically driven, or both. The pressure requirements are different according to each system. For fuel oil, the pressure may be as high as 1,000 bar, and for servo and control oil, the pressure is usually about 200 bar. The high-pressure pumps are driven by a camshaft with three lobe cams. These pumps perform several strokes with the help of the three lobe cams and a speed-increasing speed gear. For fuel oil and servo oil, the pumps are engine driven, and for control oil, it is motor driven. *The rail unit.* There are five main components that make up the rail unit:

- Fuel oil rail
- Control oil rail
- Servo oil rail
- Injection control unit (ICU)
- Valve control unit (VCU)

The rail unit is located after the pumps where the accumulated pressure from the pumps is provided to the rail. This supplies each unit when required and is located at the engine, stop platform, and just below the cylinder cover. The rail unit extends the full length of the engine and is enclosed within the engine case. Access is provided from above for maintenance and overhauling. The valve block and electronic control system. This is required for the control of the flow of the fuel oil, servo oil, control oil, and starting air from the rail to the cylinder. The valve block is managed by the electronic control unit which operates when it receives a signal indicating that the cylinder is at top dead centre (TDC). At this point, fuel must be injected, and the exhaust valve must be opened. With the help of electronic sensors, the injection can be controlled remotely by computer. For instance, if the engineers need to cut off fuel to one of the units, then the cut-off signal is given from the control system so that the appropriate valve will not open. This block is known as the ICU, and for the exhaust valve, it is known as the VCU. The control system for opening and closing the ICU and VCU is done by electro-hydraulic control. When the signal for open is given the valve for the control oil opens and the control oil pushes the valve of the ICU and VCU open. The signal for electronic control is given by a crank angle sensor, which senses each cylinder and sends signals to the system which decides whether to open or close the relevant valve. The timing of the opening of the valve can also be controlled electronically, which means that if the signal is given to open the valve early, it will open early and vice versa. There are several key advantages of the common rail system over conventional jerk pump systems. These include having the same injection pressure for the engine at all loads or rpm, which is not possible with jerk pumps, as the latter is dependent on engine speed. The injection timing can be varied during engine running, whereas with the conventional system, the engine must be stopped, and the timing settings changed manually. The design of the common rail system is simple, as there are no individual fuel pumps and the cams for each fuel pump are removed. The common rail system provides smokeless operation whereas with conventional systems, smokeless operation is only possible during high rpm running. Common rail systems require less maintenance due to the smaller number of pumps and the increased efficient combustion time between overhauls. The control of the variable opening of the exhaust valve can be achieved which is not possible with conventional systems.

TURBOCHARGER

The turbocharger is one of the most important components of the main propulsion system of the ship. However, when the turbocharger does not work properly, it can lead to surging. Turbocharger surging is a phenomenon which affects the performance of the turbocharger and reduces its efficiency. In this section, we will learn about the turbocharger and the problems associated with turbocharger surging. Turbocharger surging is defined as an audible high-pitch vibration emanating from the blower or compressor end of the turbocharger. It is frequently experienced in low-speed diesel engines. Whenever the breakdown of gas flow takes place in the turbocharger, a reversal of scavenging air occurs through the diffuser and impeller blades into the blower side. This causes surging. In other words, a large mass of oscillating airflow causes vibrations of the turbo compressor impeller and its vanes, which make the compressor unable to operate normally. This produces a high-pitch noise as a reaction, which is known as compressor surge. Other terminologies such as turbo surge or engine surge may also be used to describe this phenomenon, but the directly involved component that is surging is the compressor of the turbocharger or the turbo compressor. The turbine side or exhaust gas side of the turbocharger does not play a direct role in the surging process. It may undoubtedly affect the performance of the complete turbocharger, which may lead to the turbocharger surging. During engine operation at sea, it is common for surging to occur, as external factors such as sea state, weather, abrupt manoeuvring, and crash stopping can all lead to abnormal surges of air pressure in the turbo compressor. Such instances of compressor surge are acceptable. However, the ship's engineers must ensure that the condition of the turbocharger bearing, as well as the lube oil, is in good service condition. If the surging happens during normal engine operation and the frequency of engine surge is high, this may lead to damage of the bearing and, in some cases, results in the mechanical failure of the compressor rotor. In summary, turbocharger surging is most often the result of various engine components failing to perform in coordination with others. A worn-out engine cylinder or fuel system may lead to problems in the engine, which in turn may manifest within the turbocharger. This will result in less airflow to the compressor against the higher back pressure, making the compressor surge. Therefore, turbochargers must be matched properly with the engine air consumption rate and the pressure across the operating range of the engine. This should never be permitted to fall within the surge limits.

Surge line

As shown in Figure 4.4, the operating line of the engine should maintain the pressure and volume of the intake air at point A to maintain the equilibrium and efficient working of the turbocharger. Supposing there is an increase in the intake air volume, the pressure will decrease on the line of constant speed. To maintain the equilibrium, i.e., to remain on the operating line, the volume must decline. However, if there is a slight decrease in volume at point B (at the same pressure as A), this will result in the reduction of pressure on the constant speed line. At this stage, the compressor will not be able to maintain the required pressure and the volume will further decrease, leading to compressor surge.

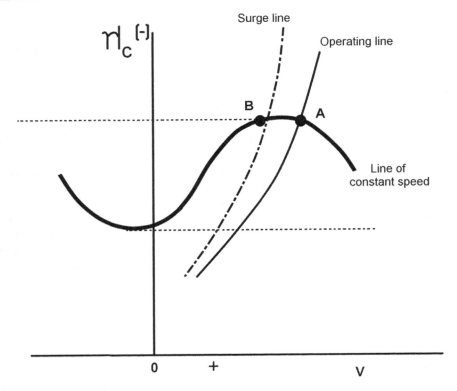

Figure 4.4 Turbocharger surge line graph.

Terms associated with turbocharger surge

- *Surge pressure dip*. The surge cycle has a specific pressure dip, and if the cycle continues without changing the operating point, the size of the pressure dip will remain.
- *Surge cycle time*. The time at which the surge starts until the operating point is changed to reach equilibrium again, i.e., the end of engine surge.
- *Surge temperature behaviour*. As the surge happens, there will be a reversal of airflow, leading to a change in the temperature of the upstream airflow.
- *Surge shaft speed variations*. The shaft of the turbocharger containing the compressor and turbine wheel will also experience a change in speed during the compressor surge. The turbochargers should therefore be matched with the engine air consumption rate and pressure across the operating range of the engine and should not fall within surge limits.

Categories of turbocharger surge

There are three categories of turbocharger surge, according to the extent and condition of the surge:

> *Mild surge*: Surges occurring under mild conditions are not significant. They may arise due to zero flow reversal and small oscillations in pressure.

Classic surge: Classic surges happen because of low-frequency oscillations combined with larger pressure oscillations.

Deep surge: This is the critical condition when the reversal of the mass flow occurs in the compressor which leads to surging.

Causes of turbocharger surge

The main causes of turbocharger surge include *inadequate power distribution*. Inadequate power distribution between the main engine cylinders may cause turbocharger surging, as one unit is producing more power and the other is producing less. Due to this, the air consumption required by both turbochargers differs, leading to a surge. *Fouled turbocharger parts*. If the inlet filter for the compressor on the turbine side is dirty, then sufficient air cannot be supplied for combustion, which leads to surging. Similarly, if the turbine side is also dirty – i.e., the nozzle, blades – then insufficient enough air cannot be introduced for combustion. Other causes may include a damaged silencer or worn-out turbocharger bearings. *Faults in the scavenge air system*. This may include a fouled air cooler and water mist catcher, insufficient water circulation inside the cooler, fouled cooling tubes, carbon deposits in the scavenge ports, and a high receiver temperature. *Faults in the exhaust system*. Highly fouled exhaust, i.e., the economiser, if fitted, may cause back pressure in the turbocharger, and thus finally lead to surging. Other exhaust problems might include the exhaust valve malfunctioning and not opening properly, a damaged or blocked protective grating before the turbocharger, pressure pulsations after turbocharger and inside the exhaust receiver, or a damaged compensator fitted on the line of turbocharger entry. *Faults in the fuel system*. If the fuel system is not operating efficiently, this can be symptomatic of low circulating or supply pump pressure, air or water in the fuel oil, low preheating temperature of the fuel, defective fuel pump suction valve, sticking fuel pump plunger and valve spindle due to carbon deposits, damaged fuel valve nozzle, or a faulty load distribution system. *Heavy seas*. In heavy seas, the engine may suddenly start racing, causing a sudden load change to take place. This happens because, during severe weather or pitching, the propeller moves in and out of the water, causing a change in load on the engine.

Preventing turbocharger surge

Discussed in the following sections are some of the methods that may be used for preventing turbocharger surge, though it should be noted that some points may vary with the design and construction of the turbocharger. Always keep the turbocharger intake filter clean, water-wash the turbine and the compressor side of the turbocharger, and always conduct proper maintenance, with periodic checks conducted for different turbocharger parts. If any issues are identified, these should be rectified as soon as practicably possible without loading the engine. Soot blowdowns should be performed from time to time wherever an economiser or exhaust boiler is fitted. Indicator cards should be taken to assess the cylinder and power distribution of individual units. Always ensure the engine auxiliaries and parts which affect the turbocharger are maintained properly. Always conduct efficient maintenance of the air-cooling system. Ensure regular cleaning and inspections of the economiser are conducted, and ensure the exhaust manifold is inspected and cleaned regularly. In addition, there are a few measures and design modifications which can be used for reducing the opportunity for surging. For instance, when a surge is about to happen, 'blow off' air from the valve located at the top of the air receiver. Bear in mind, however, this will lead to an increase in the exhaust temperatures, and care must be taken not to exceed the limiting values. There are also diverse types of turbochargers available in the market, which are designed to experience less surging phenomenon than conventional turbochargers. These include the hybrid turbocharger, which provides improved torque to the compressor turbine from the engine. This reduces the risk of turbocharger surging; the Variable Geometry Turbocharger (VGT). Variable geometry turbochargers experience almost no classic surging as it operates much closer to the surge margin which helps achieve the highest pressure possible; and the two-stage turbocharger. Two-stage turbochargers have specially designed compressors which are fitted with bypasses designed to suppress compressor surge.

ENTABLATURE

The entablature (see Figure 4.5) is one of the largest sections of the ship's main engine. It is the housing which holds the cylinder liner, along with the scavenge air space and the cooling water spaces. The entablature is usually made of cast iron. When assembling or installing an engine in the ship, the entablature can be fitted either by making castings for each cylinder of the engine and then bolting the mating surfaces, or by casting in

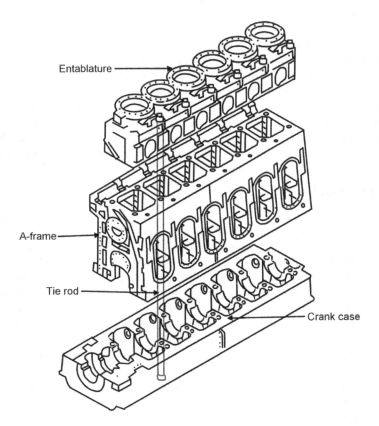

Figure 4.5 Entablature.

multi-cylinder units and then bolting them together. Fitted bolts are then used to align and bolt the 'A' frame to the entablature. It should be noted that the fitted bolts used to bolt the entablature, 'A' frames, and bedplates together are only used for alignment purposes and are not designed to resist the stresses generated from the firing forces. These stresses are instead dissipated by way of tie rods. Damage to entablature can lead to the loss of engine power and may also cause damage to the crankshaft and other critical components. The entablature of the engine comprises the cooling jacket for the cylinder liner where the elevated temperature combustion takes place. The cooling water is passed around the liner and through this jacket, which absorbs the heat of the liner and negates thermal stresses. The liner fitted in the entablature is provided with O-rings which avoid leakage of jacket water into the under-piston space. Cylinder water in the entablature jacket enters from the bottom and leaves from the top to avoid formation of air pockets or air locks. The entablature is also fitted with the liner and the cylinder head at the top. Therefore, when the head and liner is removed, thorough inspection of the entablature must be conducted.

To inspect the entablature, there are several steps which need to be followed. (1) Check the upper surface for liner face lands for cracks and deformation. (2) Check the condition of the inside of the jacket for mud formations and clean when necessary. (3) Check and clean the areas where the 'O' rings of the liner sit on the entablature jacket. (4) Check for signs of corrosion inside the water space. (5) Check for salt deposits. (6) Check the mating surface of the entablature and the 'A' frame for signs of fretting. (7) Check for evidence of cracks. (8) Check for oil or water leakages near the mating surfaces. If there is fretting between the 'A' frame and the entablature, the tightness of the tie rods must be checked. The entablature jacket inspection can be conducted without removing the liner and inspecting through the cover provided at the bottom of the jacket. Prior to opening the cover, the engine must be stopped along with the cooling water supply. The cooling water inlet and outlet supply valve should be shut and drained. The entablature must then be ventilated before any inspections can take place.

AIR BOTTLE

The main engine and auxiliary engine are the two prime components in a ship's engine room, on which, the entire operation of the vessel is dependent. However, there are several other important machineries that are necessary to support these two main components. One such piece of equipment is the air bottle or air receiver. The air bottle or air receiver is a large container which acts as a reservoir for storing the compressed air which is supplied from the main air compressor of the ship at high pressure. This compressed air is especially important as it is used for starting the main and auxiliary engines. The air bottle or air receiver also supplies the compressed control air and service air. If the quick-closing valves are air operated, safety air is supplied from the air bottle as is spring air for the exhaust valve. In effect, the air bottle or air receiver is a large metal container; however, there are various general mountings and connections which need to be inspected and maintained regularly. These include the filling valve. This is a valve fitted to the supply connection from the main air compressor to the air bottle. The outlet valve to the main engine. An outlet valve and pipe are fitted for connecting the air bottle to the main engine. It is through this outlet valve that the air bottle supplies the starter air. The outlet valve to the auxiliary engine. An outlet valve and pipe are fitted for connection between the air bottle to the auxiliary engines. This serves the same purpose as the main engines. Auxiliary connections.

66 Introduction to Ship Engine Room Systems

Other auxiliary supplies, such as service air and safety air are provided through an auxiliary connection isolating valve. Relief valve. A relief valve is fitted to the air bottle to relieve excess pressure inside the air bottle. Drain valve. A drain valve is fitted at the bottom of the air bottle to drain accumulated condensate from within the receiver. Fusible plug. A fusible plug is fitted to the air bottle with a separate connection leading out of the engine room so that in the event of fire, the plug will melt and relieve all the air within the air bottle to the outside atmosphere. Manhole door. A manhole door is fitted to the air bottle to allow internal inspections.

JACK BOLTS

Bending stress is one of the main causes of failures and cracks in the ship's main engine support girders of the bed plate, which is installed just below the main bearing. To minimise this problem, Sulzer devised a system of Jack bolts (see Figure 4.6), as opposed to holding bolts, for the main bearing. Jack bolts are hydraulically tightened long studs with an incorporated hydraulic connection which holds the upper keep of the main bearing intact. The Jack bolts do not pass through the main bearing keep housing, but instead hold the keep in place by applying hydraulic tension to the outer surface of the keep. Jack bolts are positioned at an angle of 15 degrees with the vertical plane and tightened at hydraulic pressure of around 600 bar. As the Jack bolts are mounted on the top of the upper keep,

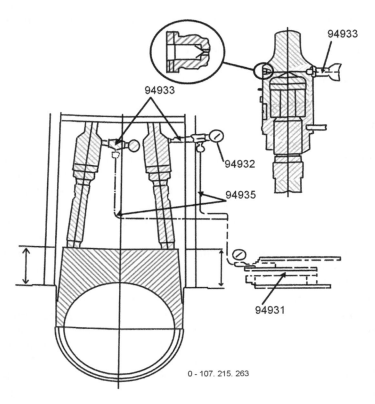

Figure 4.6 Jack bolt.

unlike conventional holding bolts, which are placed at the side of the keep, they pass through a hole provided on the housing for their stud. This allows the tie rod to be placed closer to the crankshaft centre line, which reduces the bending stresses exerted on the cross girder of the engine.

PISTON, PISTON SKIRT, AND PISTON ROD

The piston is an important part of the combustion chamber, which converts gas forces into mechanical power by way of a reciprocating motion. There are three components which make up the piston arrangement: the piston skirt, the piston rod, and the piston itself. In this section, we will discuss each of these three components. In most marine engines, two types of pistons are used. These are the crosshead piston and the trunk piston. The crosshead piston, comprising the piston crown, the piston skirt, and the piston rod (typically found in large two-stroke engines) which is connected to the crosshead functioning to transfer side thrust to the engine structure. The trunk piston comprises a piston with an elongated skirt to absorb the side thrusts. This is attached to the connecting rod by small end rotating bearings (typically used in small four-stroke marine engines).

Piston skirt

The piston skirt is fitted in both the two-stroke and four-stroke engines. It serves distinct functions for diverse types of engines. In large crosshead two-stroke engines, with uniflow scavenging, these skirts are short in length and are fitted to function as a guide to stabilise the position of the piston inside the liner. It is made of nodular cast iron, which is self-lubricating and provides superior wear resistance. The diameter of the skirt is usually kept slightly larger than that of the piston. This is done to prevent damage to the liner surface caused by the piston movement. Soft bronze alloy with lead rings is also fitted within the piston skirt. These bronze rings help during the running-in of the engine – i.e., when the engine is new – and can be replaced when necessary. In two-stroke engines which have loop or cross scavenging arrangements the skirts are slightly larger as this helps in blanking off the scavenge and the exhaust ports in the liner. In four-stroke or trunk piston engines, the skirt has an arrangement for a gudgeon pin. This gudgeon pin transmits power from the piston, via the gudgeon pin, to the top end bearing. As there are no crosshead guides in four-stroke engines, these skirts help to transfer the side thrust produced by the connecting rod to the liner walls. When inspecting the piston skirt, it is important to check for carbon deposits, signs of rubbing and wear, scuff damage to the wear ring, and cracks and or deformation of the gudgeon pin boss and trunk piston skirt.

Piston rods

Piston rods are found in large two-stroke engines. Piston rods help in transmitting the power produced in the combustion space to the crosshead and the running gear of the engine. The lengths of these rods depend on the length of the engine stroke and the manufacturer's design. The top end of the rod is flanged or attached to the underside of the piston and the bottom end is connected to the crosshead. The piston rod passes through the piston gland or stuffing box, so the rod has a smooth-running surface and low-friction coefficient. The piston rod function is such that the gas force acting on the top of the piston crown is transmitted to the piston rod by an internal mechanism, avoiding distortion of the ring belt. For cooling the piston, the rods have two through-and-through concentric holes. These holes are provided for the supply and return line of cooling oil. In most cases, the piston rods are manufactured from forged steel. Forged steel is used as it has a higher strength compared to cast steel and has a smoother surface finish. When inspecting the piston rods, there are several things to be checked including any signs of wear and or rubbing marks caused by the stuffing box gland, any scratches or dent marks caused by improper handling, the ovality of the rod at various positions, and the surface shine of the piston rod (measure surface roughness in Rs).

Trunk piston

The trunk piston is a term usually given to the pistons in four-stroke, medium-speed engines. These pistons have a composite design which consists of thin-sectioned alloy steel piston crowns with an aluminium alloy skirt. These pistons are light, strong, and rigid in construction and can resist elevated temperatures and corrosion. The piston is forged, and the space inside is provided for the arrangement of cooling spaces, which is achieved by cooling oil. The skirt consists of space for the gudgeon pin which transmits power to the connecting rod. The skirt also helps in transferring the side thrust produced by the connecting rod. The piston consists of ring grooves for fitting piston rings. The landing of the piston rings is hardened and plated with chrome to reduce wear. The top surface of the crown may be recessed to provide clearance for inlet and exhaust valves. Compression rings are fitted to the crown and are plasma coated, whereas other rings are chrome plated. An oil control ring is fitted to the top of the piston skirt. Since the piston rod is not used, the height of the engine is reduced when using a trunk piston. That said, there is no separation between the liner, piston assembly, and crankcase, which may lead to contamination in the event of blow past. The trunk piston consists of the piston crown and elongated skirt. The crown is made up of heat-resistant forged steel alloy, including chromium, nickel, and molybdenum. This alloy mix provides superior heat and corrosion resistance without compromising on strength. The skirt is made up of nodular cast iron or forged silicon aluminium alloy. This has the advantage of being light, with low inertia, which reduces the bearing loading.

TIE RODS

Marine engines are fabricated with many different components, which are held and tied together to complete the engine structure with the help of tie rods. As the name suggests, tie rods are long metal rods that are found around the periphery of the engine. The tie rod is a long strong metal rod with bolts or tie bolts at both ends. Each rod holds the three major engine components in place (i.e., the cylinder block or entablature, the 'A' frame, and the crankcase) by compression and transmits the firing load to the bed plate. Tie rods are placed as close to the centreline of the crankshaft of the engine as possible to minimise bending movements in the transverse girder. If the tie rod bolts are loosened or broken, then the marine engine will begin to exhibit various abnormal behaviours, such as vibrations from the main engine, fretting of the mating surface of the engine, and crankshaft misalignment. Prolonged conditions may lead to bearing damage, loosening of the foundation bolts and chocks, and maloperation of the turbocharger. Even a little abnormal vibration on the engine can misalign and damage the rotor or main bearings. The main causes of the tie rod breaking include not being properly tightened; the material and threading of the tie rod are underrated and not properly machined; the tie rod has exceeded its expected lifespan; the tie rod bolts are overtightened by hydraulic pressure, crossing the elasticity limit; engine overloading or peak pressures of the cylinders; previous fretting of the engine mating surface; loose foundation bolts, or damaged chocks leading to transmission vibrations in the tie rods; scavenge fires, which can loosen the tie rods as they pass from the scavenge space. The extra heat from the fire will lead to the expansion of the tie rods, weakening them; tie rods consist of quenching screws. If these screws loosen and work free, this can lead to heavy vibrations, causing cracks and splits in the rod, and, finally, in the event of heavy seas, fluctuations in the main engine load can cause the tie rods to loosen and eventually break.

DAMPERS AND DE-TUNERS

It is a well-established fact that every running machine tends to vibrate. These vibrations are caused by the various moving parts incorporated within it. When in motion, the machine develops an oscillatory motion focused on a point of equilibrium. This is the basic definition of vibration, a phenomenon common with all kinds of mechanical equipment. The main propulsion systems on board ships are some of the biggest engines ever built. It is understood, therefore, that these colossal machines generate equally colossal vibrations. The natural frequency of vibration is always present in marine engines, but the effect can be dangerous when the vibration frequency reaches elevated levels. This happens when the natural frequency of the vibration from an external source integrates with the engine vibration or when out-of-balance forces generated inside the engine start to create first- and second-order movements. Such effects can result in severe damage to the marine engine's internal moving parts, causing cracks in the structure, loosening of bolts and fastenings, and damage to bearings. Excessive vibrations are caused by axial and torsional vibrations or a combination of both. In this section, we will discuss the arrangements popularly known as dampers and de-tuners, which are used to reduce marine engine vibrations.

Dampers

As the name suggests, dampers are used to dampen or reduce the frequency of oscillation of the vibrating components. They do this by absorbing part of the energy that evolves through the equipment vibrating. There are two types of dampers used in marine engines. The first type is the axial damper. The axial damper (Figure 4.7) is fitted to the crankshaft to dampen the shaft-generated axial vibration – i.e., the oscillations of the shaft in the forward and aft directions, parallel to the shaft horizontal line. It consists of a damping flange integrated into the crankshaft and placed near the last main bearing girder, inside a cylindrical casing. The casing is filled with system oil on both sides of the flanges and is supplied via a small orifice. This oil provides the damping effect. When the crankshaft vibrates axially, the oil in the sides of the damping flange circulates inside the casing through a throttling valve provided from one side of the flange to the other. This provides the damping effect. The casing is provided with an elevated temperature alarm and pressure monitoring alarm located on both sides of the damping flanges. The alarms are designed to sound if the oil pressure on one side drops more than the values set. Low oil pressure may

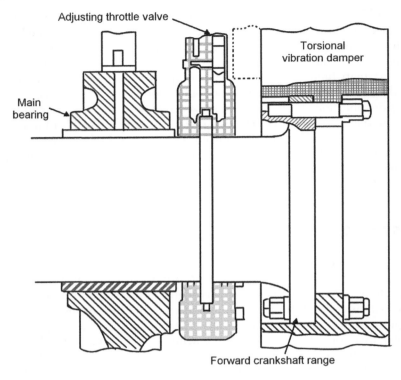

Axial damper

Figure 4.7 Axial damper.

be indicative of a low lubricating oil supply and sealing ring failure. The second type of damper used on marine engines is the torsional damper (Figure 4.8). The torsional damper works by dissipating the twisting phenomenon that evolves in the crankshaft and which spreads from one end to the other due to uneven torque pulses emanating from the different pistons. The most common type of torsional damper used on marine engines is the viscous type of damper, which consists of an inertia ring added to the crankshaft and enclosed within a thin layer of highly viscous fluid like silicon. The inertia ring is free to rotate and applies a lagging torque onto the crankshaft due to its lagging torsional motion. When the crankshaft rotates, the inertia ring tends to move in a radial direction, with a counter effect provided by the silicon fluid damping the vibrations.

De-tuners

De-tuners are used to alter the frequency of vibrating machinery by reducing the vibration of the engine. This is usually done by installing a side bracing, which is normally fitted to the top of the engine. This increases the stiffness and raises the natural frequency beyond the working range. The second method is to install flexible couplings. If the engine has a power turbine connected to its crankshaft via a reduction gear, then a flexible coupling may be used to compensate for the vibrations occurring during the motion transfer. The flexible elements are springs or special materialised rubber which help de-tune the vibration.

Key components of the main engine 71

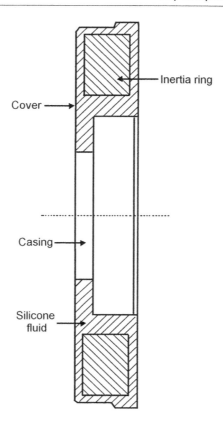

Figure 4.8 Torsional damper.

In this chapter, we have now discussed the main components that make up the marine engine, so we should have a basic understanding of what the marine engine is and what it does. We will look at propulsion systems in greater detail later, but suffice it to say, the main engine is what gives the propellers the power to move the ship through water. In the next chapter, we will begin to look at some of the main processes involved in operating the main engines, such as starting and stopping, slow turning, and performing basic maintenance.

Chapter 5

Main engine pre-start checks and monitoring

Starting the marine engines is a procedure-driven process which requires several steps to be taken into consideration. While it is necessary that none of these points should be missed, there are several important steps which must be performed. In this chapter, we will outline the pre-start checks and post-engine-starting monitoring. Before starting the main engine, it is important to (1) *check the lubrication of the main engine*. Always conduct pre-lubrication well before starting the marine engine. For the main engine, this is usually one hour, and for auxiliary four-stroke engines, at least 15 minutes in advance. (2) *Check all important parameters*. After starting the lubrication pump, check the lube oil levels and all other running pump parameters, including the cooling water pressure, fuel oil temperature and pressure, the control and starting air pressure, to ensure that all are operating within the accepted range. (3) *Open the indicator cocks and blow through*. All the indicator cocks of the marine engine must be opened to facilitate a blow through of the combustion chamber prior to starting. This helps to avoid hydraulic damage caused by water leaks. (4) *Rotate the crankshaft*. Rotate the crankshaft of the marine engine by means of the turning gear to ensure all the component parts are thoroughly lubricated prior to engaging. (5) *Manually check the turning gear*. Ensure that the turning gear is properly disengaged by checking it locally even when the remote signal is showing as 'disengaged'. Some auxiliary engines are provided with a tommy bar for rotation; in these situations, ensure that it is removed from the flywheel before the engine is started. (6) *Check the jacket cooling water temperature*. The jacket cooling water temperature of the engine should be maintained at least 60°C (140°F) for the main engine and 40°C (104°F) for the auxiliary engine (specific temperatures may vary depending upon the kW rating of the engine). (7) *Warm up the engine*. The incoming ship generator should be run at zero load for at least five minutes to allow the system to warm up. (8) *Put the load-sharing switch to manual*. When the second generator is started, it will try to come on load as soon as possible due to the autoload automation provided for sharing the equal load (if having the same rated capacity). While starting the second generator, bear in mind to put the load-sharing switch to manual. This will avoid the online generator from coming on load, giving it sufficient time for warm-up. (9) *Avoid excessive opening of the exhaust valve*. When starting the main engine with hydraulic oil-operated exhaust valves, open the spring air first and then start the hydraulic oil to the exhaust valve. This will avoid excessive opening of the exhaust valves. (10) *Examine the engine*. Engineers need to be present at or within proximity of the engine when it is started from a remote position. Auxiliary engines need to be started from a local position; avoid using remote starts wherever possible.

Remember, the smooth starting and stopping operation of the engines not only depends on following the systematic procedure set by the manufacturer, but also on proper maintenance and overhauling procedures.

ENGINE MONITORING

On ships, it is important to check the performance of the engine from time to time to ascertain the engine's working condition, and of course in the event of fault finding. Previously, the observation of marine engine performance was undertaken manually, but with the advancement of technology, automatic monitoring systems have either been retrofitted or installed as standard on new build vessels.

Engine monitoring systems

With the help of engine monitoring systems, the performance of the engine can be observed easily. This modern technology provides two types of monitoring systems. In the first system, the engine performance is monitored continuously and is thus known as online monitoring. With the second type of system, the marine engineers must manually insert the monitoring instrumentation into the cylinder head, connect the wire to the rpm sensor, and then take readings manually. These readings must then be transferred to the engine management system. On most modern ships, the main engine has an online diesel performance system, whereas diesel generators have the manual monitoring system. The type of system that is installed depends on the ship owner as well as the type of ship and the engines installed. The online system is quite costly when compared like for like with the manual system. That said, with the online system, diesel performance can be observed remotely from the engine control room as well as in the chief engineer's cabin. The system also provides various graphical representations of key operating parameters, which enables the engineers to precisely analyse the condition of the engine. These graphs are akin to the indicator cards plotted by the manual system. From the graphs obtained, various characteristics such as engine timing, compression pressure, cylinder output, can be easily analysed. They also indicate whether the engine is balanced, if some units are overloaded, and whether the timing must be adjusted. This information ensures the necessary maintenance and adjustments are conducted in a timely fashion to avoid engine failure or damage to the engine.

In most cases, the diesel performance of the main engine and of the auxiliary engines is taken once every month, with the report then analysed accordingly. A copy of the report must be retained on board, with an additional dispatched to the company's technical department together with the chief engineer's comments on the report. The technical department then checks the validity of the report, providing any feedback as required. The diesel performance reports are kept as records so that they can be compared with previous reports to identify any potential trends in operational degradation. If the reports show a consistent (or even intermittent) downward trend, this is likely to be indicative of a developing malfunction, poor maintenance, or the need for necessary parts to be replaced or adjusted. Although comparatively more expensive than the manual monitoring system, the online monitoring system provides far greater accuracy and frees up engineer time to focus on other critical tasks. This invariably helps in achieving more reliable and efficient marine engine operations. It is worth noting that good practice dictates the ship's engineers can conduct the manual monitoring procedures in the event the online monitoring system malfunctions.

As discussed earlier, the online monitoring system provides various improvements and efficiencies over the manual monitoring system. These include efficient and reliable engine operation, optimisation of Specific Fuel Oil Consumption (SFOC) monitoring, predicting necessary repairs and the prevention of engine failure through machinery malfunction, and reducing spare parts inventory through smart management, which in turn increases the time between overhauls.

Understanding the indicator diagram

Indicator diagrams are used to assess the performance of each unit of the ship's main engine. Each unit's performance is based on the indicator diagram from which the overall performance of the engine is assessed. Indicator diagrams are taken at regular intervals of time and are cross-referenced against those of the ship's sea trial diagrams to check whether there is any significant difference in performance. If any differences are identified, it is important that these are investigated and where necessary rectified before starting the engine. There are four types of indicator diagrams: (1) power card, (2) draw card, (3) compression diagram, and (4) light spring diagram. With the help of these diagrams, we can determine and interpret the following information: (1) the compression pressure inside the cylinder, (2) the peak pressure generated inside the cylinder, (3) the actual power generated by the cylinder, (4) any faulty combustion chamber parts (such as worn-out pistons, liners, rings), (5) any faulty injection parts and or wrong fuel timings, and (6) any faults in the exhaust and scavenging process. High loading is always to be prevented on main engine units lest it leads to faults developing such as bearing damage, cracking, and so forth. It is therefore critical to read these diagrams correctly, as they provide valuable information about the cylinder working pressures and loads. On older ships, it was widespread practice for the indicator diagram to be taken with the help of a mechanical indicator which was to be fitted on top of the indicator cocks. Nowadays digital pressure indicator instruments, referred to as data acquisition units, are used. These typically come as compact handheld devices. A pressure transducer is mounted on the indicator cock and then connected to the handheld unit. Once connected, the unit transmits the data directly to the engine room monitoring computer system. An incremental encoder is fitted to the engine and plugged into the data acquisition unit during the time of operation. This provides accurate data about the position of the top dead centre, or of the crankshaft angle.

The preparation and procedure for taking the indicator diagram using the data acquisition unit is as follows:

- Check the battery of data acquisition unit and change or charge if needed.
- Prepare the digital pressure indicator instrument and visually check the wires and sensors.
- Don the appropriate personal protective equipment such as high-temperature gloves and eye protection.
- Take readings of all relevant engine parameters.
- Ensure the ship, and its engine, is running at a constant speed in open sea. Readings are best taken in calm weather as swells and waves can influence engine loading.
- Use the correct tools to open the indicator cock valve.
- Connect the probe from the incremental encoder to the data acquisition unit.
- Connect the pressure transducer probe to the data acquisition unit.
- Carefully open the indicator cock of the cylinder for a few seconds and blow out the cylinder. This is done to remove any impurities such as soot and other combustion particles from inside the cock.

- Fix the pressure transducer unit to the indicator cock and open the cock to register the cylinder data.
- Repeat the procedure for all cylinders.
- On completion of the tests, disconnect the pressure transducer probe and stow it aside to cool down.
- Disconnect the incremental encoder probe from the data acquisition unit.
- Complete the required data in the digital pressure indicator software and wait for the results to be generated.

It is possible that the digital pressure indicator instrument may not be available on all ships or is not working. In these situations, a mechanical engine indicator device is available which consists of springs, drums, and a pointer to draw the diagram from the engine cylinder's pressure via the indicator cock.

- Don the appropriate personal protective equipment such as high-temperature gloves and eye protection.
- Take the reading of all the relevant engine parameters.
- Ensure the ship, and its engine, is running at a constant speed in open sea. Readings are best taken in calm weather as swells and waves can influence engine loading.
- Use the correct tool to open the indicator cock valve.
- Take the paper provided with the instrument and fix it firmly over the drum.
- Carefully open the cylinder indicator cock for a few seconds then blow out the cylinder. This is done to remove any impurities such as soot and other combustion particles from inside the cock.
- Fix the instrument onto the indicator cock so that the cord is firm.
- Draw the atmospheric line with the cock shut.
- Slowly open the indicator cock and lightly press the stylus against the paper. Make straight vertical lines as the piston moves up and down and then pull the boiler string, till the cycle is drawn on the paper.
- Close the indicator cock and remove the instrument.
- Ensure the tool is not exposed to elevated temperatures for an extended period, as this might adversely affect the mechanical components such as the springs. Elevated temperatures will also affect the accuracy of the stylus.
- Similarly, record the compression pressure line with the fuel cut-off.

The indicator diagram shown below (Figure 5.1) is a normal diagram (i.e., a diagram taken before the use of the engine). We can use this to compare the diagrams that are drawn after the engine was started to identify any deficiencies. We will now look at some of the common defects found in indicator diagrams.

When the diagram in Figure 5.1 is compared with the general graph, the compression pressure is normal, and the maximum firing pressure is too high. This can be indicative of the following:

Deficiency Type I

- Early injection
- The result of incorrect fuel timing of the cams
- Wrong variable injection timing (VIT) setting
- Leaking fuel injector

Main engine pre-start checks and monitoring 77

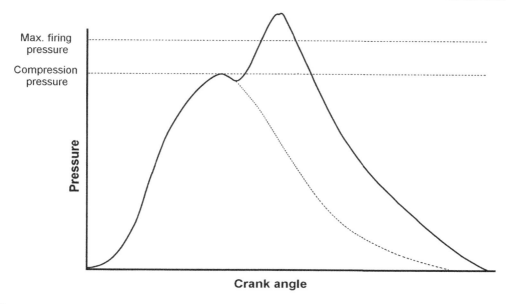

Figure 5.1 Deficiency type I.

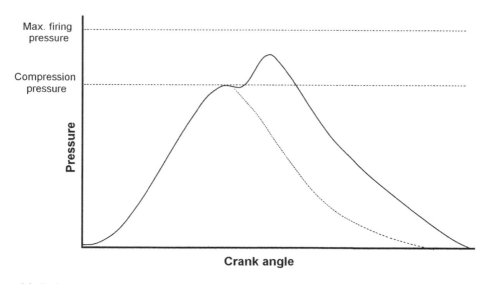

Figure 5.2 Deficiency type II.

Deficiency Type II (Figure 5.2)

- In this diagram, the compression is the same, but the peak pressure is too low. This effect can be a result of any of the following:
- Inferior quality fuel
- Blocked fuel injector nozzle
- Leaking fuel pumps
- Low fuel pressure

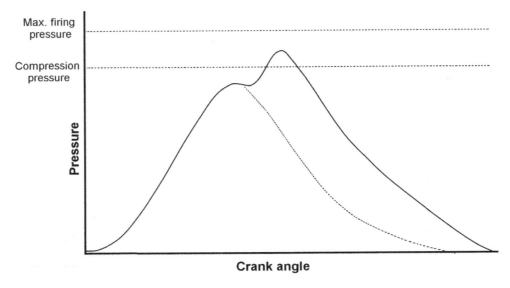

Figure 5.3 Deficiency type III.

Deficiency Type III (Figure 5.3)

- This diagram shows us that the compression pressure is low, and the peak pressure is also too low.
- This may be indicative of the following:
- Leaking exhaust valves
- Leaks through the piston rings, i.e., broken, or worn-out piston rings
- High liner wear
- Burnt piston crown
- Low scavenge pressure

Deficiency Type IV (Figure 5.4)

- This diagram shows a high compression pressure together with high peak pressure. This is usually caused by the following:
- The exhaust valve opening too late (i.e., incorrect exhaust valve timing)
- Overloading of the engine

POWER BALANCING

One of the most important considerations for ensuring maximum efficiency of the marine engine is to control the power generated by each of its cylinders. This involves a process of making fine adjustments; the aim being to achieve equal power from each cylinder. This process is referred to as power balancing. Power balancing is achieved by making minor adjustments to the fuel pumps of the individual cylinders. The quantity of fuel that

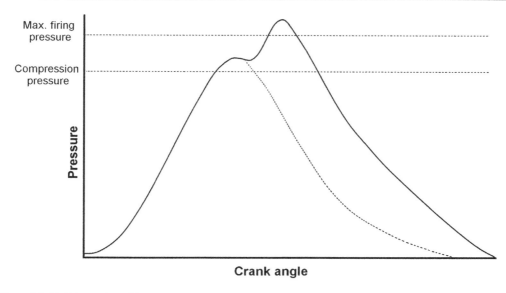

Figure 5.4 Deficiency type IV.

is injected into the cylinder is the most crucial factor in achieving power balancing. These small adjustments made to the fuel pumps should be such that the units are not overloaded, and the exhaust temperature does not exceed safe limits. It is therefore necessary to exercise extreme caution when conducting adjustments for power balancing. When preparing to adjust the fuel pumps, always ensure the fuel pump rack is in the correct position and the exhaust and cooling water return temperatures are within the acceptable range. It is important to note that not all cylinder units show equal exhaust temperatures. However, for each engine, the figures follow a certain path which can help in accessing a situation. Peak or maximum pressure of the cylinders should also be checked along with the cylinder temperatures. If proper care is not taken during power balancing, the engine can become unbalanced, leading to other serious problems including (1) overloading of bearings and running gears, (2) overheating or bearing failure; (3) piston blow past; (4) overheating or piston failure; (5) vibration followed by metal fatigue; (6) fatigue cracking in the bearings, studs, and bolts; (7) cracking in the crankshaft; and (8) failure of holding down bolts. To avoid these faults from developing, the engine room officer of the watch must maintain a regular check of the relevant temperatures and pressures (i.e., the exhaust and cooling return temperatures) and of the lubricating oil and turbocharger pressures. Any unusual noises or vibrations must be investigated immediately. Moreover, the officer of the watch must keep an eye on the exhaust for any kind of smoke. The turbocharger should be kept running smoothly without surging or panting. The fuel pump settings should be checked periodically and measurements of the clearances and timings taken when the engine is not working. It is equally important to ensure that the fuel injectors are checked at regular intervals, and certainly after cleaning and testing. Any faulty injector will not only cause loss of power but will also lead to the overloading of the other cylinders as the governor tries to maintain the normal total power output. To avoid any problems developing before, during, and after power balancing, always ensure maintenance of the engine is conducted at regular intervals and record any deviations from the normal running speed. Should any deviations be found, these must be investigated and rectified at the earliest possible opportunity to avoid further engine damage.

In this chapter, we have discussed the main pre-start checks and monitoring procedures for the main engine. It is important that these checks are carried out each time the main engine is started to avoid any problems from developing once the main engine is fully operational. In the next chapter, we will turn our attention towards operating the main engines to reduce fuel consumption through a process called slow steaming.

Chapter 6

Slow steaming and economic fuel consumption

Increasingly, companies are adopting slow steaming to save fuel costs. In the bulk carrier market, it is quite normal for ship owners and charterers to instruct the vessel to move at slow or economy speed towards their destination. This approach is often taken when shipping companies are finalising the terms of the next charter (including extensions to the existing charter). Occasionally, the charterer themselves may demand that the vessel proceeds at a slow speed, subject to a relevant clause being inserted into the charter party. Normally, ships carry a document which sets out the speed versus consumption parameters of the vessel at various rpm and in different ballast and load conditions. It is common for charterers to ask for this data prior to agreeing to the terms of the charter party. Given the rising costs of marine fuel, it is increasingly common for charterers to also request visibility of the ship's slow steaming data. Where the ship is expected to undertake an extended period of anchorage, or where the cargo to be carried is not time sensitive, it is often more economically advantageous for the charterer to run the ship as slowly as possible. This not only saves fuel but also expensive anchorage costs. Despite the demands placed on the vessel by the shipowner and or charterer, the chief engineer has a duty towards their machinery and must ensure the main engines (and indeed all equipment and machinery on board) are operated properly without compromising their safety and preventing long-term damage. Thus, it is essential that the charterer is advised of the correct and safe economical speed and rpm. Where such data does not exist (typically with older ships), then it may be necessary to hold sea trials at the start of the charter.

Depending on the vessel type and engine, there are distinct categories of slow steaming. All conventional engines (i.e., those that are not 'intelligent engines' or cam-less engines with electronic fuel injection) can be run in one of three slow steaming modes. These are (1) *low rpm with the auxiliary boiler cut-off and the auxiliary blower cut-off*. In this mode, steam demand is managed entirely (100%) by the exhaust boiler after optimising steam usage. The main engine turbocharger is set to cope with the air demand, and the oil-fired boiler is cut off. (2) *Low rpm with the auxiliary boiler firing intermittently, and the auxiliary blower is cut off*. In this mode, steam demand is managed mostly by the exhaust boiler (70%>80%), with the oil-fired boiler assisting in between and firing intermittently. The main engine turbocharger is set to cope with the air demand. (3) *Low rpm with the auxiliary boiler firing frequently and the auxiliary blower set to cut in and running*. As the exhaust temperatures have fallen, steam demand is met by the oil-fired boiler firing frequently. The main engine turbochargers cannot cope due to the reduced enthalpy of the exhaust gas, and the auxiliary blowers are running. Due to inclement weather and commercial pressures, it is sometimes necessary to run the main engine at manoeuvring rpm with the auxiliary blower cutting in. In these situations, the blowers are put into manual mode to avoid their cutting on and off. Doing so can damage the motor due to the repeated starts and the associated high starting current. Thankfully, these are rare occasions, and the chief engineer should advise the master against continuous operation unless it is unavoidable. When trying out the engine for low-load operation, it is normal to try to run the main engine at the slowest rpm at which the exhaust boiler can cope up to and the auxiliary blowers are off. It is equally important to ensure there is no sudden load change caused by a course alteration or change in weather or sea conditions. Conditions such as these will lower the load and allow the auxiliary blower to cut in intermittently. It is standard practice for charter party agreements not to allot any fuel allowance for the firing of the boiler. To avoid falling foul of the charter party, it is necessary to submit these rpm data readings to the charterer.

To provide an example, the following (see Table 6.1) are the main engine characteristics recorded during an actual sea trial conducted at sea on a fair-weather day. The purpose of the trials was to collate and report the economy rpm/speed to the vessel's charterers. The vessel in question is a 73,000 mega tonne deadweight Panamax bulk carrier fitted out with a six-cylinder MAN B&W 6S50MC-C main engine with a Maximum Continuous Rating (MCR) of 14,100 bhp at 119 rpm. The vessel has a rateable fuel consumption of 35 tonnes per day at a sea speed of 13 Kt at 85% MCR (equal to 11,990 bhp at 113 rpm). In this instance, the engine turbocharger exhaust outlet temperature is a limiting factor, as at 330°C (626°F), the elevated temperature alarm is generated and resets at 324°C (615°F). The other limiting factors to be aware of are the high cylinder exhaust temperatures, low scavenge pressure, and boiler pressure. The oil-fired boiler fires up when the pressure drops below 5.5 bar.

The following remarks were recorded in the sea trials report: (1) the main engine can be run in a stable condition at 99 rpm onwards, over a load index of 55; (2) the main engine can run at 98 and 97 rpm if the load index is 55 and above; (3) the main engine can be run

Table 6.1 Main engine characteristics based on actual sea trials

Main engine rpm	Load	Turbocharger exhaust out temperature	Turbocharger exhaust temperature alarm	Scavenge pressure	Boiler pressure	Turbo-charger rpm	Mean exhaust temperature	Highest exhaust temperature	Auxiliary blower status	Ship speed
99	55	322	Off	1.31	6.1	9,000	365	379	Off	13.4
98	54.5	324	On/Off	1.18	6.1	8,800	362	379	Off	13.3
97	54	328	On/Off	1.07	6.1	8,500	358	380	Off	13.2
96	53	331	On	0.97	6.1	8,400	350	381	Off	13.1
95	51	334	On	0.90	6.1	8,200	365	382	Off	12.8
94	50	339	On	0.87	6.2	8,000	366	388	Off	12.7
93	49.5	344	On	0.73	6.2	7,800	367	384	Off	12.6
92	49	344	On	0.71	5.9	7,500	368	384	Off	12.5
91	48	347	On	0.70	6.1	7,400	369	384	Off	12.4
90	47.5	345	On	0.68	6.1	7,200	365	378	On	12.2
89	46	332	On	0.75	5.9	7,350	350	363	On	12.2
88	45	326	On	0.70	6.1	7,000	345	358	On	12.1
87	44.5	324	On	0.67	6.1	6,900	344	356	On	12.0
86	43	324	Off	0.63	6.1	6,700	343	353	On	11.8

continuously at or below 86 rpm with the auxiliary blowers and two generators running; (4) the main engine cannot run between 86 to 98 rpm corresponding to a load index of 43 to 55; (5) the auxiliary blower starts at around 90 rpm when the load falls to 47 at 0.68 bar scavenge pressure; (6) the boiler pressure is stable up to 87 rpm below when it starts falling; (7) at 99 rpm, the fuel oil consumption will fall between 28 to 29 tonnes per day, weather depending; (8) the actual ship's speed may be less than that recorded as measurements were done for short intervals only; (9) the actual speed is weather dependent; and (10) the actual weather conditions at the time of the trials were:

- Heading   140°
- Wind   Westerly 4
- Swell   Westerly 4
- Current   Northwest 0.4

From the information provided above and in Table 6.1, we can begin to infer certain data characteristics. First, we can deduce that where the load is above 55% MCR, irrespective of rpm, the main engine can be run safely. Second, between a load of 54% and 44% MCR, the main engine cannot run as the turbocharger exhaust outlet temperature becomes dangerously high. Third, the main engine can be run at and below 43% load corresponding to an approximate 86 rpm with the auxiliary blowers continuously on and two generators running. This extra fuel consumption (equating to 2 tonnes per day) slightly offsets the savings made. Thus, on analysing the previous data, we can infer that, after considering the advice of the manufacturer and the associated risks and maintenance involved in slow steaming, the engines can be run at 99 rpm, with a representative fuel consumption of 28 tonnes per day. This equals an approximate saving of 7 tonnes of fuel per day, amounting to $6,062 (€5,746, £4,912, as at Rotterdam, April 2022).

PREPARING TO SLOW STEAM

Traditionally, main engines are designed to run between 75% to 85% load range during continuous operation. However, to run the ship's engine for slow steaming, several precautions need to be taken to run the marine engine at low loads. In this section, we will discuss the various checks and precautions that need to be taken for preparing the marine engine for slow steaming. With traditional marine engines (except intelligent engines) there are several checks and procedures that are necessary before commencing slow steaming. For example, it is necessary to carry out frequent scavenge inspections and under-piston-area checks; inspect the piston rings for signs of breakage, fouling, and lack of springiness; inspect and clean the exhaust boiler (consider using high-pressure jet machines for effective cleaning); check the cylinder lubrication rate and inspect the liners and pistons for over and under lubrication and scuffing; check the turbocharger rpm as well as the scavenge air pressure (any drop in rpm or scavenge air pressure at the same load may indicate fouling of the turbocharger); check the temperature difference of the exhaust gas between the turbocharger exhaust inlet and outlet (any reduction in the difference may indicate fouling of the turbine); and check the funnel stack temperature after the exhaust gas boiler, as any gradual increase in temperature at the same load and any decrease in steam pressure may indicate fouling of the exhaust boiler tubes. Any sudden increase may indicate a minor fire; frequently conduct indicator card checks and inspections of main engine performance, and conduct frequent drainage of the air cooler of water.

It is a well-established fact that most breakdowns related to slow steaming occur not during the slow steaming itself but when the engine is expected to return to its normal range. To avoid engine breakdowns following a period of slow steaming, it is crucial that certain precautions and routines are conducted diligently during the period of slow steaming. It is strongly advised to maintain the jacket cooling water at an optimum temperature of between 80°C and 85°C (176°F–185°F), unless stipulated otherwise by the engine manufacturer, and where possible, to avoid large fluctuations in the cooling water temperature. Doing so helps reduce thermal stresses on the liner and cold corrosion. In load-dependent cylinder lubricators, slow steaming may lead to reduced feed rates, hence a more suitable higher base number (BN) cylinder oil may be needed to protect against corrosion. Commonly, a reduction of engine load from 90% to 30% increases the residence time (i.e., the time spent inside the cylinder for each charge of cylinder oil) by as much as a factor of three. This means higher BN cylinder oil must be used. Where a lower BN oil is used, a higher feed rate must compensate. Where the ship is expected to ultra-slow steam, this must be conducted with the auxiliary blowers running. In so doing, extra electric motors must be supplied and kept on board. Freshwater generation will fall due to the reduced heat load. This means considering preheating the jacket water to generate sufficient cooling media and avoid water purchase costs. The exhaust gas temperature after the exhaust gas boiler should not be allowed to fall below 220°C (428°F). This is required to keep the exhaust above the dew point for sulphuric acid. Regular engines load up should be done at least every second day to around 80% to 85% of MCR, to prevent the fouling of the exhaust gas boiler and the exhaust manifold. Doing so will also burn away any unburned fuel and oil residues left over in the exhaust manifold.

Dry washing of the turbine wheel and washing out of the compressor must be conducted during the load up procedure. Soot blowing of the exhaust gas boiler must also be conducted during this period. Avoid the build-up of water condensation in the air coolers and aim to maintain the scavenge air temperature around 40°C to 45°C (104°F–113°F). Maintain the hot well temperature by cooling water control of the condenser and directly allowing some condensate to flow through the hot well bypass valve. Always use the correct cylinder oil feed rate as per the manufacturer's recommendations. Only ever use cylinder oil having the correct BN as recommended by the manufacturer. Thorough maintenance must be conducted on the fuel injectors and, where necessary, revised maintenance intervals should be issued. This is due to the increased incidence of fouling and dripping during slow steaming. Cold corrosion can be caused by low exhaust temperatures during exceptionally low-load operations. Care should be taken to avoid exhaust temperatures after the cylinder drops to below 250°C (482°F). This figure is particularly important, as the temperature will drop further after the extraction of heat in the exhaust boiler. Frequent washing of the exhaust gas boiler and extra soot-blowing routines are required to ensure the exhaust gas boiler functions at optimum condition. The main injection viscosity of the fuel oil should be maintained between 12 to 13 kinematic viscosity (CST). Maintaining a higher low temperature (for example, in the central cooling plants) will facilitate optimum scavenge temperatures and jacket cooling water temperatures. The freshwater generator may need to be bypassed to maintain the jacket water temperature on some ships. Keep the auxiliary blower continuously on (in manual mode) to avoid elevated exhaust temperatures after the cut-off and before the cut-in period. Exhaust temperatures above 450°C (842°F) can cause hot corrosion and burning of the exhaust valves. Finally, it should be acknowledged that low-load operation can cause unburned fuel and cylinder oil to accumulate in the exhaust manifold. If left unattended, this may suddenly burn, causing subsequent overspeeding and damage to the turbocharger when the load is increased. To avoid this from occurring, conduct frequent exhaust manifold inspections and remove all oil residues.

OPTIMISING THE MAIN ENGINE FOR SLOW STEAMING

In the previous section, we discussed several concerns that marine engineers have regarding slow steaming. As low-speed marine engines are not traditionally suited for prolonged slow steaming, various precautions are needed to be taken in the event slow steaming operations are adopted without engine modification. In this section, we shall discuss the checks to be done, the additional maintenance required, and the precautions to be taken to prevent long-term damage to the ship's machinery. When considered together, these procedures help to optimise the marine engine for slow steaming operations.

Optimisation of the ship's main engine

Traditionally, main engines are designed to run at between 70% to 85% load range during continuous operation. The matching and designing of all the auxiliaries are based on this load range operation. The exhaust boiler size (i.e., surface area) is determined based on the exhaust temperature, volume of exhaust gas flow, and the waste heat recovery within this range. Low-load operations make this waste heat recovery system ineffective as there is lower production of steam, which in turn increases the load on the oil-fired boiler. The air cooler size (again, the surface area) is selected based on the heat load of the air within this operating range. During low-load operations, the cooling water to the air cooler needs to be controlled by bypassing the cooler and throttling the water valves to maintain optimum scavenge air temperature. Too much throttling of the water valves reduces the flow velocity of the cooling water thereby increasing the deposit rates of precipitants, leading to fouling and contamination of the tubes. The turbocharger selection and matching to the main engine are based on the enthalpy of the exhaust gas that needs to be extracted. The other selection criteria are the quantity of the scavenge air that needs to be supplied to the cylinders for optimum combustion. The turbocharger is selected for normal running load range of between 70% to 85%. Low-load operations on the main engine lead to a lower running rpm of the turbocharger and reduced generation of scavenge air. This leads to incomplete combustion, causing increased fouling, and renders cleaning measures such as turbine dry grit cleaning ineffective. The propeller is designed to provide maximum efficiency for the rpm within this range. Due to the lower rpm, propeller efficiency may become affected. Additionally, specific fuel oil consumption is also optimised for running within this range. Even though the fuel consumption is lower in totality, the specific fuel oil consumption is higher at part loads as fuel injection and combustion are not complete.

Other factors that need to be considered when optimising the main engines for slow steaming include consideration of the fuel injectors and fuel pumps. The fuel injectors and fuel pumps are designed for this range thus atomisation and penetration may be affected at low-load operation. The engine's operating parameters and their alarms and monitoring systems are designed for this range. Hydrodynamic lubrication is rpm dependent, and the grade of oil and its properties, such as viscosity, are selected for this range. The shaft generators are designed and selected based on this range. Low-load operation may render shaft generators unusable. Thus, running the main engine below its normal operating range of 70% to 85% MCR means the entire system is not optimised. If engine modifications and retrofitting is conducted on the main engine, then the vessel is safe for slow steaming as well as ultra-slow streaming. That said, we have limited ourselves to discussing slow steaming without any engine modifications in this section. In summary, slow steaming up to 50% to 55% load can be done on most engines without causing long-term harm provided certain precautions are taken. That is the point above which the auxiliary blowers cut in.

Now that we have covered the main principles of slow steaming, we can turn our attention to economical fuel consumption. It should be obvious that whereas slow steaming does provide distinct advantages in terms of saving fuel costs, it comes at the price of potentially causing long-term damage to the ship's machinery. An alternative method of reducing expenditure on bunkerage, without causing potential long-term damage to the ship's machinery, is to consider economical fuel consumption. This is done using performance curves, which we will now discuss.

DEFINING ECONOMICAL FUEL CONSUMPTION

After the construction of the ship is complete and prior to handing her over to her new owners, sea trials are conducted to determine whether the ship can deliver the contractual speeds. The primary purpose of the sea trials is to determine the speed of the ship with reference to the rpm in accordance with the power produced. In addition to evaluating the hull of the ship, the important machineries of the engine room such as the boilers, auxiliary engines, and the main engine are also thoroughly evaluated. Machineries have a test record which is developed separate from the sea trials data. This is collated within the manufacturing plant and is called as test bed data. In most instances, the main engines, generators, motors, and pumps will all have their own test bed data. Once the sea trials are completed, the data collated from the trials are compared to the test bed data. Any discrepancies are then investigated. Assuming the test bed data and sea trials data corroborate each other, a set of performance curves are drawn up. It is these performance curves which enable the chief engineer to run the ship's machinery safely and economically. As mentioned previously, it is standard practice within a charter party to agree and fix the ship's speed and fuel consumption amongst other things. There is often extraordinarily little margin for error, and if the ship cannot perform at the agreed speed, this may give rise to a speed claim. Moreover, if there is overconsumption of fuel to maintain the agreed speed, this too may lead to a fuel claim. Both the speed and fuel claim are tantamount to a breach of contract, which may be remedied by the payment of compensation. Where bunkerage costs are high, any such claim could effectively wipe out any profit from the charter. To prevent this from happening, it is essential that the main engine performs satisfactorily and gives the rated power at the rated rpm within narrow but permissible limits (usually in terms of temperature and pressure at the correct specific fuel oil consumption). In addition, lubricating oil and cylinder oil consumption must also be kept within allowable consumptive limits.

Performance curves

During the test bed or shop trials, the performance curves of the engine are developed. The performance curves (see Figure 6.1) are a set of graphs demonstrating different parameters set on an x-axis plotted against the engine power or load on the y-axis. These different

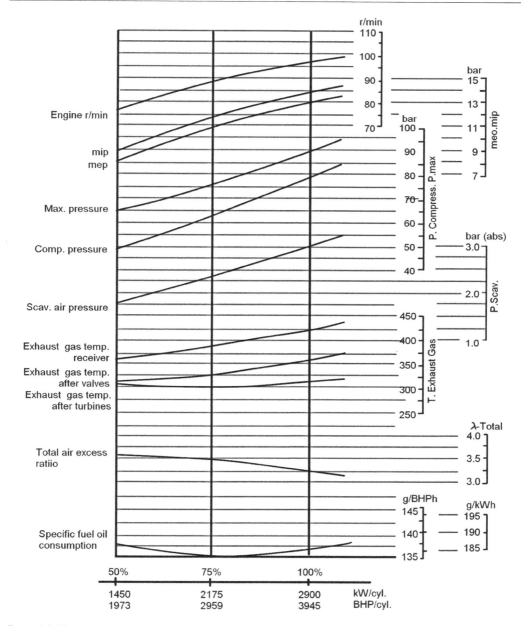

Figure 6.1 Typical example performance curve for a slow-speed, two-stroke diesel marine engine.

plotted curves are as follows: (1) *Engine rpm vs. load*. This curve helps in ascertaining whether the main engine is overloaded or not. A higher power generated at a lower rpm indicates an overloaded main engine. (2) *Mean effective pressure vs. load*. Mean effective pressure is used to calculate horsepower; hence, these two values should co-relate. In the event they do not, then there may be some error in the calculation or the instrumentation. (3) *Maximum pressure vs. load*. This curve helps in knowing the condition of the fuel injection equipment, the injection timing, and the compression within the cylinder. (4)

Compression pressure vs. load. This curve indicates the condition of the parts responsible for maintaining compression such as the piston, piston rings, and exhaust valves. (5) *Scavenge air pressure vs. load.* This curve indicates the condition of the turbocharger and its associated equipment. (6) *Exhaust gas temperature in the receiver vs. load.* This curve indicates the enthalpy of the exhaust gas prior to entry into turbocharger. This value, when compared with the value after the turbocharger, provides the temperature drop across the turbocharger. This serves as an indicator of turbocharger efficiency. (7) *Exhaust gas temperature after the exhaust valve vs. load.* This curve demonstrates quality of combustion, fuel injection, timing, compression. A higher temperature may be caused by after burning. (8) *Exhaust gas temperature after the turbocharger vs. load.* This curve is especially useful as it indicates the enthalpy captured from the exhaust by the turbocharger and hence its condition. In the event the receiver temperature is within range, but the outlet temperature is higher than expected, this may be indicative of fouling of the turbocharger and hence the associated lower scavenge air pressure and high exhaust gas temperature. (9) *Total excess air ratio vs. load.* This curve is scarcely used by ship staff and is useful for design engineers. The curve demonstrates scavenging and the turbocharger capacity and condition. It shows that as the power increases, the excess air decreases due to consumption. (10) *Specific fuel oil consumption vs. load.* This curve helps to counter-check whether the engine is consuming fuel oil as expected in accordance with the load. In addition, other parameters may be required and or listed as per the manufacturer's instructions.

Economical fuel consumption

A ship's main engine will run economically if the engine is well maintained and is run at the rated economic rating where the specific fuel oil consumption is at the lowest margin. An engine is said to be performing well or is well maintained if it can be safely run at the rated rpm at the rated load. For instance, if an engine is having a continuous service rating of 15,000 bhp at 104 rpm but cannot reach the rated rpm and is developing 15,000 bhp prematurely at 98 rpm, this is indicative of a loss of the ship's speed. The shipowner is subsequently liable for a speed claim. It also tells us that there is a problem – i.e., the ship cannot give speed, it is overconsuming fuel, and the engine is overloaded. This points to either hull fouling, a damaged propeller, or a faulty prime mover. In such cases, the careful study of the sea trial data, engine shop trial data, and performance curves will help to determine the cause of the problem. For troubleshooting, first, the main engine performance must be determined on a mild weather day when the engine load is steady. The main engine must be run to its rated power. Thereafter the data found must be superimposed on the performance curves. After superimposing the measured parameters on the performance curves, we will know whether the parameters are normal or abnormal. A complete study of the parameters will help us to pinpoint the probable cause of the problem. An example of the performance data superimposed on the performance curve is given in Figure 6.2.

From the diagram in Figure 6.2, the following points may be inferred: (1) at 75% MCR, the rpm attained is lower than during the sea trials; (2) the average maximum cylinder pressure P max is lower than during the sea trials; (3) the compression pressure P comp is the same as the sea trials, confirming that the running gears (such as the piston, piston rings, and exhaust valves) are functioning normally; (4) the scavenge pressure is almost normal, suggesting that the turbocharger is in a satisfactory condition, and the enthalpy of the exhaust gas is higher than normal for this rpm; (5) the exhaust temperatures are all

Slow steaming and economic fuel consumption   89

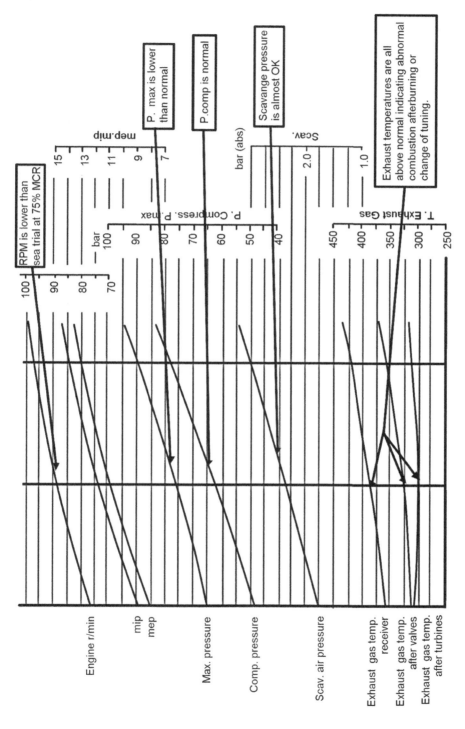

Figure 6.2 Example superimposed performance curve for a slow-speed, two-stroke diesel marine engine.

increased, suggesting either abnormal combustion, after burning, or a change of timing. It may also indicate faulty fuel injection equipment.

Hopefully, the example will help us to understand the use and benefit of performance curves. After the main engine performance has been taken and plotted on the original performance curves from the sea trial data, any problems can be identified and the specific fuel oil consumption restored to normal values. By doing so, at any stage during the lifetime of the ship, we can determine why she is not performing as expected by comparing the test bed data against the parameters plotted on the performance curves.

In the next chapter, we will examine the exhaust gas system and a relatively recent introduction, the scrubber.

Chapter 7

Exhaust gas system and scrubbers

The work performed by the marine engines to keep the plant running for propelling a ship requires the burning of fuel. The energy converted inside the cylinder of the engine is not a 100% efficient conversion as part of it is lost in the form of exhaust gases. The exhaust gas system of modern marine engines is designed in such a way that the unused gases coming out of the cylinders are redirected to the turbocharger and exhaust gas boiler (EGB) to recover this waste energy. To use the maximum energy from the waste gases, the exhaust system consists of various components, including the exhaust gas pipes, EGB, the silencer, the spark arrester, and a series of expansion joints. In this chapter, we will briefly discuss the role and function of each of these main components.

EXHAUST GAS PIPING

The exhaust gas piping system (see Figures 7.1 and 7.2) conveys the exhaust gases from the outlet of the turbocharger to the atmosphere. When designing the exhaust piping system, the following critical parameters must be taken into consideration:

- The exhaust gas flow rate
- The maximum back force from the exhaust piping on the turbochargers
- The exhaust gas temperature at the turbocharger outlet
- The maximum pressure drops within the exhaust gas system
- The maximum noise levels at the gas outlet to the atmosphere
- Sufficient axial and lateral elongation ability of the expansion joints
- Utilisation of heat energy of the exhaust gases

The exhaust gas from the cylinder unit is sent to the exhaust gas receiver where the fluctuating pressure generated from the different cylinders is equalised. From here, the gases which are at constant pressure are sent to the turbocharger where waste heat is recovered to provide additional scavenge air to the engine. The most important thing to consider when designing the exhaust piping system is the back pressure on the turbocharger. The back pressure in the exhaust gas system at specified maximum continuous rating (MCR) of the engine depends on the gas velocity and is inversely proportional to the pipe diameter to the fourth power. It is general ship practice to avoid excessive pressure loss within the exhaust pipes. The exhaust gas velocity is maintained about 35 mps to 50 mps at the specified MCR. The other factors which affect the gas pressure are the installation of the EGB and the spark arrestor, which are typically installed in the path of the exhaust gas travel. At the specified

92 Introduction to Ship Engine Room Systems

Figure 7.1 Ship's exhaust piping system.

Figure 7.2 Ship's funnel stack.

MCR of the engine, the total back pressure in the exhaust gas system after the turbocharger (as indicated by the static pressure measured in the piping after the turbocharger) must not exceed 350 mm WC (0.035 bar). To have a back pressure margin for the final system, it is recommended at the design stage to initially use a value of about 300 mm WC (0.030 bar).

EGB

The maritime EGB is one of the most efficient waste heat recovery systems developed for ships. When the ship's propulsion plant is running at its rated load, the auxiliary boiler may be switched off as the EGB can generate the required steam for various ship systems. The exhaust gas passes through the EGB, which is usually placed near the engine top or in the funnel. The efficiency of the EGB is affected by pressure loss of the gases across the boiler and the parameters governing the pressure loss (for example, exhaust gas temperature and flow rate) are affected by the ambient conditions. The recommended exhaust pressure loss across the exhaust gas boiler is considered at 150mm WC at the specified MCR. If the exhaust system is not provided with additional equipment (i.e., spark arrester or silencer), the pressure loss value may be a little bit higher than the previously stated value (150 mm WC at the specified MCR).

SILENCER

The engine room is the single biggest contributor to noise levels in the ship's accommodation, which is made worse by the fact the exhaust gas piping system is located within proximity to the ship's accommodation block. Acceptable noise levels are governed by the Maritime Labour Convention (MLC) and must not exceed the mandated levels. Where the exhaust gas system is not fitted with a silencer, the noise levels must be continuously determined. This is done by taking soundings approximately 1 m from the exhaust gas pipe outlet at an angle of 30 degrees. If the noise level recorded is within acceptable limits, no further action need be taken by the ship's engineers. If, however, the noise level recorded is out of the acceptable range, this must be addressed immediately as it contravenes the MLC. One method for reducing the noise level is to fit a silencer after the EGB. This then reduces the noise level in the exhaust gas manifold. Conventional silencers consist of absorptive and reactive chambers. They are constructed for a gas velocity of 35 mps. The reactive chamber is only effective at one frequency, although the latest silencer designs consist of three chambers to overcome this limitation. The three elements are composed of a reactive element for attenuation of lower frequencies, a resistive element-absorptive silencer to tackle higher frequencies, and a combination element of both reactive and resistive elements. This setup reduces the noise effectively without increasing back pressure on the turbocharger by tuning the elements to match the engine over the noise range.

SPARK ARRESTOR

Low-load operations on marine engines tend to produce partially burnt carbon deposits and soot within the exhaust gas piping system. As the exhaust gases produced after combustion are rich with oxygen, these partially burnt carbon particles are discharged from the exhaust funnel as a combustible spark. To prevent ignitions within the exhaust gas system, a spark arrester may be fitted within the exhaust piping system to prevent sparks from the exhaust gas being spread over the top decks. The spark arrester is placed at the end of the exhaust gas system of the engine. Newer designs help the gases to create rotatory movements by forcing them to pass through a fixed number of angled-positioned blades. The heavy carbon particles are then smoothly collected in the soot box, which is emptied and cleaned as required. The spark arrester can often be combined with the silencer as one unit to save both space and installation costs. The main disadvantage of the spark arrester, however, is a considerable drop in pressure. For the main engine, it is recommended that the combined pressure loss across the silencer and or spark arrester should not be allowed to exceed 100 mm WC at the specified MCR.

EXPANSION JOINTS

The exhaust gas system typically experiences huge temperature variations. It is not possible to construct the entire exhaust piping system using one single piece, therefore multiple sections are joined together to complete the system. When the engine is at a standstill, the temperature of the exhaust pipe may vary from 10°C to 40°C (50°F–104°F) depending on the surrounding environment or geographical location of the ship. When the engine is up and running, the exhaust system temperature crosses 200°C (392°F). This major temperature variation requires the need of special joints to safely absorb the heat-induced expansions and contractions of the pipes and tubing systems. For this purpose, bellows and expansion joints are used. These are specially designed to ensure they can withstand the stresses caused by temperature fluctuations and avoid cracks brought about by the continuous change in temperature. In accordance with Boyle's Law,[1] when tubing is subjected to high-temperature fluids, the bar pressure also increases. This means expansion joints are needed to bear the extra force that accumulates within the piping. Expansion joints are used in the tubing and piping systems, and bellows are used to connect the exhaust gas pipes to the funnel. The expansion joints are chosen with an elasticity that limits the forces and the moments of the exhaust gas outlet flange of the turbocharger as stated by each turbocharger manufacturer. The expansion joints are placed at various locations throughout the exhaust gas piping system.

EXHAUST GAS SCRUBBERS

Scrubbers or exhaust gas cleaning systems (EGCS) are used to remove particulate matter and harmful components, such as sulphur oxides (SOx) and nitrogen oxides (NOx) from the exhaust gases generated because of the combustion processes in marine engines, to implement pollution control. These scrubbing systems have been developed and employed to treat exhaust from engines, auxiliary engines, and boilers, onshore and onboard marine vessels, to ensure that no damage is done to human life and the environment by toxic chemicals. Sulphur emissions to the atmosphere by oceangoing vessels are limited by international regulations which came into effect starting 01 January 2020 under MARPOL. The IMO regulations mandate that the sulphur content in fuels, which is carried by merchant vessels, must be limited to 0.50% globally and 0.10 % m/m in emission control areas. Before this, the maximum sulphur cap in fuels was kept at 3.5% m/m. Compliance with the new regulations requires that vessels either use expensive low sulphur content fuel or else clean the exhaust gases by using exhaust scrubbing systems. Exhaust gas scrubbers are hence being installed on a substantial number of ships to comply with the 2020 international regulations.

The operating principle of the scrubber system is that exhaust gas streams are pushed through the scrubber where an alkaline scrubbing material is used to neutralise the acidic nature of the exhaust gases and to remove any particulate matter from the exhaust. The neutralised scrubbing material is then collected with wash water which may be stored or disposed of immediately as treated effluent. The cleaned exhaust is passed out of the system and into the atmosphere. The scrubbing material is chosen such that specific impurities like SOx or NOx can be removed through inert chemical reactions. For de-sulphurisation purposes, marine scrubbers use lime or caustic soda, where after sulphur-based salts are produced after treatment. These can be easily discharged overboard as they do not pose a threat to the marine environment. Scrubbers may use seawater, freshwater with added calcium or sodium sorbents or pellets of hydrated lime as the scrubbing medium, due to their alkaline nature. To increase the contact time between the scrubbing material and the exhaust gases, packed beds consisting of gas-pollutant removal reagents (such as limestone) are placed inside the scrubbers. These packed beds slow down the vertical flow of water inside the scrubbers and intensify the exhaust gas cooling and acidic water neutralisation process. Based on their operation, marine scrubbers are categorised as dry and wet scrubbers. Dry scrubbers employ solid lime as the alkaline scrubbing material which removes sulphur dioxide from the exhaust gases. Wet scrubbers use water, which is sprayed into the exhaust gas to achieve the same objective. Wet scrubbers are further classified into closed-loop or open-loop scrubbers. In closed-looped scrubbers, freshwater or seawater is used as the scrubbing liquid. When freshwater is used in a closed-loop scrubber, the quality of water surrounding the ship has no effect on the performance and the effluent emissions of the scrubber. Open-loop scrubbers, however, consume seawater in the scrubbing process.

Wet scrubbers

Inside the wet scrubber, the scrubbing liquid used may be seawater or freshwater with chemical additives. The most common additives are caustic soda (NaOH) and limestone ($CaCO_3$). Scrubbing liquid is sprayed into the exhaust gas stream through nozzles to distribute it evenly. With most scrubbers the design is such that the scrubbing liquid flows downstream, however, some scrubbers may be designed with an upstream movement instead. The exhaust inlet of the scrubber can be made in the form of a venturi, as shown in Figures 7.3 through 7.9. In this design, the exhaust gas enters at the top, and water is sprayed into the high exhaust gas speed areas at the neck or above the neck in the form of a spray.

The exhaust intake is positioned either on the side or at the bottom of the tower. The design ensures that the SOx present in the exhaust gas is passed through the scrubbing liquid; here it reacts to form sulphuric acid. When diluted with alkaline seawater, the sulphuric acid, which is highly corrosive, is neutralised. The wash water is discharged into the open sea after being treated in a separator to remove any sludge and the cleaned exhaust passes out of the system. Mist eliminators are used in the scrubbing tower to remove any acid mist that forms in the chamber by separating droplets that are present in the inlet gas from the outlet gas stream.

96 Introduction to Ship Engine Room Systems

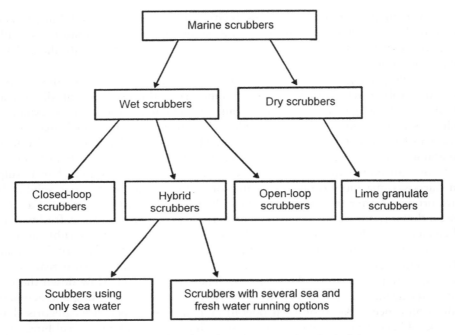

Figure 7.3 Classification of marine scrubbers based on their operational principles.

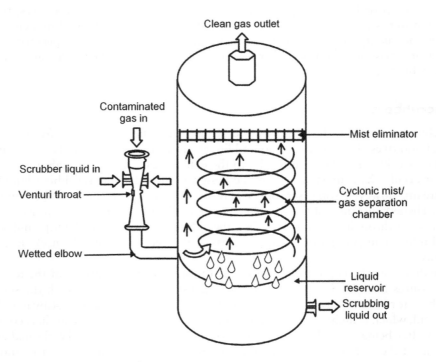

Figure 7.4 High-energy venturi scrubber.

Exhaust gas system and scrubbers 97

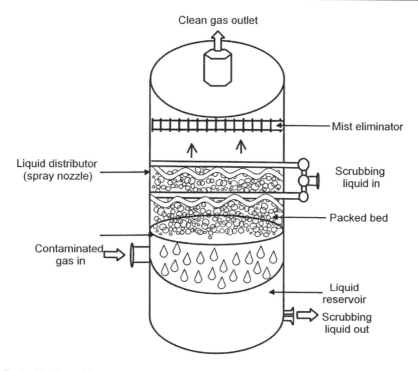

Figure 7.5 Packed bed scrubber.

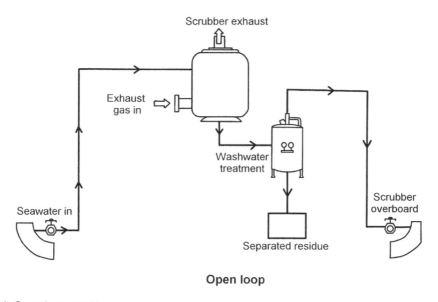

Figure 7.6 Open-loop scrubber system.

98 Introduction to Ship Engine Room Systems

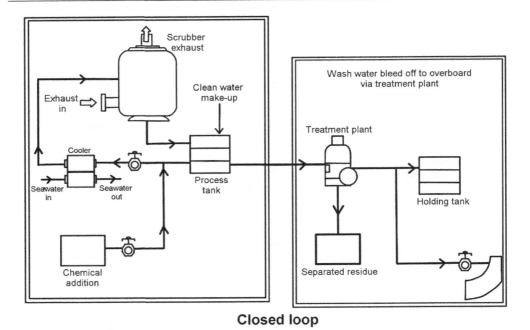

Figure 7.7 Closed-loop scrubber system.

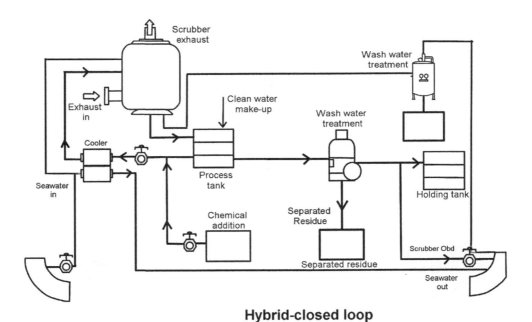

Figure 7.8 Hybrid scrubber system.

Figure 7.9 Exhaust gas scrubber.

MARPOL regulations require that the wash water must be monitored before being discharged out to sea to ensure that its pH value is neither too high nor too low. Since the alkalinity of seawater varies according to a number of reasons, such as the distance from land, volcanic activity, and the marine life present in the water body, wet scrubbers are accordingly divided into two types: the open-loop and closed-loop systems. Both these systems have been combined into a hybrid system, which can employ the most suitable scrubbing action depending upon the conditions of the voyage.

Open-loop scrubber system

The scrubbing system uses seawater as the scrubbing and neutralising medium, which means no other chemicals are required for desulphurisation. The exhaust stream from the engine or boiler passes into the scrubber and is treated with alkaline only seawater. The volume of this seawater depends on the size of the engine and its power output. The system is extremely effective but requires large pumping capacities as the volume of seawater required is significant. An open-loop system works perfectly satisfactorily when the seawater used for scrubbing has sufficient alkalinity. However, seawater which is at a high ambient temperature, freshwater, and even brackish water are not as effective and cannot, therefore, be used. An open-loop scrubber for these reasons is not considered suitable technology for areas such as the Baltic Sea where salinity levels are not sufficiently high.

Reactions involved in the open-loop scrubber system:

SO^2 (gas) + H_2O + ½O^2 → SO^4 2– + ^2H+
(sulphate ion + hydrogen ion)

HCO^{3-} + H+ → CO_2 + H_2O
(carbon dioxide + Water)

The advantages of this system include (1) it has very few moving parts, the design is simple and easy to install on board; (2) other than de-fouling and operational checks, the system requires minimal maintenance; (3) the system does not require storage for waste materials. The disadvantages of this system are (1) cooling of the exhaust gas is a problem faced by wet scrubber systems; (2) the operation of the system depends on the alkalinity of the water available and is not suitable to be employed in all marine conditions; (3) a very large volume of seawater is required to obtain efficient cleaning; this means the system consumes has significant power demands; (4) in emission control area (ECA) zones and ports, expensive and cleaner fuels must be used.

Closed-loop scrubber system

The closed-loop scrubber system works on similar principles to the open-loop system; it uses freshwater treated with a chemical (usually sodium hydroxide) instead of seawater as the scrubbing media. The SOx from the exhaust gas stream is converted into harmless sodium sulphate. Before being recirculated for use, the wash water from the closed-loop scrubber system is passed through a process tank where it is cleaned. The process tank is also needed for the operation of a circulation pump that prevents pump suction pressure from sinking too low. Ships can either carry freshwater in tanks or generate the required freshwater via the freshwater generators present on board. Small volumes of wash water are removed at regular intervals to the holding tanks where freshwater can be added to avoid the build-up of sodium sulphate in the system. A closed-loop system requires almost half the volume of wash water compared to the open-loop version; however, more tanks are required. These include a process tank or buffer tank, a holding tank from which discharge directly to the sea is prohibited, and a storage tank capable of regulating its temperature between 20°C and 50°C (68°F and 122°F) for the sodium hydroxide, which is usually used as a 50% aqueous solution. Dry sodium hydroxide also requires large storage space. Some hybrid systems use a combination of both wet types that can operate as an open-loop system when water conditions and discharge regulations allow and as a closed-loop system at all other times. Hybrid systems are hence proving to be the most popular choice due to their ability to cope with different conditions.

The reactions involved in the closed-loop system are:

2NaOH + SO2 → Na2SO3 + H_2O (sodium Sulphite)

Na2SO3 + SO2 + H_2O → 2NaHSO3 (Sodium hydrogen sulphite)

SO2 (gas) + H_2O + ½O2 → SO42 + 2H + NaOH + H2SO4 → NaHSO4 + H_2O
(sodium hydrogen sulphate)

$$2NaOH + H_2SO_4 \rightarrow Na_2SO_4 + 2H_2O \text{ (sodium sulphate)}$$

The advantages of this system include (1) minimal maintenance requirements and (2) independence of the operating environment of the vessel. The disadvantages are, however, (1) the system requires substantial storage space (i.e., buffer tanks) to hold wastewater until it can be discharged; (2) selective catalytic reduction systems must operate before the wet scrubbers are put into action; and (3) fitting the system together, especially for dual-fuel engines can be quite complex and therefore expensive and time-consuming.

Hybrid scrubber system

The hybrid scrubber system offers a simple solution for retrofitting vessels with scrubbers that are capable of operating on both open-loop and closed-loop configurations. These systems run on an open-loop mode at sea and closed-loop mode in ECA zones and ports. The system can be switched over relatively easily. As the system can run on lower costing fuels for longer periods of time and almost anywhere around the world, the long-term advantages of the system can overcome the high initial installation cost. The advantages of the hybrid scrubber system include (1) being suitable for long and short passages worldwide; (2) ships with hybrid scrubbing systems can spend more time in ECA zones and in port than those with open-loop systems; (3) ships can use lower costing heavy fuel oil all of the time. The main disadvantages of the hybrid scrubber are (1) the need for more structural modifications, (2) large storage space requirements for chemicals and additives, and (3) high installation times and costs.

Dry scrubbers

With these types of scrubbers, water is not used as a scrubbing material at all; instead, pellets of hydrated lime are used to remove the sulphur. The scrubbers are operated at higher temperatures than their wet counterparts, and this has the additional benefit of the scrubber burning off any soot and oily residues left in the system. The calcium present in caustic lime granulates reacts with the sulphur dioxide in the exhaust gas to form calcium sulphite. Calcium sulphite is then air-oxidised to form calcium sulphate dehydrate which, when mixed with water, forms gypsum. The used pellets are stored on board for discharge onshore at ports; however, they are not considered a waste product, as the gypsum formed can be reused as a fertiliser and as construction material. Dry scrubber systems consume less power than wet systems as they do not require circulation pumps. However, they weigh considerably more than wet systems. The reactions involved in the dry scrubber process is:

$$SO_2 + Ca(OH)_2 \rightarrow CaSO_3 + H_2O \text{ (calcium sulphite)}$$

$$CaSO_3 + \tfrac{1}{2}O_2 \rightarrow CaSO_4 \text{ (calcium sulphate)}$$

$$SO_3 + Ca(OH)_2 \rightarrow CaSO_4 + H_2O \text{ (gypsum)}$$

The main advantages of the dry scrubber system over the wet types of scrubber system are (1) there is efficient removal of nitrogen and sulphur oxides; (2) this type of system does not result in the production of liquid effluent which must be disposed of overboard;

(3) the gypsum obtained after the exhaust gas cleaning process can be sold for use in various industrial applications. The disadvantage of this system includes (1) the requirement for significant onboard storage to handle the dry bulk reactants and products associated with the process; (2) there must be a readily available supply of reactants; (3) the reactants used are costly, especially the urea needed for NOx abatement, and calcium hydroxide for SOx abatement. In summary, for a shipping company to select the most suitable kind of scrubber system to be installed on board, there are many factors it must consider. These include the installation spaces available on board, the area of operation and the chartering schedule of the ship, the power and output of the engine and boiler, the availability of freshwater, and the available power on board to run the system in different conditions, amongst many more.

The importance of reducing exhaust emissions cannot be overstated. Historically, the maritime industry has been one of the biggest contributors of NOx and SOx emissions, which are proven to affect human health and cause environmental damage. That said, over the past 20 years or so, the industry has made significant strides in reducing emissions from ships. Sadly, we are still far off from entirely eliminating ship emissions. In the next chapter, we will look at the various lubrication systems used throughout the engine room and how marine engineers should assess the quality of lube oil to ensure the ship's machinery is kept in good working order.

NOTES

1. Boyle's law, also known as the Boyle–Mariotte law, or Mariotte's law in France, is a gas law that describes how the pressure of a gas tends to decrease as the volume of the container increases.

Chapter 8

Engine room lubrication systems

Lubrication is essential for any kind of machinery onboard ships. For example, the lubrication system of the main engine is responsible for lubricating and cooling the internal parts, which, when acting relative to each other, create immense friction and heat, which can result in the overheating of parts. Lubrication not only provides cooling but also the removal of any debris or impurities. There are several types of lubrication systems available including hydrodynamic lubrication, hydrostatic lubrication, boundary lubrication, and elasto-hydrodynamic lubrication. In this chapter, we will very briefly discuss the role and function of each of these lubrication systems. *Hydrodynamic lubrication*. With this type of lubrication, the oil forms a continuous oil film of adequate thickness between the moving surfaces. The film is formed due to the motion of the moving parts and the self-generated pressure. For example, the journal bearings of the main engines have hydrodynamic lubrication. A film is formed between the main bearing and the journal of the crankshaft with the help of a wedge formed by the rotating shaft. Thrust bearings with a tilted pad design also have this type of lubrication as they form a converging wedge to obtain hydrodynamic lubrication. *Hydrostatic lubrication*. Where the oil film cannot be formed due to the motion of moving parts, the oil pressure has to be supplied externally. These types of lubrication are referred to as hydrostatic lubrication. For slow-moving heavy parts, their relative motion is not enough to provide self-generated pressure for lubrication, and hence pressure is provided externally with the help of a pump. For example, many crosshead bearings require an additional crosshead lubrication pump to boost the pressure for crosshead bearing lubrication as the pressure cannot be self-generated. *Boundary lubrication*. In this type, there is a thin film between two rubbing surfaces, which may experience surface contact. Boundary lubrication is used where there are relatively slow speeds, high contact pressure, and rough surfaces. For example, boundary lubrication in the main engine occurs during starting and stopping due to the aforementioned conditions. *Elasto-hydrodynamic lubrication*. With this type of lubrication, the lubricating film thickness considerably changes with elastic deformation of surfaces. This is seen in line or at the point of contact between rolling or sliding surfaces, for example, with rolling contact bearings and meshing gear teeth. Elastic deformation of metal occurs, and there is an effect of high pressure on the lubricant.

LUBE OIL SYSTEMS

Main engine lubricating system

The main engine has three separate lubricating oil systems:

1. Main lubricating oil system

2. Cylinder oil system
3. Turbocharger lubricating oil system

The main or crankcase lubrication system is supplied by one of two pumps, one of which will be operating and the other on standby, set for automatic cut-in should there be a lubricating oil pressure reduction or primary pump failure. The main lube oil pumps take their suction from the main engine sump tank and discharge oil via the main lube oil cooler, which takes away the heat. An automatic backflushing filter unit with a magnetic core helps to remove any metal debris. The plate-type lube oil cooler is cooled from the low-temperature central cooling freshwater system. The supply pressure in the main lubrication system depends on the design and requirement and is generally around 4.5 kg/cm². Lube oil supply to the cooler is via a three-way valve which enables some oil to bypass the cooler. The three-way valve maintains a temperature of 45°C (113°F) at the lubricating oil inlet to the engine. The main lube oil system supplies oil to main bearings, camshaft, and camshaft drive. A branch of lube oil goes to an articulated arm or a telescopic pipe to the crosshead from where it performs three functions:

1. Some oil travels up the piston rod to cool the piston and then comes down.
2. Some oil lubricates the crosshead bearing and the shoe guides.
3. The remaining oil passes through a hole drilled in the rod connecting to the bottom end bearing. A branch of lube oil is led to the hydraulic power supply unit for actuation of exhaust valves, to the thrust bearings, to the moment compensator, and to the torsional vibration damper. The cooling effect of the oil at the vibration dampers is important.

For the operation of the main engine lubricating oil system, it is assumed that the engine is stopped but is being prepared for starting: (1) check the level of oil in the main engine sump tank and replenish, if necessary; (b) ensure that the low-temperature central cooling system is operating and that freshwater is circulating through the main lube oil cooler; (3) ensure all pressure gauge and instrumentation valves are open and that instruments are reading correctly; (4) ensure that the steam heating is applied to the main lube oil sump tank if the temperature of the lube oil is low; (5) set the line and make sure all right valves are open. Normally, it is assumed that the main engine lubricating valves are left open; (6) select one main lube oil pump as the master (duty) pump and the other as the standby pump (Note: The main lube oil pumps have large motors and are generally fitted for autotransformer starting; after a start, the autotransformer must be allowed to cool down for 20 minutes before another start is attempted. Restarting is inhibited for 20 minutes between starts); (7) keep the lube oil system circulating and allow the temperature of the system to gradually increase to normal operating temperature; (8) check the outlet flows from the individual units; check that temperatures are similar and that all pressure gauges are reading correctly; and (9) when lubricating system temperatures and pressures are stable, the engine may be started. The main engine lubrication system is replenished from the main lube oil storage tank.

Main engine lube oil purifier. The main engine lube oil purifier takes suction from the main engine lube oil sump and purifies the oil. Its feed temperature is maintained around 90°C (194°F) (as maximum density difference is achieved at that temperature) to allow efficient separation. The main engine lube oil must be tested frequently in order to determine whether or not it is fit for further service. Samples should be taken from the circulating oil and not directly from the sump tank. The main engine lubrication system also has a subsystem (depending on whether the main engine is cam-less or has a camshaft). *Hydraulic power supply unit.* In cam-less engines, a branch from the lube oil inlet to the main engine is provided to the hydraulic power supply (HPS) unit. The function of the HPS unit is to control the fuel injection and exhaust valve actuators hydraulically and drive the cylinder lubrication units. In the main engine with a camshaft, a lubrication system feeds to camshaft roller guides and bearings, which actuates the exhaust valves and fuel pump. *Main engine lube oil sump tank.* It is located under the engine in the double bottom and is surrounded by cofferdams. A sounding pipe to know the level of lube oil in the sump is provided, along with a sounding pipe for cofferdam to know if there is any leakage. *Cofferdam.* The cofferdam needs to be inspected on a regular basis to know any signs of leakages. The main engine lube oil sump consists of a level gauge, sounding pipe, air vent pipe, heating steam coil, manholes, suction pipe, and valves for lube oil pump and lube oil purifiers.

Turbocharger lubricating oil system

The turbocharger bearing lubricating system can be separate from the main engine lubricating system or can be fed through the main engine lubricating system, depending on the design. It is essential to have a separate filter for turbocharger lubrication which is generally a duplex filter. From the duplex filter outlet, the turbocharger lube oil flows to the inlet manifold supplying turbochargers. The outlet of lube oil from turbochargers has a sight glass to make sure the flow is continuous. Under normal circumstances, a lube oil supply is always maintained to the turbochargers to ensure that they are always available for service and to prevent damage. A lube oil supply must be maintained when the engine is stopped, as natural draught through the turbocharger will cause the rotor to turn. Hence, the bearings must be lubricated.

Cylinder lubrication system

The load-dependent lubrication of the cylinders is performed by a separate cylinder lubrication system. Cylinder lubrication is required to lubricate the piston rings to reduce friction between the rings and liner, to provide a seal between the rings and the liner, and to reduce corrosive wear by neutralising the acidity of the products of combustion. The alkalinity of the cylinder lubricating oil should match the sulphur content of the heavy fuel oil supplied to the engine. If the engine is to be run on low-sulphur fuel oil for a prolonged period, advice must be sought from the cylinder oil supplier and the engine builder as to the most suitable cylinder oil to use. The ability of an oil to react with an acidic reagent, which indicates the alkalinity, is expressed as total base number. It should correspond to the sulphur percentage of fuel oil to neutralise the acidic effect of combustion. When high sulphur fuel oil is used for main engines, a high total base number grade of cylinder oil needs to be used. When the main engine is 'changed over' to low-sulphur fuel oil or low-sulphur marine gas oil (LSMGO), a low total base number of cylinder oil needs to be used.

There are two important systems used in modern lubrication systems:

1. Accumulation and quill system (for Sulzer engines)
2. Cylinder lubricating units pumping to orifices in the liner (MAN B&W engines)

The cylinder lubricating oil is pumped from the cylinder oil storage tank to the cylinder oil measuring tank which should contain sufficient lube oil for two days' cylinder lubricating oil consumption. Cylinder lubricating oil is fed to the cylinder lubrication system by gravity from the measuring tank; a heater is in the gravity line and pipe; pipes are electrically 'trace heated' – i.e., the outer surface of the pipe is maintained at a certain temperature. The heater and trace heating maintains a temperature of 45°C (113°F) at the lubricating unit. Before starting the main engine, it is necessary to pre-lubricate the liners. Pre-lubrication before the start can be made manually or by a sequence in the bridge manoeuvring system. The following criteria determine the control:

- The cylinder oil dosage must be proportional to the sulphur content of the fuel.
- The cylinder oil dosage must be proportional to the engine load, i.e., the cylinder fuel supply.

The quantity of cylinder oil injected at the individual injection points is controlled by the cylinder lubrication control system. Each cylinder lube oil injector (quill) is effectively a non-return valve that is opened by the pressure oil directed to it by the lubricator control system. Cylinder oil feed rates can be adjusted, but adjustments must only be made by authorised personnel. Correct cylinder lubrication is essential for efficient engine operation, to minimise lubricating oil costs, and to optimise maintenance costs. It is essential that the cylinder lubricators are correctly set and that the correct cylinder lubricating oil is used for the fuel being burned. No adjustment should be made to the engine cylinder lubrication system without the permission of the chief engineer. The cylinder oil measuring tank is replenished from the cylinder oil storage tank using the cylinder oil shifting pump. In the event of failure of the electrically driven cylinder oil shifting pump, a hand-operated pump is provided. The electrically driven cylinder oil shifting pump is started manually, but a high-level switch in the cylinder oil measuring tank stops the pump when the tank level reaches a high value. The tank is fitted with a low-level alarm. A separate cylinder oil storage tank for use with low-sulphur heavy fuel is also fitted, and the cylinder oil from this tank must be used when the main engine is changed to low-sulphur heavy fuel oil operation. The cylinder oil measuring tank has an overflow system via a sight glass; the overflow line has a three-way valve which must be set to direct the overflow oil to whichever cylinder oil storage tank is in operation.

Piston rod stuffing, the box, and scavenge space drainage system

The piston rod gland or stuffing box provides a seal for the piston rod as it passes through the separating plate between the crankcase and the scavenge airspace. The stuffing box has two sets of segmented rings that are in contact with the piston rod; the upper set of rings scrapes crankcase oil from the piston rod, and the lower set of rings prevents oily deposits in the scavenge space from entering the crankcase. In the middle of the stuffing box, there is a 'dead space' which should normally be dry if the rings are working effectively. Any oil or scavenge space material that enters this space is drained directly to the oily bilge drain tank.

LUBE OIL PROPERTIES

Lube oil is one of the essential elements for operating any kind of machinery on board a ship. Lube oil is responsible for lubrication and cooling of the parts which are operating relative to each other, giving rise to frictional and other types of stresses on the machinery. Without the use of lube oil, we cannot imagine any machinery operation on the ship. Several types and grades of lube oils are available for machinery, depending upon the working condition, operation, and requirements of the machinery itself. When it comes to marine engines, it is critical to select the best grade of lube oil that can be used as crankcase oil or cylinder oil. The lube oil is selected based on the properties which will improve the engine operation and reduce the wear-down rate and hence the maintenance cost of the machine. The following properties are considered vital for superior quality marine lube oils: *Alkalinity*. Lube oil alkalinity plays an important part in maintaining marine engines. When fuel burns, the fumes carry sulphuric acid, which causes acidic corrosion. For a trunk piston engine or four-stroke engines, the main lube oil is responsible for providing piston and liner lubrication; hence, it comes directly in contact with the combustible fuel. For two-stroke engines, separate grades of lube oil are used as cylinder oil. Its alkalinity depends on the engine fuel grade (i.e., whether heavy-sulphur or low-sulphur fuel oil). *Oxidation resistance*. Lube oil is always in contact with air and thus oxygen presence in lube oil is inevitable. Moreover, at elevated oil temperatures, oxidation rates increase. After 85°C (185°F), any increase of 10°C (50°F) the oil oxidation rate doubles leading to the formation of sludge, acid production, and bearing corrosion. To minimise these effects, additives are added to maintain the properties of the lube oil. Lube oil temperature is controlled by passing it through the lube oil cooler. *Load-carrying capacity*. This is one of the most important characteristics of lube oil as it influences the viscosity of the oil. The load subjected to different internal parts of the marine engine is extremely high; therefore, the load-carrying capacity must be sufficient to withstand the pressure inside the engine. If this is not achieved, then oil will be forced out resulting in metal-on-metal contact. This in turn will wear down the machinery faster. *Thermal conductivity*. The internal parts of the marine engine are always producing heat energy. This heat energy must be dispersed; otherwise, it can lead to wear down due to thermal stresses. The lube oil must cool down the internal parts of the engine to avoid such situations, which means the lube oil must have good thermal conductivity. *Detergency*. The detergency of the oil is obtained by adding metallic-based additives which prevent the build-up of small deposits on the metal surface. In two-stroke engines, cylinder oil detergency is very important, as it removes deposits from the ring pack area and keeps the combustion space as clean as possible. *Disparency*. Disparency is the property of lube oil which prevents impurities from mixing with the oil and instead keeps them suspended on the surface. This makes it easy for the separator or clarifier to remove the impurities from the oil. *High flash point*. The flash point is the minimum temperature at which the oil vapourises to provide an ignitable mixture of air. The flash point should always be on the higher side so that in the event of an increase in oil temperature, the risk of fire can be avoided. Normally for marine engine lube oils, the flash point is higher than 220°C (428°F). *Low demulsification number*. It is not impossible to completely avoid the contamination of oil with water. A low demulsification number helps to separation the water from the oil in the separator or when stored in the settling tank.

ASSESSING THE LUBE OIL

Almost all machinery on board ships requires some form of lubrication to ensure smooth and efficient operation. This is achieved by using different grades and types of lubricating oils which are stored in designated lube oil tanks, drums, or receptacles. As for

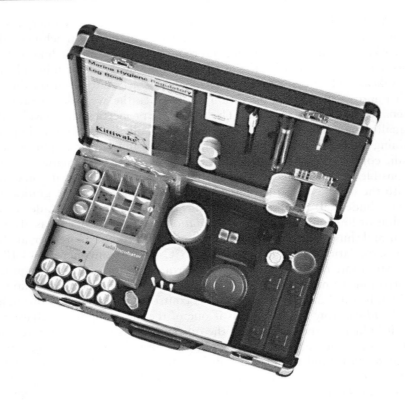

Figure 8.1 Lubricating oil test kit.

machinery spares, lube oil also has a limited period of operation after which it must be renewed. The renewal period is determined by the properties of the lube oil, the type of machinery it services, and the type of conditions in which the oil and machinery are used. Apart from renewal, lube oil must be checked and assessed for quality and purity (see Figure 8.1). As discussed in the previous section on lube oil properties, to maintain these properties while in operation, the lube oil must be checked both onboard and at an onshore laboratory.

Onboard lube oil tests vary according to the preventative maintenance schedule, but in most cases, it ranges from one every 15 days to once a month. Lube oil samples must be sent ashore every three months for special laboratory tests which include spectroanalysis. *Taking samples for tests.* The onboard lube oil tests are conducted by taking samples from the sampling point (see Figure 8.2), which should be located after the system, with the system in running condition. Before taking the sample, the oil must be drained so that any stagnant oil at the sampling point is removed. The sample must be kept within the control room to cool down to normal atmospheric temperatures. When doing this, the lid on the sample bottle must be half open; otherwise, the vapours will condense during the cooling process and alter the results of the assessment.

Onboard lube oil tests. For all types of lube oils on ships, the following lube oil tests are conducted: (1) *water content test*. A 5ml sample is taken by a digital water content

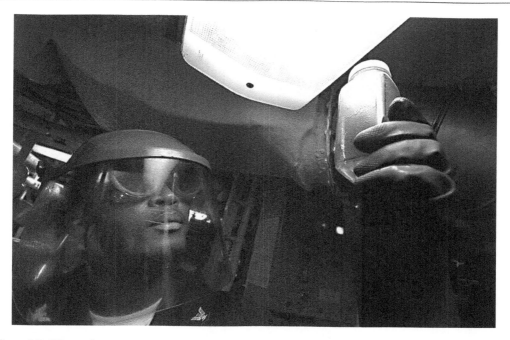

Figure 8.2 Oil samples.

metre and is mixed with 15ml of reagent containing paraffin or toluene. Before closing the lid of the digital metre, a sealed sachet containing calcium hydride is added and the container is closed tight. The metre is shaken by hand and the pressure rise due to the chemical reaction in the test container is shown as a water percentage on the digital display. (2) *PH test*. This is done using a pH paper which changes colour once in contact with oil. It is then compared with standard values. This test determines the reserve alkalinity of the oil sample. (3) *Viscosity test*. This test is performed using a flow stick, in which two paths are provided for the flow of oil, side by side. Fresh oil is filled into one path, and in the other path, sample oil is filled. Now the flow stick is tilted, allowing the oil from both paths to flow in the direction of the tilt due to gravity. A finish point is provided together with reference points along the flow stick. The position of the used oil is checked when the fresh oil reaches the finish point. This method demonstrates the contamination of lube oil which may be caused by diesel oil, heavy fuel oil, or sludge resulting in a change of viscosity. (4) *Spot test*. In this test, a drop of lube oil is placed on blotter paper and then dried for several hours. The dry spot is then compared with a standard spot. This determines any insoluble components present in the lube oil. (5) *Flash point test*. This test is performed by using a *Pensky Martin* closed cup apparatus which determines the temperature at which the vapour will flash up when an external ignitable source is provided (see Figure 8.3). (6) *Water crackle test*. This is another method for determining the presence of water in the lube oil. Here, the oil sample drops are heated in an aluminium container over a flame. If water is present, this will result in an audible crackling sound.

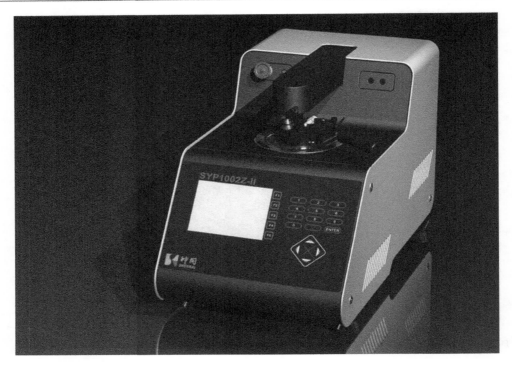

Figure 8.3 Flashpoint test.

In this chapter, we have learnt about the importance of using superior quality lubricating oils, as opposed to cheaper alternatives, and the significance of testing and assessing lube oil for impurities and contamination. In the next chapter, we will build on this knowledge by examining essential engine room machinery maintenance and troubleshooting techniques and procedures.

Chapter 9

Basic engine room machinery maintenance and troubleshooting

To ensure the marine engine runs smoothly, periodic routines and maintenance jobs are conducted by the engineering department. For marine engineers working on ships, troubleshooting problems related to engine room machinery is the most important task they must deal with daily.

PRINCIPLES OF TROUBLESHOOTING

Technically, there are several aspects that play important roles in solving an engine room machinery problem. Though marine engineering training inculcates the basics of maritime concepts, it is only through first-hand experience that marine engineers can really learn and put into practice the techniques for good engine room operation and maintenance. The process of troubleshooting the ship's machinery involves three essential elements, which are as follows: (1) the requirement, (2) the approach, and (3) continuous learning. (1) *The requirement*. Though the art of troubleshooting cannot be learnt solely from reading marine engineering books, knowledge is the foundation of all experience. This means marine engineers must have the right technical background, which is usually best achieved by taking a marine engineering course at university or a specialised training institution. Good marine engineering courses provide their students with a balance between learning engineering principles from books and other texts and applying their learning through simulation training and activity-based learning. Doing so not only enhances practical learning but is more engaging and leads to improved problem-solving capabilities. (2) *The approach*. Though it is mandatory for a person to fulfil the aforementioned 'requirements' to seek employment on ships, these requirements are not by themselves sufficient. Indeed, fulfilling the requirements is merely the first step towards successful troubleshooting. The second element of good troubleshooting is seeking the right approach. This means knowing each piece of machinery inside-out. Learning the starting and stopping procedures and reading and rereading the manufacturer's operations manuals to the extent they become second nature. Only after the engineer has absorbed as much information and data as is humanly possible, are they ready to begin the troubleshooting process. This process begins by understanding what makes the machinery perform at its most efficient (e.g., electric power, oil, water, air, temperature, and so forth). Once this has been ascertained, it is possible to identify the main causes of malfunction. By comparing the current operational parameters with previous records, it should become obvious where the fault lies. For example, if the machinery is working fine, but the indicators are showing some abnormality, it is likely the indicators are faulty rather than the machinery itself. Where it is not immediately possible to identify the cause of the fault, it may be necessary to perform a reverse diagnosis. Rather than searching for faults from the beginning, it is sometimes easier and more efficient to start at the end and work backwards. Divide the system into sections and investigate the root cause logically. Even if it is unlikely that the cause of the fault is rooted in one piece of equipment or section, it is always worthwhile double-checking for hidden faults or abnormalities that may surface later. (3) *Continuous learning*. Seafarers often forget that troubleshooting is a continuous learning process. There is no such stage as 'know-it-all'. There is always something new to learn from every day, even if one has years of sailing experience. Even if we are unable to solve the problem at hand, never be disappointed. Learn from the issue. Learning from the last problem always adds to our bank of experience which can be used for future troubleshooting situations. Once the fault is resolved, share the problem – and the solution – with the other departmental members, as they may encounter the same or similar issues in the future. Always keep accurate and timely records of each fault and the way it was resolved. This serves as both an official log, but also provides a living source of information for the future.

Now that we have covered the three essential elements of good troubleshooting, we can begin to look at some of the main types of faults and malfunctions that occur within the engine room.

COMMON FAULTS AND MALFUNCTIONS

The marine engine is the main propulsion power source and by far the biggest machine on any ship. A substantial amount of effort, resources, and time is spent to ensure that the engine runs smoothly and efficiently, taking the ship from one port to another without breaking down. Yet, no matter how many precautions are taken, faults and breakdowns do occur. This is usually no fault of the engineers but is consequent to the substantial number and complexity of parts that make up the marine engine.

> *Clogged or stuck fuel rack*. A stuck fuel rack is one of the most common problems engineers encounter with the oil-fired, two-stroke marine engine. The governor controls the fuel pump delivery through a fuel rack, which is a combination of mechanical links. Sometimes the fuel rack can become clogged, leading to a lack of fuel supply. This results in either fluctuation in the engine rpm if running or else the engine will not start if at standstill. To prevent and or resolve a stuck fuel rack, the mechanical links of the fuel rack must be well lubricated and greased before starting the main engine. If after starting the main engine, the engine rpm continues to fluctuate constantly, even at lower speeds in calm weather, inspect each fuel rack as one or more will still be clogged.
>
> *Starting air valve leak*. Any leakage from the starting air valve will lead to hot gases returning to the engine air line, which may contain thin oil films. This mixture of oil and film can lead to a starting air-line explosion. Fortunately, this kind of explosion is not common today due to the various safety features incorporated into the air line (for example, bursting disk in the MAN B&W engine and relief valves in the Wärtsilä Sulzer engines). That said, it is important not to overlook the possibility of these devices malfunctioning, which can lead to a starting air-line explosion.

Normally, there is no remote monitoring of temperature for the air line supplying air to starting air valve. The best way to determine any faults is to check the temperature of the air line manually during manoeuvring. This problem is more likely to occur when the engine is started frequently rather than when the engine is left running continuously.

Fuel leak and or fuel valve malfunction. Problems with the fuel system are commonly observed in the main engine. When there is a deviation in the temperature of one unit, the fuel system, and in particular, the fuel valve, these need to be checked immediately. Overhauling and pressure evaluating the fuel valve must be done as per the vessel's preventative maintenance schedule. If the engine is fed with heavy diesel oil, there are good chances of leaks developing from the pump seals. Also, if the fuel treatment is improperly managed, and the fuel temperature is not maintained, this can lead to the propagation of cracks and leaks in the high-pressure fuel pipe. Any leaks in the main engine fuel oil system can be determined from the 'high-pressure leak off tank' level and alarm.

Sparks in the main engine exhaust. The engineering department often gets calls from the bridge informing them about sparks coming from out of the funnel, which is the main engine exhaust. Sparks from the funnel occur most often when slow steaming and during frequent manoeuvrings, during which unburnt soot deposits build up in the EGB boiler path. Frequent cleaning (at least monthly) of the exhaust gas boiler should avoid or eradicate the discharge of sparks from the funnel.

Starting air leaks. This is also one of the most underrated, yet common, problems related to marine engines. The control air supplies air to various parts and systems of the main engine. It is always in open condition when the engine is in use. Small leaks are normal and can be rectified only by tightening or replacing the pipes or joints. When the engine room machinery is in working condition, it is difficult to hear any air leaks. The best way is to trace each of the airlines by feeling the connections and or joints by hand. The easiest way to find air leaks is when there is an intentional blackout. At this moment, all the machinery will be placed in a 'stop' position whereby leakage sounds (hissing noises) will become loud and clear.

Stuck air distributor. The air distributor is responsible for maintaining the air supply which opens the starting air valve in the engine cylinders. Since it is a mechanical part, it is prone to malfunction. The main engine will not start if the air distributor does not supply air to open the starting air valves, as no air will be present in the cylinder to commence the fuel combustion. Many engine designs, such as the MAN B&W, have their air distributor located at the end, with an inspection cover, which can be opened when the engine is not running. From here it is possible to conduct inspections of the air distributor and lubricate if necessary.

Malfunctioning gauges. It is especially important to have local parameter gauges on various systems of the main engine. When recording readings in the logbook, it is always recommended to take local readings rather than remote readings. Often engineers find that one or more gauges (such as the pyrometer, pressure gauges, manometers) are not working or are in a dilapidated condition. The main reasons are due to loose parts and connections caused by vibrations. The easiest solution to this type of fault is to replace the faulty parameter gauges with new ones as early as possible.

Faulty alarms and sensors. The main engine is fitted with various sensors, which measure and transmit factual data to the central alarm console. Contributing factors such as vibrations, elevated temperatures, high humidity, dust, and so forth, can cause these sensors to malfunction, leading to false alarms. To prevent this from happening, routine checks need to be performed on all engine room sensors and alarms.

Preventative maintenance is one of the most important tasks conducted by marine engineers on board a ship. It involves conducting routine maintenance tasks on the marine engine and associated machineries. Unlike preventative maintenance, which is done to prevent machineries from breaking down, corrective maintenance is conducted in response to faults and malfunctions. Ideally, all maintenance should be preventative, but as we have already discussed, machinery is liable to break down despite the most initiative-taking maintenance schedules. Marine engine parts need to be checked on a regular basis to avoid breakdowns or heavy losses caused by the ship going off charter. Marine engine repairs are conducted by marine engineers as per their basic understanding of the machinery, sound troubleshooting knowledge, and correct techniques used for testing and overhauling. Moreover, there are several agencies around the world that provide services for marine engine repairs, many of which cannot be conducted by the ship's marine engineers on account of the lack of special equipment, expertise, or contractual constraints. Some examples of heavy maintenance which cannot often be performed by the ship's staff include metal stitching or metal locking, reconditioning of the pistons, honing of liners, and so forth.

When we talk about marine engine repairs, we are not just speaking about maintenance and repair work on the mechanical parts of the engine but also repairs on the various electrical equipment. Thus, marine engine repair is categorised into two parts – mechanical and electrical. For the effective performance of the marine engine and to prevent breakdowns proper procedures must be followed as described in the manufacturer's manuals. Marine engine repairs must be done at specific running hours as described in accordance with the preventative maintenance system. On board the ship, there is a team of marine engineers, along with engineering ratings such as the motorman, oiler, and fitter, whose expansive responsibilities are to conduct the work of marine engine repairs. The team of engineers includes the chief engineer, the second engineer, third engineer, and fourth engineer. The chief and second engineers are management-level officers, whereas the third and fourth engineers are operational-level engineers. The chief engineer looks after the many different surveys that are to be conducted on the marine engine and plans when they are to be conducted. The second engineer supports the chief engineer by scheduling repair work that is pending or soon due. The second engineer also looks after the main engine and the various pumps in the engine room. The third engineer manages the boiler and auxiliary engines, whereas the fourth engineer looks after the compressors and purifiers. For electrical equipment, the repairs are conducted by a separate resolute electrical engineer, who looks after the various motors, batteries, and print card electronics. For marine engine repair, the most critical issue is to make available several sets of spare parts on board the ship. If there is a shortage of any of these parts, then they need to be ordered by the respective engineer, who is looking after that specific machinery. Some special considerations may also need to be given to emergency, safety, and life-saving equipment. The marine engineers must also ensure that all the ship's equipment is in good working order. External agencies such as Port State Control and Flag state authorities are entitled to inspect and detain any vessel that has defective equipment. This not only includes within the engine room but also the emergency generators, lifeboat engines, firefighting appliances, systems and equipment, and the navigational equipment on the bridge. Port State Control and Flag state authorities have the power to fine individual officers where they have been found negligent in their duties.

CRANKSHAFT FAULTS AND MALFUNCTIONS

The crankshaft is the intermediate part of the marine engine, which transfers the power of a firing cylinder from the reciprocating piston to the rotating propeller (or alternator in the case of a generator). The function of other components, such as the camshaft, depend upon the correct rotation of the crankshaft. A failure in any single part of the crankshaft can stall the engine as well as the entire ship. Some of the most common faults and malfunctions to affect the crankshaft are as follows:

Fatigue failure. The majority of steel crankshaft failure occurs because of fatigue failure, which may originate at the change of the cross-section such as at the lip of the oil hole bored in the crankpin.

Failure due to vibration. If the engine is running with heavy vibration, especially torsional vibration, this may lead to cracks developing in the crankpin and journal.

Failure due to insufficient lubrication. If the lubrication of the bearings in the crankshaft is starved, this may lead to wipe out of the bearing and failure of the crankshaft.

Over-pressurisation of the cylinder. It may happen that there is a hydraulic block (for example, caused by water leakage) inside the liner. Due to extreme pressure, the crankshaft may slip or even bend if the safety valve of the unit is not working properly.

Cracks in the crankshaft. Cracks can develop at the fillet between the journal and the web, particularly between the position corresponding to ten o'clock and two o'clock when the piston is at top dead centre.

Crankshaft misalignment. The crankshaft is a massive component when fully put together in the engine. Initially, the complete crankshaft is aligned in a straight line (the connection drawn from the centre of the crankshaft makes a straight line) before setting on the top of the main bearings. With time, due to numerous factors, the straight line may deviate and misalign. A degree of misalignment is acceptable within limits, but where the value goes beyond that rated by the manufacturer, this may lead to damage or even crankshaft breakage. There are many reasons and causes for the crankshaft to suffer misalignment; with the main causes listed here: (1) damage or wipe out of the main bearing, (2) loose engine foundation bolts leading to vibrations, (3) deformation of the ship's hull, (4) cracks in the bearing saddle, (5) loose main bearing bolts leading to damage of the main bearing, (6) excessively high bending moments on the crankshaft caused by excessive force from the piston assembly, (7) grounding of the ship, (8) crankcase explosion or fire, (9) defective or worn-out stern tube or intermediate shaft bearings, (10) loose or broken chokes in the foundation, (11) cracked bearing pockets, (13) deformation of the bed plate and/or damage to the transverse girder, (14) slack or broken tie bolts, and (15) weakening of the engine structure due to corrosion. It is thus advisable to regularly inspect the crankcase and crankshaft deflection to check for misalignment.

Crankcase inspections

The crankcase contains some of the most sensitive components of the main engine. The crankcase lubricating oil needs to be maintained in good. If not maintained and checked periodically, the crankcase lubricating oil can damage the bearings and other parts of the engine. As well as being entirely avoidable, crankcase repairs are extremely costly and time-consuming. Moreover, if the damage is extensive, the ship may need to go off charter. To avoid sustaining damage to the crankcase, it is imperative that the lube oil is evaluated at least weekly by way of the lube oil water test. This test is conducted to ascertain that there are no leaks in the crankcase. If the water content of the lube oil is 2% or less of the total volume, this is acceptable and can be reduced through purification. If, however, the water volume in the lube oil is above 2%, this is indicative of a serious leak within the crankcase. In these situations, the cause of the leak must be investigated immediately and any remedial actions conducted as soon as possible. Only once the cause of the leak has been found, and appropriate remedial actions conducted, should the oil in the crankcase be replaced completely. Other weekly checks to be conducted on the crankcase include tests to ascertain the total base number and viscosity of the oil. The crankcase must be topped up or a complete oil change conducted as per the manufacturer's instructions. An oil sample must be sent ashore for laboratory analysis every three months. The spectrographic analysis helps determine the amount of metal wear through the determination of fine particles. Where the analysis shows the number of fine particles is above the acceptable limits, the laboratory report will recommend appropriate procedures or precautions.

For larger low-speed engines, a full crankcase inspection must be conducted once a month, whenever the ship is in port and there is sufficient time for inspection. This thorough inspection is required to evaluate the conditions inside the crankcase compartment and to assess any damage to the bearings. Before conducting an inspection of the crankcase, the following procedures must be followed. First, permission must be sought from the port authorities to ensure there are no prohibitions on conducting crankcase inspections. This immobilisation permission is needed, as the vessel will effectively become dead in the water. Once permission has been received, the chief engineer is required to complete the crankcase inspection checklist. This checklist contains a list of actions and safety precautions to be implemented and followed throughout the crankcase inspection. Any safety-related issues must be discussed with the members of the ship's staff and any shore-based personnel involved in the inspection. When the engine is placed into an isolated or 'stopped' condition, the lubricating oil pump and crosshead oil pump must be stopped. The breaker should be removed to prevent the pumps from inadvertently operating. Safety signs should be prominently displayed in the immediate area around the crankcase, throughout the engine room, immediately outside the engine room, and on the bridge. As the engine crankcase is an enclosed space, a permit or work order must be applied for from the ship's master, duly completed and signed. After stopping the engine and the pumps, the crankcase doors should be opened to allow thorough ventilation and cooling of the crankcase space. It is important to remember the environment inside the crankcase will be extremely hot and deprived of air. Once the interior of the crankcase space has cooled down and is thoroughly ventilated, the individual entering the space must be donned in appropriate personal protective equipment, including boiler suit, safety harness, and anti-slip shoes or boots. It is important to ensure no jewellery, loose tools, or pens can fall into the crankcase space, as these may cause damage to the bearing and machine parts. Finally, before entering the crankcase space, the inspector should be thoroughly briefed on the purpose of the inspection and what parts and components need to be inspected.

During the inspection, it is typical for the following checks to be made: (1) checks on the overall quality of the oil, i.e., whether it is clean or dirty with carbon particles; (2) checks for any distinguishing smells; if present, this is usually indicative of some form of bacterial contamination of the oil and has a rancid rotten egg odour; (3) checks for any metal particles near the grating in the crankcase; (4) checks on the condition and damage to the grating; (5) checks on the slip marks of the web; these should be in the same line; if slippage is found, this should be reported to the company's technical department and the ship's classification society; (6) checks for any bluish dark patches, which are indicative of hot spots caused by insufficient lubrication friction; (7) checks for crosshead damage; (8) crosshead guides for damages and marks; (9) checks of the bed plate for signs of weld cracks and fissures; (10) checks for any naked metal seen near or at the bearings, which may be caused by wiping; (11) checks of the piping and any loose connections; (12) checks of the locking wires and locking washers on the stuffing box bolts; and (13) any other checks specified by the shipowner's technical department. When exiting the crankcase space, always check and then double-check for any items which may have fallen or been left inside.

MAIN BEARING FAULTS AND MALFUNCTIONS

The bearings of a marine engine are said to be worn down when the lining (Babbitt or tin-aluminium) is worn away through lining scuffing, wiping of the linings due to excessive loading, abrasive wear caused lube oil contamination, melting out or extensive fatigue of the lining (Babbitt) due to lack of supply of oil or bearing high temperatures, and, in the worst case, from direct steel-to-steel contact. The bearings can be inspected or surveyed in one of three ways: (1) manually through a crankcase inspection, (2) manually through crankcase deflection, or (3) through automatic monitoring. The crankcase inspection forms part of the Preventative Maintenance System. Crankshaft deflection is a procedure conducted to measure the misalignment of the shaft at various levels when compared to the original or last reading of the measured deflection. Normally, this procedure is performed in conjunction with a crankcase inspection. The recorded readings are interpreted in a graph and compared to the original to determine, whether there are any worn-out or defective bearings which are causing crankshaft misalignment. Some, if not most, modern ship engines have an automatic monitoring system which monitors the state and play of the main bearings. By having this system installed on board, the ship's engineers are advised well ahead of any steel-to-steel contact situation developing. As well as being inexpensive and easy to install on new builds, the monitoring system can be retrofitted to existing marine engine infrastructures. The advantage of the system is it senses and provides continuous condition-based data. This reduces the need for scheduled open inspections, decreases engine downtime, and saves man-hours. It is also reducing the potential for cross-contamination of internal parts, such as the sump oil, which is always a risk when the crankcase chamber is opened. Within the automatic monitoring system, there are three main parameters that are of primary interest. These are the *bearing wear monitor* (BWM), the *bearing temperature monitor* (BTM), and the *water in oil monitor* (WIOM). The purpose of the BWM, BTM, and WIOM systems is to prevent bearing damage well ahead of time. It should be noted, however, that monitoring does not in itself protect the bearing shells but avoids consequential damage to the crankshaft and bed plate in the event of catastrophic bearing failure. A separate device, the *Propeller Shaft Earthing Device*, is designed to protect the bearing from spark erosion.

> *Bearing wear monitoring.* The principle of the bearing wear monitoring system is to measure the vertical position of the crosshead in bottom dead centre (BDC). The general bearing wear monitoring system monitors all three principal crank train bearings using two proximity sensors located forward and aft per cylinder unit, inside

the frame box. The sensor continuously monitors the guide shoe bottom ends and measures the distance to the crosshead in BDC. These sensors send the continuously monitored data to an alarm system in the ship's computer monitoring system. Hence, when there is notable wear in the main bearing, the crank pin, or the crosshead bearing, this monitored vertical position will shift, and the same will be reflected in the monitoring system of the ship. If this shift reaches the set alarm values for one or more units, the engineer will be alerted about the situation. This monitoring system is usually connected to the safety system of the engine, which may cause the engine to slow down automatically.

Bearing temperature monitoring. A rise in temperature is often considered an important sign of an abnormality developing in the engine bearings. If the temperature of the bearing can be monitored and considered before it rises to a dangerous level, a major breakdown of the engine crankshaft and bearing arrangement can be prevented. The measurement of bearing temperature is done in two ways: first, through *direct measurement* using temperature sensors normally fitted at the rear side of the bearing shell, and secondly, through *indirect measurement*, whereby detection of the temperature is taken as readings of the return oil from each bearing in the crankcase. The temperature monitoring system continuously monitors the temperature of the bearings. In the event a specified temperature is recorded (either a bearing shell temperature or bearing oil outlet temperature), an alarm is raised. For shell temperatures in the main bearing, crankpin, and crosshead bearings, two high-temperature alarm levels apply. The first level alarm is indicated in the alarm panel, whilst the second level activates a slowdown command. For oil outlet temperatures in the main bearing, crankpin, and crosshead bearings, two high-temperature alarm levels – including a deviation alarm – apply. The first level of the elevated temperature/deviation alarm is indicated in the alarm panel, while the second level activates a slowdown command.

Water-in-oil monitoring. The water content in the lubricating oil is a crucial factor for maintaining a good bearing condition in the main engine. A significant increase in the water content (typically max. 0.2 vol.%; for a brief period up to 0.5 vol.%) can be extremely damaging to the engine bearings. Any increase in water content will have the following effects on different bearings:

- Excessive water content will cause the lead overlay, which acts as a running layer in the crosshead bearings (tin-aluminium lined) to corrode away.
- Main and crankpin bearings lined with Babbitt or tin-aluminium may also suffer irreparable damage caused by water contamination.

This damage can be easily prevented if the lubricating oil of the engine is continuously monitored for water contamination. There are two methods for monitoring water contamination. The first is the manual method. This involves checking for water contamination on a weekly basis using a water-in-oil test kit provided by the engine manufacturers. The main disadvantage of this method is the time gap and discontinuity in the monitoring of water content process. The second method, or automatic method, uses an installed water-in-oil monitoring system in the engine lube oil system. This works by continuously measuring the relative humidity of the system oil. A probe in the oil piping system is installed which transmits the signal to a unit which calculates the humidity as water activity. The major advantage of this system is the continuity in monitoring the water content in oil, which allows early intervention. Moreover, this system is independent of oil type, temperature, or age.

The system is directly connected to the engine room alarm system, which activates the alarm should the water content reach the set value.

Propeller shaft earthing device. On the average merchant ship, different metals are used in the construction of the propeller, hull, bed plate, crankshaft, and bearings. The current from the cathodic protection system is present in these parts, which eventually creates the perfect environment for spark erosion. When two currents, carrying dissimilar metals are in contact, a spark travels at the point of contact, which erodes away the metal, creating a cavity. To suppress the effect of galvanic corrosion, especially at the stern part of the ship where the propeller is present, an Impressed Current Cathodic Protection system is used. The propeller shafting is earthed to achieve a continuous circuit. This circuit usually exists when the propeller is at a rest, where metal-to-metal contact is made between the shaft and the stern tube liners or the main engine bearings and journals. However, when the shaft is turning, the bearing oil film creates an intermittent high resistance which effectively insulates the propeller from the hull structure.

Bearing inspections and surveys

When conducting inspections or surveys of the main bearings, the following process should be followed. *Prior to opening the bearing for inspection or survey*, ensure to check the previous bearing opening and survey report. Check the details of the records and clearances. Check all bearing-related service letters from the manufacturer. Check the laboratory lubricating oil analysis report. Conduct an onboard lubricating oil test and record the results. Inspect the work done report or logbook for any critical issues relating to the bearings (e.g., grinding of the pins, under or oversized bearings). Inspect all photographs from the last inspection or survey. Conduct a risk assessment of the job, and perform a toolbox talk for all personnel involved in the inspection or survey. Apply for immobilisation permission from the Port Authority. *During the bearing inspection or survey*, ensure the crankcase is properly ventilated and don appropriate personal protective equipment, including helmet and safety harness. Check and record the clearance of the bearing. Check the condition of the bearing metal. Check for signs of squeezing, scoring, cracking, and pitting. Check the surface shine of the pin; the pin should be shiny in appearance. If there is evidence of scoring, pitting, or cracks, in the pin, this should be polished, ground, or reconditioned. If deemed necessary, the replacement of the bearings must be conducted in accordance with the manufacturer's instructions. The pin and bearing should be thoroughly cleaned and lubricating oil applied before refitting. Ensure sufficient photographs are taken to record the condition of the bearing, pin, and internal environment of the crankcase chamber. *After the inspection or repair*, ensure the bearing and other parts are secured as per the manufacturer's instructions. The tightening value of the hydraulic bolts should be cross-checked and conducted in the presence of a senior engineer officer. The engine should be turned via the turning gear for a minimum of ten minutes with the lubricating oil pump on and the oil pressure recorded. The turning gear current should be observed and any abnormalities recorded. Once the engine is closed and ready for operation, running-in should be performed as per the manufacturer's instructions. During the running-in procedure, ensure to record all parameters. Once the inspection or repair is complete, prepare the maintenance survey report. File the report with the ship's record and send the complete work with photographic evidence to the ship owner's technical department. This report can be used as a reference during continuous machinery surveys, ensuring the relevant bearing need not be opened.

FUEL VALVE OVERHAULING

The fuel valve on the marine engine plays an especially significant role in determining the quality of combustion inside the engine cylinder. The penetration and atomisation of fuel, governed by the condition of the fuel valve, ensures efficient combustion. Fuel valves are an integral part of the ship's preventative maintenance system and are therefore overhauled and cleaned at regular intervals. Maintenance work on the fuel valves needs to be conducted with utmost care to prevent damage to the valve and to ensure thorough cleaning. The following tasks should be completed when performing fuel valve overhauling: (1) *Check the valves*. Before opening the fuel valve for overhauling, ensure that the engine is isolated; the fuel inlet and outlet valves are shut, and the fuel valve cooling water line is isolated. (2) *Cover the cylinder head*. When the fuel valve is removed from the cylinder head, ensure to cover the opening with a large cloth to avoid any tools or parts falling inside the cylinder. (3) *Check the testing machines*. Before assessing the fuel valve, ensure to check the diesel oil, hydraulic oil, and other electrical systems of the fuel valve testing machine (depending on the type of valve testing machine. It may be either electrical, manual, or hydraulic). Also, check the rated opening pressure of the valve in the manufacturer's operating manual. (4) *Loosen the pressure setting screw*. Before opening the fuel valve for overhauling, tightly hold the valve body in a vice and loosen the pressure setting screw before attempting to open the needle holder. (5) *Stow all overhauled parts in diesel or kerosene*. Ensure to keep and clean all opened parts in diesel or kerosene. Clean the fuel valve body with diesel and remove all heavy fuel oil deposits from the inside. (6) *Inspect the spring*. Check the spring for its length, elasticity, or signs of any breakage or cracks. Also, check the spring quarter and holder for breaks. (7) *Check the surface of the valve needle*. Check the surface of the valve needle holder and lap it with fine lapping paste if pitting or scratch marks are visible. Do this until the surface is smooth. Measure the face width. If it is less than required, replace the holder. (8) *Check the needle movement*. Check the surface of the needle tip and the movement of the needle inside the holder guide. (9) *Clean the nozzle holder atomisation holes*. Clean the nozzle holder atomisation holes with microwire (or gas torch hole cleaning wires). (10) *Check the valve properly after overhauling*. Once the fuel valve is overhauled and reassembled, check the opening pressure, atomisation, and dripping of fuel in the fuel valve testing machine. Ensure to use protective glasses and gloves when performing the tests. Check the opening pressure in the manufacturer's operating manual and then increase or decrease the opening pressure accordingly with the help of the pressure setting screw provided on the top of the valve. Some marine engine fuel valves with a zero-sac volume are not recommended for pressure testing and instead should be used directly in the engine after overhauling.

FUEL CHANGEOVER PROCEDURES

Emission control area (ECA) zones are designated coastal regions where SOx and NOx emissions are regulated by the laws laid down under MARPOL annex VI – Prevention of Air Pollution by Ships. Some local laws regarding air pollution are more stringent than those laid down by MARPOL. For instance, in Europe, whilst the ship is at the port, all running machinery consuming fuel must use only fuel with less than 0.1% sulphur content. As the SOx emission is purely dependent on the quality and sulphur content of the fuel, when entering ECAs zones, the ship is required to switch over to a lower sulphur content fuel. This means flushing out any residual fuel from the system with a sulphur content of more than 1.0% sulphur prior to entering the ECAs zones. This process is called fuel oil changeover. Considering that most ships today run off high-sulphur fuel oil, changing over fuel at the right time is critical. Moreover, in consideration of today's economic conditions, it is even more imperative to change over fuel from high to low sulphur at the correct time, as an early changeover will lead to unnecessary consumption of low-sulphur oil, which is prohibitively expensive compared to standard high-sulphur fuels. If the ship changes over too late, however, the ship may be liable to prosecution for violating the MARPOL annex VI regulations.

Fuel changeover procedures for the main engine. Most ships in operation today are equipped with one service tank and one (two) settling tank, which can result in the mixing of two different grades of oils when performing a changeover of fuel. Every ship is provided with a changeover low-sulphur fuel oil calculator which advises the correct changeover time at which the system should be running on low-sulphur fuel oil before entering the ECAs zone. This system requires the input of four key data parameters: (1) the sulphur content of the high-sulphur fuel oil currently in the system; (2) the sulphur content of the low-sulphur fuel; (3) the fuel capacities of the main engine system, including the settling tank, service tank, main engine piping, and transfer piping from the service tank to the main engine; and (4) the capacity of transfer equipment – i.e., the fuel oil transfer pump and fuel oil separators.

Once the changeover time has been calculated, which also accounts for the time of intermixing the two different sulphur grades (typically around 48 hours), the following procedures should be carried out accordingly: (1) first, ensure that no further transfers of high-sulphur fuel to the settling tank are carried out; (2) ensure that the low-sulphur bunker tank steam is open for transfer and purification of the fuel is operating as expected; (3) if two separate settling tanks are present, one can be dedicated to the low-sulphur fuel oil; this will reduce the changeover time; (4) keep the separator running until the settling tank level reaches the minimum level; (5) if filling of the service tank with high-sulphur fuel oil increases the calculated time of changeover, stop the separator and drain the settling tank; (6) the settling tank can be first drained into the fuel oil overflow tank; the oil drained can then be transferred to the bunker tanks containing the same grade of oil; (7) once the settling tank is drained of high-sulphur fuel oil, fill the settling tank with low-sulphur fuel oil via the transfer pump; (8) as the separator is stopped, the service tank oil will be consumed by the main engine system; (9) remember not to lower the level of the service tank below which the fuel pumps cannot take suction; (10) start the separators from the settling to the service tank; these will now be filling with low-sulphur fuel oil; (11) fill the low-sulphur fuel oil into the settling and service tanks as per the quantity required to cross the ECAs zone as calculated by the chief engineer in accordance with the passage plan. When changing over from high-sulphur fuel oil to low-sulphur fuel oil, and again from low-sulphur fuel oil to high-sulphur fuel oil, it is important to maintain an accurate record in the oil record book as the oil record book is likely to be inspected by Port State Control. The records that must be made include the fuel tank levels when at the point the changeover procedure commenced (for example, 48 hours prior); and the date, time, and position of the vessel when the changeover from high to low sulphur was started, together with the quantity of low-sulphur oil in the settling and service tanks. It is considered good practice to record this same information in the engine logbook.

Fuel changeover procedures for the boiler. Some ports have regulations pertaining to the use of gas oil boilers while the vessel is in port (for example, European ports). This requires the boiler to changeover from high-sulphur fuel oil to low-sulphur fuel oil (i.e., diesel oil with a sulphur content less than 0.1%). To perform this procedure: (1) shut off the steam to the boiler's fuel oil heaters; (2) when the temperature drops below 90°C (194°F), open the diesel oil service tank valve leading to the boiler system; (3) shut the heavy oil valve for the boiler system slowly and observe the pressure of the supply pump; (4) check the quality of the flame and combustion in the boiler; (5) keep the heavy oil outlet open and keep the diesel oil outlet shut; this is to ensure no heavy oil contaminates the diesel oil system; (6) when the line is flushed with diesel oil, open the diesel outlet valve, and shut the heavy oil outlet valve.

Fuel oil changeover for the generators and auxiliary engine. The generators must be changed over from one grade to another while at load, as this improves the flushing out of the system. If only one generator is being changed over, keep the other generator(s) running as a backup in the event of an emergency. Start by (1) shutting off the steam to the fuel oil heaters in the boiler; (2) when the temperature drops below 90°C (194°F), open the diesel oil service tank valve that feeds the generator system; (3) open the local diesel inlet valve and shut the heavy oil inlet valve simultaneously and slowly; (4) keep an eye on the fuel pressure; (5) change only one generator into diesel by way of a separate diesel pump; (6) keep the heavy oil outlet open and the diesel oil outlet shut until the system is thoroughly flushed; (7) allow some time for the system to settle before opening the diesel oil outlet and shutting the heavy oil outlet; (8) if the complete system is to be changed into diesel oil, open the diesel oil inlet valve to the generator supply pump simultaneously whilst closing the heavy oil inlet valve; (9) if the return line is provided to the diesel service tank, open after a few moments whilst simultaneously closing the heavy oil return – this should only be done after the system is flushed properly.

Once the changeover procedure is complete, change the human-machine interface (HMI) setting of the cylinder oil lubricator system (alpha lubrication) or change over the cylinder oil tank suitable for low-sulphur operation.

OVERHAULING THE CYLINDER LINER

The cylinder liner is an integral part of the combustion chamber, through which power is generated onboard. Like all other machinery and engine parts on ships, it must be overhauled according to specific intervals set by the engine manufacturer. In this section, we will discuss the liner removing procedure for a typical two-stroke marine engine.

Checking wear in the liner and when to overhaul. The liner is an enclosed area and a part of the combustion chamber where the fuel is burnt. Heat energy is transformed into kinetic energy by way of the pistons, crossheads, bearings, and crankshaft. For marine engineers, it is important to know the various methods for checking the condition of the liner to ensure the combustion chamber is producing the required pressure efficiently. There are four main methods for inspecting the condition of the cylinder, which are scavenge inspections, routine liner ovality checks, piston overhauls, and rectifying problems involving the liner.

Scavenge inspection. Scavenge inspections are performed every time the scavenge space of the engine is cleaned of sludge and deposits. After the cleaning is done, the second engineer must enter the scavenge space to check the general liner condition. This is done using the following method: (1) to inspect a larger area of the cylinder liner and piston, it is often expedient to enter the scavenge air receiver and make observations from the 'exhaust side'; (2) dismount the small covers on the scavenge air boxes and clean the openings; (3) when the piston has been turned below the level of the scavenge ports, inspect the cylinder liner walls and the piston crown.

Routine liner ovality check. Readings are taken at the port and starboard positions at various levels to calculate the change in the ovality of the liner.

Piston overhaul. When the piston is overhauled as per the preventative maintenance schedule, or in response to engine breakdown, the liner ovality must be checked and the liner surface inspected for the presence of defects.

Rectifying problems with the liner. Issues with the liner, such as a leaking water 'O' ring, cracked liner, and blow-past from the piston, require immediate remedial actions, such as checking and changing the piston and liner. The normal overhauling schedule of the liner is dependent on the efficient operation of the engine, the operator, the type of fuel used in the engine, and how the engine parameters (i.e., temperature and pressure) are maintained by the operator. If the liner is working fine, the preventative maintenance schedule must be followed, with the liner checked and gauged after a certain number of running hours as prescribed by the engine manufacturer. In most cases, this ranges from between 12,000 to 16,000 hours. The general lifespan of a cylinder liner will again depend on the way the engine is operated and the type of fuel oil used for combustion. The typical lifespan of a liner varies from 40,000 running hours to 90,000 running hours. The size of the liner bore is related to the lifespan of the liner; therefore, liners with small bores will have a shorter lifespan compared to liners with large bores.

When preparing to carry out a liner inspection or replacement, the following procedures must be followed (common to all liner types and makes): (1) inform the company and request permission to immobilise the vessel; (2) request permission from the port authorities to immobilise the vessel and wait for the issue of a Port State Control immobilisation certificate; (3) review the operations manual and hold a toolbox talk with the members of the ship's staff involved in the liner operation; (4) prepare the tools and spares needed for overhauling the liner as provided in the manufacturer's manual; (5) prepare a risk assessment and ensure all personal safety equipment is issued and donned; (6) shut the starting air for the main engine and display placards or warning signs around the engine room and on the bridge; (7) engage the turning gear; (8) open the indicator cocks for all cylinders; (9) stop the main lube oil pump and switch off the breaker; (10) once the engine jacket temperature has cooled, shut off the inlet water valve for the unit to be overhauled; (11) keep the other units in the jacket preheating system to maintain the jacket temperature; (12) drain the jacket water of the relevant unit from the exhaust valve and liner; (13) shut off the fuel oil to the unit; (14) dismount the cylinder head; (15) discard the sealing ring from the top of the cylinder liner; (16) turn the piston down far enough to make it possible to grind away the wear ridges at the top of the liner with a hand grinder; (17) dismount the piston in accordance with the manufacturer's instructions. If the cylinder liner is stuck in the cylinder, apply hydraulic pressure at the bottom of the cylinder liner. Once the liner has been dislodged, it may be lifted out by the engine room crane. After removing the liner from the engine, place the cylinder liner vertically on a wooden plank. Clean the interior of the cylinder frame, paying special attention to the contact surfaces of the cylinder liner at the top of the cylinder frame. Discard the O-ring on the cooling water pipe and clean the pipe carefully. Ensure to inspect the liner for cracks and other defects.

Liner removal (MAN B&W MC and ME engines). To remove the liner from a MAN B&W MC and or ME engine, the following steps should be followed. Ensure the liner lifting tool is well maintained. Two lifting screws are used with a lifting hook connected via a chain. Ensure the chain, screw, and lifting hook are fastened together with no deformations. Ensure the safety strap in the lifting hook is working properly. Tighten the two lifting tool screws in the liner as per the rated torque given in the manufacturer's manual on both sides. Ensure there is no gap between the liner surface and the screw landing surface after tightening. Use a 0.05mm feeler gauge. Disconnect the cylinder oil pipe connections and unscrew the non-return valves. Dismount the four cooling water pipes between the cooling jacket and cylinder cover. Clean carefully. Remove the screws from the cooling water inlet pipe. Attach the crossbar to the engine room crane; this completes the lifting arrangement for the cylinder liner. Hook the chain from the lifting cross bar on the lifting screws and lift the cylinder liner with the cooling jacket out of the cylinder frame.

Liner removal (MAN B&W RTA and RT Flex engines). Ensure the liner lifting tool is well maintained and the suspension bridge beam does not have any loose connections or distortions. Remove the screw which fastens the supporting ring to the liner. Disconnect the cylinder oil pipe connections and unscrew the non-return valves. Dismount the cooling water pipes between the cooling jacket and cylinder cover. Clean thoroughly. Remove the screws from the cooling water inlet pipe. Remove the passages for the lubricating quills and their protecting bushes. Mount the suspension bridge beam over the top landing surface of the liner. Fasten the suspension bridge beam to each side of the liner and tighten to the rated torque stated in the manufacturer's manual. Attach the engine room crane to the lifting tools. Lift the cylinder liner with the cooling jacket out of the cylinder frame.

TIMING CHAIN TIGHTENING AND ADJUSTMENTS

Conventional marine engines comprise a crankshaft and camshaft, whose combined effect is to produce power either to drive a propeller or to generate electrical power for the ship. The camshaft unit is used to drive the fuel pump and valve unit of the marine engine. For this, a chain drive is used to transmit motion from the crankshaft to the camshaft. This drive is called the timing chain and is responsible for the rotation of the camshaft, which governs the fuel pump and exhaust valve timings. In a typical two-stroke engine, the rotational speed camshaft is the same as that of the crankshaft rotational speed, and in a four-stroke engine, it is half of that of the crankshaft. For two-stroke engines, two methods are used to transmit the crankshaft rotation to the camshaft: first, by the timing gear or reduction gear, or second, by the timing chain. In this section, we will discuss the tightening procedure of the timing chain drive which is used in MAN B&W engines (see Figures 9.1 to 9.6). As always, it is necessary to inform the company and request permission to immobilise the vessel. Once permission is granted, inform the Port State Control and request an immobilisation certificate. Review the operations manual and hold a toolbox talk with the

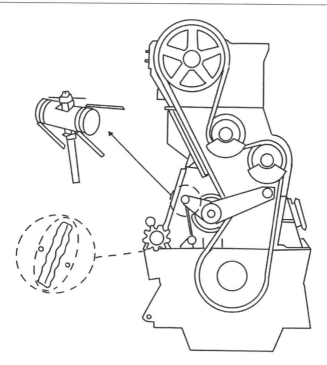

Figure 9.1 MAN B&W chain-tightening procedure (a).

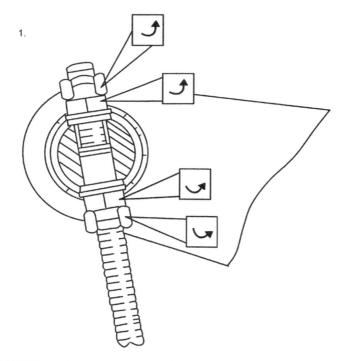

Figure 9.2 MAN B&W chain-tightening procedure (b).

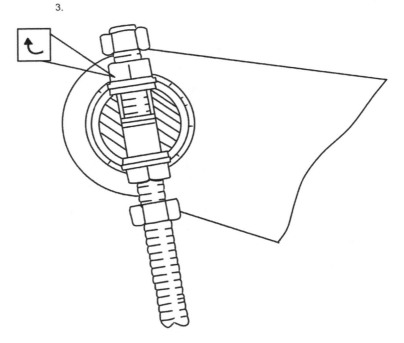

Figure 9.3 MAN B&W chain-tightening procedure (c).

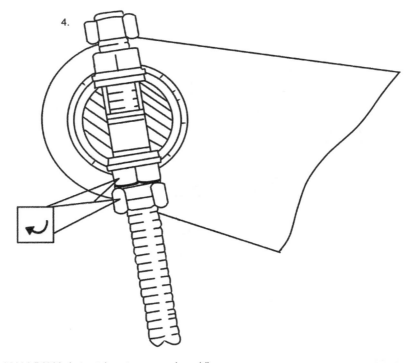

Figure 9.4 MAN B&W chain-tightening procedure (d).

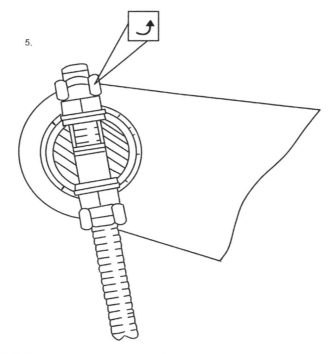

Figure 9.5 MAN B&W chain-tightening procedure (e).

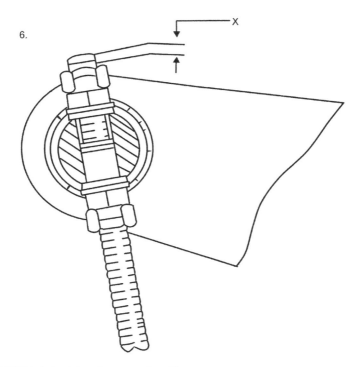

Figure 9.6 MAN B&W chain-tightening procedure (f).

members of the ship's staff involved in the chain-tightening operation. Prepare the tools and spares needed. Prepare a risk assessment and ensure all personal safety equipment is issued and donned. Shut off the starting air for the main engine. Engage the turning gear. Open the indicator cocks. After stopping and cooling down the engine, stop the main lube oil pump. Open the crankcase doors for the forward and aft unit and the door incorporating the chain drive. Turn on the blower and ventilate the crankcase chamber thoroughly. Prepare the enclosed space entry checklist, which involves checking the internal atmosphere for oxygen and hydrocarbon gases. After sufficient ventilation, don personal protective equipment and enter the crankcase chamber. Open the tab washer for nut A–B and C–D. Loosen nuts A, B, C, and D to free the chain-tightener bolt.

Turn the engine so that the slack side of the chain is on the same side as the tightener wheel. Ensure the balance weights are hanging downwards. Tighten nut B as it is loose on the chain-tightener bolt (keep measuring with a feeler gauge) until there is a minimum clearance of 0.1mm between the shaft and the nut.

Tighten nut B as stated in D-2 (tightening angle – 720 degrees equals 12 hexagons). Tighten nut C hard against the contact face of the shaft.

Tighten nut D and then lock nuts C and D with a tab washer. Tighten nut A and lock nuts A and B with a tab washer.

Measure distance 'X'. If the chain is worn, i.e., 'X' >165mm repeat the entire tightening procedure but tighten nut B on a reduced tightening angle, i.e., reduce the tightening angle to 600 degrees (10 hexagons).

EXCESSIVE WATER LOSS FROM THE FRESHWATER EXPANSION TANK

The main reason for this excessive makeup is leakage from the freshwater tank. These leaks may be caused by several issues, some of which we will now briefly discuss. *Leaks from the cylinder head 'O' rings*. This happens mainly because of insufficient preheating (below 45°C (113°F)) but stops when the engine is running, and the jacket cooling water outlet temperature is 80°C–82°C (176°F–180°F) due to thermal expansion. Regular maintenance together with the use of the correct size and type of O-ring and good cleaning of surfaces is key to avoiding this problem. In some engines, there is an intermediate cylindrical piece, which forms part of the jacket. If this is not correctly fitted (assuming the dowel pin and rubber ring are not oversized), this piece may crack. To cut off this unit, we may need to close the inlet and outlet valves of the cylinder. The jacket cooling water inlet and outlet valves of the main engine must be overhauled on all units during dry docking. Engineers also need to practice how to cut off the fuel to a particular cylinder in the correct manner, as trying to figure this out at the last moment is never a clever idea. *Leaks from the cylinder liner 'O' rings*. During cylinder overhauling, the marine engineers should try to pull out the liner and renew the 'O' rings after good cleaning of the landing surface. This process requires time and immobilisation of the ship. Even so, it is important to conduct this work whenever possible. *Leaks from the main engine turbocharger water-cooled casing*. The turbocharger casing should be cleaned chemically on the waterside (never hard scrape or hammer the casing) after a minimum of ten years of operation. Ultrasonic gauging of the casing at the top (near the air vent) and at the bottom (where mud collects inhibiting circulation) is required. If, unfortunately, the casing develops a crack, it is exceedingly difficult to trace and equally difficult to repair.

Rigging of air cooling may be needed to ensure the oil temperature does not exceed 120°C (248°F). *Leaks from the pump gland.* With improvements in pump designs and the use of mechanical seals, leaks from pump glands are quite rare in modern engines. However, on older engines, it is necessary to renew the pump sleeve using the correctly sized gland packing. *Leaks in the freshwater cooler.* The main engine's freshwater cooler for jacket cooling water should be regularly checked, cleaned, and pressure evaluated as per the ship's preventative maintenance schedule. Any leaking tubes must be plugged as per the manufacturer's instructions. *Degraded cooling water properties.* Maintaining cooling water quality is of prime importance. Once every six months, the engineers should send a cooling water sample for analysis. Always try to keep the pH of the water between 8.0 and 8.5 through regular chemical dosing. *Improper maintenance and overhaul.* The ship's engineers often overhaul the exhaust valves but do not pay minute attention to the cooling water side by removing the plugs. This means the cylinder heads may develop cracks around the air starting valve area without the engine room staff being aware. This can lead to time-consuming and costly repairs.

OVERSPEEDING AND PREVENTION

Marine diesel engines are designed to cope with the mechanical stresses associated with the centripetal and centrifugal forces of the moving parts inside within specified operational ranges. Centripetal force is directly proportional to the square of the rotational speed. This means stress increases rapidly with each increase in speed. Mechanical connection strength can be overcome by the exceeding stresses due to the increase in operational speed. This can result in damage to the rotating parts or damage to the machinery itself. Overspeed is thus a serious safety hazard and can lead to fatal situations developing quite rapidly. An overspeed trip is a safety feature provided on the diesel engine of the ship to restrict uncontrolled acceleration of the engine. Left unchecked, this can lead to mechanical failure. Due to sudden changes in the load on the diesel engine, the speed of the engine may vary. Though a governor is provided to control the speed of the engine, the speed might go out of control, causing damage to the engine.

Preventing overspeeding of engine

Reducing the likelihood of an uncontrolled and catastrophic overspeed is essential and can be achieved through one of two methods. The first is by use of a mechanical overspeed trip, and the second uses an electronic overspeed trip. In this section, we will briefly discuss the electronic overspeed trip. The electronic overspeed trip consists of (1) a *flywheel-mounted speed sensor*. Magnetic speed sensors are preferred in generator engines. Due to the discontinuity of the actuator surface (i.e., the gear tooth of the flywheel) voltage is excited in the pick-off coil of the sensor, producing an electric analogue wave. This cyclic wave as created by the flywheel is read by the sensor. (2) *Signal condition unit.* This unit acts as a receiver for the speed sensor. The basic function of the signal condition unit is to convert one type of electronic signal which may be difficult to read into another type which is more readily analysed. This can be achieved through amplification, excitation, and linearisation of the electrical signal. (3) *Detection and comparison unit.* There is a set value which is normally 10% above the rated speed. This serves as the base value for this unit. Signal condition unit output is continuously detected and compared against the set value. (4) *Trip signal unit*. If the difference between the set value and detected value is above the allowable limit, this unit gives a trip signal which in turn shuts down the generator.

In this chapter, we have discussed some of the basic engine room machinery maintenance routines and common troubleshooting procedures. We have covered crankcase inspections, faults and malfunctions, main bearing faults and malfunctions, fuel valve overhauling, fuel changeover procedures when entering ECAs, overhauling the cylinder liner, tightening the timing chain, and problems associated with the main engine freshwater expansion tank. Finally, we looked at the problems associated with overspeeding and how to prevent main engine overspeed. Before conducting any maintenance or repairs on the main engine, or indeed any equipment or machinery in the engine room, be sure to consult the manufacturer's operations manual first. A few minutes spent reading the correct procedures is time better spent than going in half-cocked and causing more damage than necessary. Remember: failing to prepare, is preparing to fail!

Chapter 10

Mechanical measuring tools and gauges

The machinery on board ships requires regular care and maintenance so that their working life and efficiency can be increased, and the cost of operation, which includes unnecessary breakdowns and spares, can be reduced. For different types of machinery and systems, various measuring tools, instruments, and gauges are used. Measuring instruments and gauges are used to measure various parameters, such as clearances, diameters, depths, ovality, and trueness. These are critical engineering parameters which describe the condition of the working machinery. In this chapter, we will briefly discuss the main mechanical measuring instruments and mechanical gauges which are used in the ship's engine room, and indeed throughout the ship:

1. *Ruler and scales*. These simple tools are used to measure length and other geometrical parameters. This tool is one of the most used measuring instruments in mechanical engineering. They can be a single steel plate or a flexible tape-type tool. They are usually available in the measuring scale of inches or centimetres. They are used for the quick measurement of parts and always kept with other measuring gauges or tools in the workshop for quick access. The ruler and scales are not used where precise measurement is required. It is made from stainless steel which is durable and will not rust or corrode.
2. *Callipers*. These are usually of two types: the inside and outside calliper (see Figures 10.1 to 10.3). They are used to measure internal and external sizes (e.g., the diameter) of an object. It requires an external scale to compare the measured value. This tool is used on surfaces where a straight ruler scale cannot be used. After measuring the body or part, the opening of the calliper mouth is kept against the ruler to measure the length or diameter. Some callipers are integrated with a measuring scale; hence, there is no need for other measuring instruments to check the measured length. Other types of callipers found on the bridge include the odd leg and divider calliper.
3. *Vernier callipers*. Counted amongst the list of most used measuring instruments, the vernier calliper is used to measure small parameters with high accuracy. It has two different jaws to measure the outside and inside dimensions of an object. It can be a scale, dial, or digital-type calliper.
4. *Micrometre*. The micrometre is an excellent precision tool which is used to measure small parameters and is much more accurate than the vernier calliper. The micrometre size can vary from small to large. The large micrometre calliper is used to measure large outside diameters or distances, for example, large micrometres are

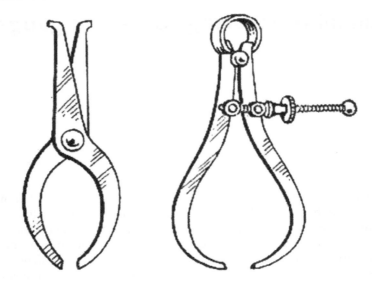

Figure 10.1 Typical callipers.

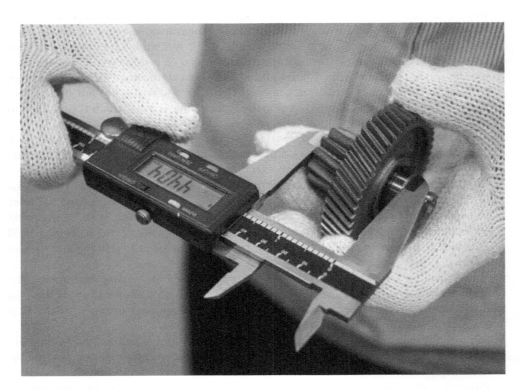

Figure 10.2 Typical vernier callipers.

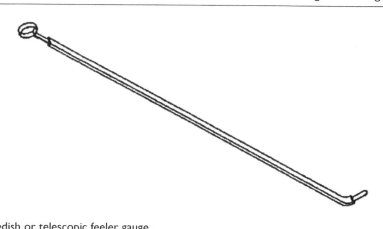

Figure 10.3 Swedish or telescopic feeler gauge.

used as a special mechanical measuring tool for the main engine to record the outer diameter of the piston rod.

5. *Feeler gauge*. Feeler gauges are a bunch of fine thickened steel strips of different thicknesses bundled together. The thickness of each strip is marked on the surface of the strip. The feeler gauge is used to measure the clearance or gap width between surfaces and bearings such as piston ring clearance, engine bearing clearances, or tappet clearances.

6. *Swedish or telescopic feeler gauge*. Similar in functionality to the feeler gauge, this type of gauge is also known as a tongue gauge, as it consists of a long feeler gauge inside a cover with a tongue or curved edge. The long feeler strips protrude out of the cover like a telescope so that the gauge can be inserted into remote places where feeler gauge access is not possible – for example, to measure the bearing clearance of the top shell. It is essential that after the use of the telescopic gauge, the strip is cleaned and retracted back into its housing; otherwise, the feeler strip may become damaged or bent out of shape.

7. *Poker gauge*. The poker gauge is one unique tool amongst many different tools used for measuring. It is available as a mechanical or digital type. The poker gauge has only one purpose: to measure the propeller stern shaft clearance, also known as 'propeller wear down'. It is a type of depth measurement instrument, whose reading indicates the wear down of the stern shaft. A special access point or plate is provided which can be either opened, bolted, secured, or welded, depending on the ship design. The poker gauge is inserted into this access point to measure the propeller drop. Because of its importance, the poker gauge is always kept locked under the custody of the chief engineer.

8. *Bridge gauge*. As the name suggests, the bridge gauge looks like a bridge carrying the measuring instrument at the centre of the bridge. It is used to measure the amount of wear on the main engine bearing. Typically, the upper bearing keep is removed, and the clearance is measured for the journal. A feeler gauge or depth gauge can be used to complete the process.

9. *Liner measurement tool*. The liner measurement tool is a special tool for marine engines which comes in a set of straight rods of varying marked lengths. These can be assembled to form a measuring tool of the required length. It is used to measure

the wear down or increase in diameter of the engine liner. The liner measurement tool is considered one of the critical tools of the engine room and as such is usually kept separate under the chief engineer or second engineer's supervision.

10. *American wire gauge.* The American wire gauge or AWG is a standard type of engine room tool which is circular and has various slots of different diameters in its circumference. It is used to measure the cross-section of an electric cable or wire. This tool is usually kept in the electrical workshop of the ship, which the electrical officer uses for measuring wire thickness.

11. *Bore gauge.* The bore gauge is a tool used to accurately measure the diameter of any hole. It can be a scale-, dial-, or digital-type instrument. The most common type which is used on the ship is the dial-type bore gauge, which comes with a dial gauge attached to the shaft and replacement rods, also known as measuring sleds, of different sizes to measure various hole dimensions. It is usually calibrated to 0.001 in (0.0025 cm) or 0.0001 in (0.00025 cm).

12. *Depth gauge.* The depth gauge is used to measure the depth of a slot, hole, or another surface of an object. It can be a scale, dial, or digital type. The depth gauge can be a micrometre style type, a dial indicator type, or a modified vernier type tool, which means the measuring base is fitted on the reading scale of a micrometre, dial indicator, or vernier scale.

13. *Angle plate or tool.* As the name suggests, this is a tool comprising two flat plates which are at right angles to each other and is used to measure the exact right angle of an object or two objects joined together. This tool is usually kept in the workshop away from other tools or chemicals which may roughen the surface of the angle plate

14. *Flat plate.* The flat plate or a surface plate is a precision flat surface used to measure the flatness of an object when it is kept over the flat plate acting as a reference. The flat plate is also kept in a workshop in a secure location, and a wooden piece is usually held on the top of the flat surface as the protective cover to safeguard the surface. Regular visual inspection and calibration need to be done to check for wear, scoring, etc., on the surface

15. *Dial gauge.* The dial gauge is used by many of the aforementioned tools and can be used separately to measure the trueness of a circular object. It consists of an indicator with a dial, which is connected to the plunger carrying the contact point. Once the contact point is put into contact with an object (to be measured), any unevenness or jumping will cause the plunger to move. The plunger is connected to the pointer in the dial. The dial is attached so that it does not retract but swings in an arc around its hinge point to show the reading in the indicator.

16. *Lead wire.* The lead wire is a conventional method that uses a soft lead wire or lead balls to measure the wear down or clearance between two mating surfaces. The lead wire or balls, which are of a fixed dimension (which is usually larger than the expected clearance), are kept between two surfaces, and both are tightened against each other just as in a normal condition. The change in the width of the lead wire or ball will show the clearance or wear down.

17. *Oil gauging tapes.* Also known as sounding tapes, these are special types of gauges which are only used to measure the level of fluids (such as heavy fuel oil, diesel oil, lube oil, or freshwater) inside the ship's tanks. The sounding tapes can be of a mechanical type where the tape is retracted into a coil and connected to a heavy bob at the end. Mechanical tapes are the most commonly used on dry ships; however,

on tankers, electronic-type sounding gauges are typically used (such as electrically powered servo-type gauges and ultrasonic types).
18. *Seawater hydrometer*. The seawater hydrometer is a small glass instrument used for measuring the density and saturation of seawater. This is an essential tool for deck officers, as the draught survey will be determined using the water density to calculate the cargo weight for loading. It is also used for ensuring compliance with the load line survey.
19. *Crankshaft deflection gauge*. This is a form of dial gauge specifically designed to measure the crankshaft deflection of the main engine. The working principle is similar to the dial gauge; the only difference is the construction, which lets this tool hang between two webs, allowing it to measure the deflection when the crankshaft rotates.
20. *Engine peak indicator*. A measuring instrument for the marine engine with a pressure indicator dial used to measure the peak pressure generated inside the engine cylinder. The pressure indicator dial is connected to the blowdown valve located on the top of the cylinder. There is a check valve provided before the indicator. When this opens, the pressurised gases continually flow inside the indicator until they reach the maximum value in the dial. Once the pressure is measured, an exhaust valve provided on the side of the valve is opened which releases the pressurised gas from the instrument. It is an oil-filled pressure gauge instrument which helps in resisting vibration and also acts as good heat resistant.
21. *Engine indicator diagram tool*. This is a cylindrical device containing the indicator piston with spring and needle, used to draw the indicator diagram for a particular cylinder when it is fixed on the indicator cock of the unit. The internal pressure changes in the cylinder are transferred to the indicator piston which is balanced with the spring. The displacement in the piston is magnified and transformed into an indicator diagram by using a precision link mechanism connected to a metal stylus.
22. *Planimeter*. An instrument which is used to measure areas of irregularly shaped areas of an arbitrary two-dimensional shape on plans or drawings.

These are the most used tools and gauge types in the engine room. In the next section of this chapter, we will look at some of the main workshop processes marine engineers are typically expected to carry out during their time on board.

WORKSHOP PROCESSES

Discussed here are some of the main workshop processes marine engineers carry out daily in the engine room:

1. *Welding*. This is the process by which metals are joined by heating and melting the metals and simultaneously adding filler material. This forms a weld pool and makes a strong joint when cooled down. It functions on the principle of coalescence. Welding is widely used for fabrication and maintenance operations. The different types of welding are electric arc, laser, electronic beam, etc., but the most common of these is electric arc welding.
2. *Brazing*. Brazing is the process of joining metals by heating base metals at a temperature of 426°C (800°F) after which a nonferrous filler metal with a melting point

well below the base metal is added to form a strong joint by capillary action. When brazing is done, flux is used, as it prevents oxide formation when the metal is heated.
3. *Gas cutting.* Gas cutting is the process of cutting metals by the application of high-temperature flame or torch produced through the combination of two gases: oxygen and acetylene. It is the most common method used on board ships. Other metal-cutting procedures are carbon air cutting, plasma arc cutting, etc.
4. *Annealing.* Annealing is a heat treatment process done to induce ductility in metal. The metal is heated above its recrystallisation temperature and then cooled. This action relieves its internal stresses and refines the structure.
5. *Riveting.* Riveting is a process of fastening one metal to another metal by the use of a riveting machine and small cylindrical shaft with a head on one end. It is not as strong as annealing or welding but is still useful throughout different parts of the ship.
6. *Lathe practice.* A lathe machine is one of the most important parts of the ship's workshop, as it is used for various purposes such as manufacturing, cutting, shaping, and checking different spares and parts of the ship. Many different operations can be performed on the lathe such as machining, surface finishing, thread making, gear making, and knurling.
7. *Drilling.* Drilling is a process of cutting or enlarging a cylindrical hole in a solid material. This is done by applying rotational pressure on top of the metal through a strong drill bit. The drill bit is a drilling tool made from a higher-strength metal like high-tensile steel or cobalt steel alloy.

Figure 10.4 Example of a crew member welding on deck.

8. *Grinding.* This process is used to smoothly cut metal and to remove edges from the metal. In this process, a grinding machine is used which rotates a highly abrasive grinding wheel which acts as a cutting tool. The grain on the wheel cuts off the shards of metal by shear deformation.
9. *Buffing.* Buffing is the process of cleaning and removing debris and hard deposits like carbon and sludge from the surface of metals. A buffing wheel or buffing tool, which is a metal wire wheel, is attached to a portable hand-driven buffing machine or a fixed buffing wheel.
10. *Tapping.* Tapping is the process of making threads in metal. Worn-out threads are restructured by using taps and drills. Tapping tools are used in series to perfect the thread. The tools are plug tap, intermediate tap, and taper tap.
11. *Thread extraction.* This is the process of removing or extracting a broken part of a bolt or metal which is threaded in a hole. Extracting tools are fitted after drilling a hole in the metal or bolt to be removed. It is a reverse tap and turns the thread in the direction of the drawn pitch (Figure 10.4).

This is just a brief overview of the main tools and processes that take place in the engine room workshop. In the next chapter, we will look at the role and function of the marine diesel generator.

Part II

Power generation

Chapter 11

Marine diesel generators

Marine electricity or marine electrical power is a vital part of a ship's operation. Without electrical power, ships would not be able to run any of their machinery. We cannot define the term 'marine electricity' as a whole. To understand its meaning, first, we need to understand the words separately. In this context, 'marine' refers to ships, ports, dry docks, and any other structures which cater to the shipping of cargo by sea. '*Electricity*' is a type of energy resulting from the existence of charged particles (such as electrons or protons), either statically as an aggregation of charge or strong as flowing current. The electricity which is produced, supplied, and distributed onboard ship, port, dry dock, and shipyard for running or repair of the cargo and passenger ships is referred to as marine electricity. Marine electricity generation can be achieved onboard ships by using diesel, shaft, or steam-driven generators. For ports, shipyards, and structures located inland, marine electricity is provided by land-based power generation plants. Unlike on land, the ship's generator has insulated neutral points, i.e., the neutral is not grounded or connected to the ship's hull. This is done to ensure all essential machinery is operational even in the event of an earth fault. Ships plying in international waters have a three-phase DC electrical supply with a 440V insulated neutral system. Vessels such as passenger ferries and cruise ships, which have large electrical load requirements, are usually provided with high-voltage operating generator sets in the range of 3kV to 11kV. By comparison, on land, the frequency of the power supplied can be 50 or 60 Hz depending on where the supply is. Ships have adopted a similar practice where 60hz frequency is standard, as this allows the hundreds of motors on the ship to run at higher speeds even where they are of smaller size. The supply which is at 440V is stepped down, using a transformer, to 220V or 110V for lights and low-power signal equipment. The electrical equipment onboard ships are like on land, however, by necessity, they are upgraded to withstand the harsh conditions of the sea including humidity, frequent temperature fluctuations, salty and corrosive atmospheres, vibration, and so forth.

SUMMARY OF THE MARINE ELECTRICAL SYSTEM ON SHIPS

The marine electrical system onboard ships can be divided into four specific systems: the generator system, the main switchboard system, the emergency switchboard system, and the distribution system.

1. *Generator system.* The generator system (Figure 11.1) consists of an alternator and driver for the alternator which can be either a diesel-driven or steam-driven engine.

Figure 11.1 Ship's generator.

Many ships are equipped with a shaft generator wherein the rotation of the main engine is used to operate the alternator to generate additional electricity.

The power produced by the generators is transported to the main switchboard using busbars. There are no electrical wire connections inside the main and emergency switchboards on ships for connecting the power supply from the generators to these switchboards, as all high-voltage and high-current systems are connected by these busbars.

2. *Main switchboard system.* The main switchboard is the distribution hub of the ship's electrical system. It takes power from the generator and distributes it to the power consumer spread all over the ship. It provides a power supply of 440V. A part of the main switchboard is provided with a 220V supply via a step-down transformer. This supplies the bridge equipment, navigation lights, radio communication equipment, and so forth. The power from the auxiliary switchboard is used to charge the battery which is used for the emergency lights.
3. *Emergency switchboard.* An emergency generator is required to always remain operational in case the main generator fails. This emergency generator must start automatically to provide power to the emergency switchboard. The emergency equipment supply is connected to the emergency switchboard. Like the main switchboard, the emergency switchboard is divided into two sections: a 440V section and a 220V section.
4. *Distribution system.* The distribution system is located after the switchboard and consists of the following components: (a) *Distribution boxes.* These boxes are enclosed

Marine diesel generators 143

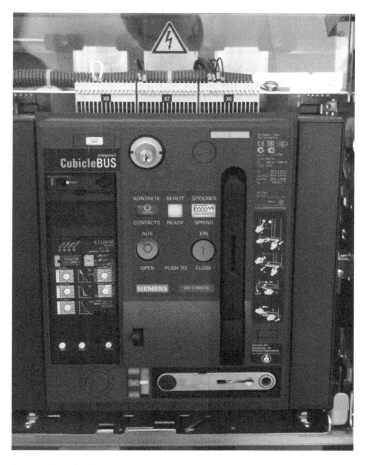

Figure 11.2 Typical circuit breaker box.

and made from metal to supply power to localised parts of the ship's machinery. (b) *Motor starter boxes*. There are hundreds of motors operating several mechanical machinery onboard ships. Each group of motors is provided with a motor starter box containing 'ON' and 'OFF' switches, together with safety devices. Local gauges for amperage and temperature are fitted to the starter panel (see Figure 11.2). (c) *Shore connection boxes*. When the ship is in a port where emission control requirements are in place, or during dry docking where the ship generator cannot run, shore power is supplied instead. The shore panel is usually located near the accommodation entry or near the bunker station to allow an easy connection of the shore supply cable. (d) *Lighting distribution panel*. The lighting distribution panel supplies power to the ship's lighting systems, accommodation systems, small heating appliances, circuits, and motors of 1/4 hp or less. (e) *Emergency switch-off panel*. For safeguarding the ship's machinery and personnel, various emergency switch-off panels are provided at various locations. These are used for shutting down machinery and equipment in emergency situations. In summary, the main objective of the distribution system is to have an operational, alarm, and safety console for individuals or groups of machinery. Power is then supplied through circuit breakers to the large auxiliary machineries at high voltage. For smaller machinery and equipment, a supply fuse and miniature circuit breaker are provided instead.

WORKING PRINCIPLES OF THE MARINE GENERATOR

Shipboard power is generated using a prime mover and an alternator working together. For this, an alternating current generator is used on board. The marine generator works on the principle that when a magnetic field around a conductor varies, a current is induced in the conductor. The generator consists of a stationary set of conductors wound in coils on an iron core. This is known as the *stator*. A rotating magnet called the *rotor* turns inside this stator producing a magnetic field. This field cuts across the conductor, generating an induced electromagnetic force (EMF) as the mechanical input causes the rotor to turn. The magnetic field is generated by induction (in a brushless alternator) and by a rotor winding energised by DC current through a series of slip rings and brushes. There are a few points worth noting: first, AC, three-phase power is universally preferred over DC as it provides more power for the same size; and second, three phase is preferred over single phase as it draws more power and in the event of failure of one phase, the other two will remain operational.

Marine power distribution

The power distributed on board a ship needs to be supplied efficiently throughout the ship. For this, the power distribution system is used. A shipboard distribution system consists of different components for the distribution and safe operation of the system. These are the ship's generator consisting of the prime mover and alternator; the main switchboard which is a metal enclosure taking power from the diesel generator and supplying it to the ship's machinery; busbars, which act as the carrier and allow the transfer of load from one point to another; circuit breakers, which act as a switch and in unsafe condition can be tripped to avoid electrical overloads; and fuses, which act as a safety device for individual machineries. Transformers step up or step down the voltage. For example, when supply is to be given to the lighting system, a step-down transformer is used to reduce the electrical current supplied by the distribution system. As we have already noted, the power distribution system supplies voltage at 440V. There are some large installations however where the voltage is as high as 6,600V. In these instances, a step-up transformer is used to increase the voltage. Power is supplied through circuit breakers to large auxiliary machineries which operate at high voltage. For smaller machineries, fuses and miniature circuit breakers are used instead. The distribution system consists of three wires which are neutrally insulated or earthed. In most cases, an insulated system is preferred to the earthed system as during an earth fault essential machinery such as the steering gear can be lost.

Marine emergency power

In the event of a failure of the main power generation system, an emergency power system or a standby system is needed. The emergency power supply ensures that essential machineries and systems continue to operate unimpeded. Emergency power can be supplied by batteries or an emergency generator, or both. The rating of the emergency power supply should be such that it is able to provide sufficient power to the ship's essential systems including the steering gear system, the emergency bilge and fire water pumps, watertight doors, the firefighting system, the ship's navigational lights and emergency lights, and the ship's communications and alarms systems. The emergency generator is normally located outside the machinery space of the ship. This is done to avoid situations where access to the engine room is not possible. A switchboard in the emergency generator room supplies separate power to the ship's essential machinery.

ESTIMATING THE POWER REQUIREMENT FOR THE SHIP

One of the most important stages of the ship design process is the estimation, calculation, and optimisation of the ship's power requirements. A ship with higher power requirements will automatically demand larger amounts of fuel for each passage, resulting in significantly higher running costs for the ship operator. Moreover, in response to attempts by the shipping industry to improve its environmental credentials, ships are now rated in terms of their overall efficiency using the Energy Efficiency Design Index (EEDI). The lower a vessel rates on the EEDI, the more efficient the ship is from an environmental perspective. As the rating of the EEDI is proportional to the ship's power requirement, naval architects and ship designers are increasingly obliged to reduce the power requirements of their vessels in every way possible. In reducing the power requirement, the EEDI rating decreases (i.e., improves), which in turn diminishes the ship's carbon footprint. Given that we now know the importance of determining the ship's power requirement, we can begin to look at how this power requirement is determined in the first place. The step in this process is to calculate the resistance of the ship. To calculate the resistance of the ship, we must first conduct a towing tank test. In the case of new hull forms, a towing tank test is always preferred. If, however, the hull form of the ship in design has already been tank tested, we need not repeat it and instead can follow the scaling method (which we will discuss later). In a towing tank test, the resistance of the model scale can be obtained through the computation of the carriage. This is then scaled up to the ship's scale using a set of steps as recommended by International Towing Tank Conference (ITTC). It is worth noting the towing tank only provides the bare hull resistance of the ship. Air resistance, resistance due to appendages, and correlation allowances must be added to obtain the actual total resistance of the ship. This total resistance, when multiplied by the ship's velocity, provides the effective power of the ship (PE). The second step is to decide on the type of propulsion system. This is one of the most important decisions to be made in the ship design process. Selecting the wrong type of propulsion system can cause long-term complications and even render the vessel operationally uneconomical. Diesel mechanical propulsion is preferred in most cargo ships which require low-speed operations and lower operating costs. If we recall from Part I, the operation cost for heavy-sulphur fuel oil as used in marine diesel engines is cheaper than the cost of operating diesel-electric type propulsion systems. As we noted in Chapter 6, slow steaming has become the accepted method for countering the effects of increased fuel costs. Subsequently, diesel propulsion is by far the most preferred propulsion system for large and heavy vessels such as bulk carriers, oil tankers, and container ships. Diesel-electric propulsion, on the other hand, is preferred for ships which require more electrical power (for example, cruise ships require more electrical power to run their passenger and hospitality facilities, and drill ships require electrical power for the operation of dynamic positioning systems) and ships that require constant operations

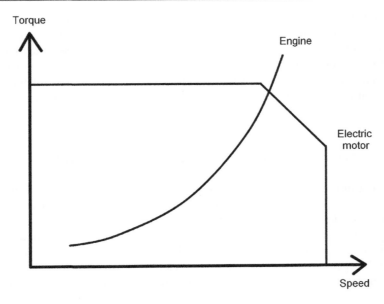

Figure 11.3 Variable torque operations for diesel engines and electric propulsion.

with varied torque demands (such as tugs). This is one of the most notable advantages of diesel-electric propulsion. Where diesel mechanical propulsion systems do not offer high efficiencies at all torques, diesel-electric propulsion systems can operate at high efficiencies at all torque variations.

In Figure 11.3 we can see it is evident that the torque supplied by the diesel engine varies with speed. That is, higher torque can only be obtained at higher operating speeds. But if higher torque is required at lower speeds (which is most often the case for tugs and dynamic positioning systems), an electric motor (which is used in the diesel-electric propulsion system) is needed.

The third step is to estimate the engine or diesel alternator ratings. *Diesel mechanical propulsion*. The resistance calculated from the towing tank tests only provides the bare hull resistance – that is, the effect of the propeller is not considered. This is important as when a propeller operates behind the ship, the resistance of the ship increases from the value calculated in bare hull condition. The propeller must operate at a torque that is sufficient to overcome this augmentation in resistance, as well as enable the ship to overcome its bare hull resistance. Hence, due to the loss in power induced by the operation of the propeller, the power delivered to the propeller (PD) at the shaft output should be more than the effective power (PE). The ratio of the effective power to the delivered power is called the *Quasi Propulsive Coefficient (QPC)*. QPC usually ranges from 0.55 to 0.65. The power of the engine output (i.e., the shaft input) is not fully obtained at the shaft output. This is because of the frictional and heat losses that occur along the length of the shaft. These are referred to as 'shaft losses'. Shaft loss margins are usually accepted to be as much as 2%. In the case of smaller ships where high rpm engines are used, reduction gearboxes either reduce the shaft rpm or the engine is operated at variable rpms. The losses induced by the gearbox are called 'gearbox losses'. Gearbox losses range from 5% to 5%. The resistance estimated during the design phase does not consider the effect of waves.

Table 11.1 Engine-rated power calculations

	Factor	Power	
Effective power		200	KW
QPC	0.55	363.6364	KW
Shaft losses	0.98	371.0575	KW
Gearbox losses	0.95	390.5869	KW
Sea margin	0.85	411.1441	KW
85% MCR operation	0.85	463.6989	KW
Total engine BP		483.6989	KW
		648.6509	BHP
BP per engine		241.8494	KW
		324.3255	BHP

Due to wave actions, the actual resistance which acts on the ship is higher than that in calm water conditions. Hence, a margin of 15% is used as the 'sea margin'. This means the engine power must be rated to enable it to overcome the sea margin. It is always desirable to keep SFOC as low as possible. For marine diesel engines, SFOC is the minimum where the rpm of the engine corresponds to 85% of MCR. This means that the design speed should be calculated not at the rated MCR but at 85% of the MCR. To obtain the MCR, the corresponding factor of 0.85 is used. Table 11.1 illustrates the calculation that is used to obtain the rated engine power from the effective power of a twin-engine ship using the aforementioned factors:

Diesel-electric propulsion. In this section, we will discuss the basic components of a diesel-electric propulsion system just to the extent that makes it possible for the reader to understand what we will be discussing regarding the estimation of power rating a diesel-electric propulsion system. The basic components of a diesel-electric propulsion system are (1) the diesel generators, (2) transformers, (3) electric motors, and (4) loads. Now, the loads on the system may be an electric motor-driven propulsion pod, a bow thruster, or indeed any component of the ship's hotel load (such as lighting, HVAC). In this instance, the electric propulsion motors, propellers, and other loads together form the load of the entire power plant. Remember though that not all loads will be in operation in every condition. For example, when the vessel is in port, the propulsion loads will be absent, whereas the hotel loads will be operational. Where the vessel has dynamic positioning systems installed, both hotel load and propulsion units will be in operation. In this latter case, the load on the diesel generators will be at its maximum. With this information in mind, we need to calculate the total power requirement before deciding on the number of diesel generators required to meet all load conditions. Once the total power requirement has been determined, the number of diesel generators needed can be decided upon based on certain principles that we shall discuss later. In the first instance, to calculate the total power requirement, the ship designer needs to prepare a load chart which lists all the electrical loads of the ship. As the load chart is prepared, the designer must consider the following three operating conditions: sailing, harbour, and manoeuvring. In the load chart, the power requirements of each electrical load on the ship are calculated by multiplying the maximum rated power (MRP) of the component with two factors: (1) the load factor, which is the ratio of the operating power to the maximum power rating of the component,

Table 11.2 Utility factor for a steering gear assembly

EQUIPMENT	Nos Installed	Nos in Use	Maximum Power of Each in kW	Installed Power in kW	SAILING			HARBOUR			MANOEUVRING		
					LF	UF	Power [kW]	LF	UF	Power [kW]	LF	UF	Power [kW]
Steering Gear	2	1	24.00	26.67	0.8	0.8	17.07	0.8	0.0	0.00	0.8	0.8	17.07

and (2) the utility factor, which is a factor that determines the extent of the operation of the particular component in a particular condition. As an example, refer to Table 11.2.

Note that the utility factor is '0.8' for sailing and manoeuvring conditions, but '0' in harbour condition, as in a harbour condition the steering gear is not used. This means the contribution of the steering gear equipment to the total power requirement in a sailing condition will be zero. In the analogous manner as illustrated earlier, the load chart is prepared for all the electrical components on the ship. A sample of which would look like Table 11.3.

Once the load chart is prepared, the total power requirement for each of the three conditions (sailing, harbouring, and manoeuvring) can be calculated by adding up the power requirement for each component for each of the conditions (see the following). Once these calculations have been established, we can understand how the total number of diesel generators is determined:

1. Sailing Condition

   At SAILING total sea load in kW = 1,258.28

   2 × 80 0 kW – Generators
   % Load on each generator = 63.41%

2. Manoeuvring Condition

   At MANOEUVRING total load in kW = 1,669.48

   3 × 800 kW – Generators
   % Load on each generator = 55.65%

3. Harbour Condition

   At HARBOUR total load in kW = 658.73

   1 × 800 kW – Generator
   % Load on generator = 65.87%

The two rules to be followed in deciding the number of generators are (1) if more than one generator is operating in any condition, both generators should share an equal amount of load, and (2) the load on each generator in any of the three conditions should not be more than 70% of the rated power of the generator (or the maximum rating of each generator

Table 11.3 Electrical load chart

EQUIPMENT	Nos Installed	Nos in Use	Maximum Power of Each in kW	Installed Power in kW	SAILING			HARBOUR			MANOEUVRING		
					LF	UF	Power [kW]	LF	UF	Power [kW]	LF	UF	Power [kW]
Steering gear	2	1	24.00	26.67	0.8	0.8	17.07	0.8	0	0.00	0.8	0.8	17.07
Windlass	1	1	37.00	41.11	0.8	0	0.00	0.8	0	0.00	0.8	0.7	23.02
Baggage crane	2	1	14.00	15.56	0.8	0	0.00	0.8	0.6	7.47	0.8	0	0.00
Mooring winch	2	1	20.00	22.22	0.8	0	0.00	0.8	0	0.00	0.8	0.7	12.44
Engine room crane	1	1	4.00	4.71	0.8	0.1	0.38	0.8	0.2	0.75	0.8	0	0.00
Provision davit	2	2	5.00	5.88	0.8	0.1	0.94	0.8	0.5	4.71	0.8	0.1	0.94
Galley equipment	1	1	489.47	543.86	0.8	0.2	87.02	0.8	0.1	43.51	0.8	0.2	87.02
Laundry equipment	1	1	85.10	94.56	0.8	0.2	15.13	0.8	0	0.00	0.8	0.2	15.13
Ventilation system	1	1	109.76	121.96	0.8	0.8	78.05	0.8	0.4	39.03	0.8	0.8	78.05
Side thruster	2	2	250.00	277.79	0.8	0	0.00	0.8	0	0.00	0.8	0.7	268.89
Incinerator	1	1	14.00	16.477	0.8	0.2	2.64	0.8	0	0.00	0.8	0.2	2.64
Workshop equipment	1	1	10.00	11.11	0.8	0.2	1.78	0.8	0.2	1.78	0.8	0.2	1.78
Welding equipment	1	1	32.00	35.56	0.8	0.1	2.84	0.8	0.1	2.84	0.8	0.1	2.84
Starting air compressor	2	2	8.60	10.12	0.8	0.2	3.24	0.8	0.3	4.80	0.8	0.3	4.86
Control air compressor	1	1	2.90	3.41	0.8	0.4	1.09	0.8	0.3	0.82	0.8	0.4	1.09
Control air drier	1	1	0.30	0.35	0.8	0.4	0.11	0.8	0.3	0.08	0.8	0.4	0.11

is calculated based on the condition that 70% of the maximum rating is more than the load on the generator in any one of the three conditions). One additional generator should always be included, which is for emergency standby. Note that this standby generator will not share the load in any of the aforementioned three conditions unless any of the working generators are taken out of service. Therefore, the standby generator is not included in the earlier calculations, though it will usually have the same rating as the other generators. This process is iterated by varying power ratings and varying numbers of generators until the earlier first two conditions are satisfied, and a situation like the one outlined on page 000 is obtained.

STARTING AND STOPPING THE GENERATOR

Unlike the conventional generators that are used on shore, a ship's generator requires a special procedure for starting and stopping. Though not particularly complex, the process demands good diligence and a step-by-step system to be followed. Missing just one step can lead to the generator failing to start or stop. This can lead to a blackout situation, a situation best avoided at all costs (we will cover blackouts in more detail in the next chapter). In this section, we will cover the normal step-by-step procedure for starting and stopping a typical marine generator (bear in mind some generators and ships may have their own specific operating procedures).

Generator starting procedure – automatic start

This method is only possible if there is sufficient starting air available. The air valves and interlocks are operated in the same manner as turning gear. In this method, the operator has nothing to do, as the generator starts itself depending on the load requirement. During manoeuvring, however, and in restricted areas, the operator must start the process by going into the computer-based power management system (PMS). Once inside the system, the operator goes to the generator page and clicks start. In the PMS, the automation follows the sequence of starting, matching the voltage and frequency of the incoming generator. Once the correct parameters are met, the generator will come on load automatically. In the event of a blackout or dead ship condition, the engineer may have to start the generator manually.

Generator starting procedure – manual start

Before we start, it is worth noting that this manual starting procedure is not followed on unmanned machinery space (UMS) ships. Furthermore, in engine rooms which have water mist firefighting systems installed, this procedure is not followed, as when the engine is given a manual kick with open indicator cocks, a small amount of smoke is emitted from the heads. This can lead to a false alarm, resulting in the release of water mist.

The manual process is different from the automatic start system. First, we must check that all the necessary valves and lines are open, and no interlock is active on the generator before operating. Before starting the generator, the indicator cocks are opened, and a small air kick is given by way of the starting lever. After this, the lever is brought back to the zero position, which ensures there is no water leakage into the generator. The leakage may come from the cylinder head, the liner, or from the turbocharger. The generator controller is then set to the local position; this ensures the generator can be started locally. In the event any water is found, this must be reported to a senior engineering officer or the chief engineer immediately. Before any further steps can be taken, the water must be drained, and the cause of the leak identified and rectified. Having checked for leakages, the indicator cocks are closed, and the generator is started again from the local panel. The generator is then allowed to run on a zero-load condition for about five minutes. After this interval, the generator control is changed to remote mode. Once the generator controller is put into remote mode, the generator will come on load automatically having checked the voltage and frequency parameters. If this does not happen automatically, then the engineer must go to the generator panel in the engine control room and check the voltage and frequency parameters of the incoming generator. Where necessary, the frequency can be increased or decreased by the frequency controller or governor control on the panel. The incoming generator is checked in synchroscope to determine whether it is running fast or slow. This means the frequency is too high or too low (again, we will cover this later in more detail). In synchroscope, the needle is checked to see whether it moves in a clockwise and anticlockwise direction. A clockwise direction means it is running fast and an anticlockwise direction means it is running slow. In this case, the breaker is pressed when the needle moves in a clockwise direction very slowly until it settles at the eleven o'clock position. It is important that this process is performed under the supervision of an experienced officer, especially for the first time. Any mistakes in the manual procedure can lead to a blackout condition, which as stated earlier, must be avoided at all reasonable costs. Once the procedure is complete, assuming the parameters are correct, the generator load will be shared equally by the number of generators running. It is then necessary to monitor the operation of the generators against any abnormalities.

Generator stopping procedure – automatic stop

In this procedure, the generator is stopped by going into the PMS on the engine room control system and selecting the stop button to bring the generator to a stop. This procedure must be conducted only when two or more generators are running. Even if we try to stop the only running generator, the safety system will inhibit us from doing so. This prevents potential blackout conditions. When the stop button is pressed, the load is gradually reduced by the PMS until the generator is brought to a standstill. This completes the automatic stopping procedure.

Generator stopping procedure – manual stop

In this procedure, the generator to be stopped is taken off load from the generator panel in the engine control room. The load is reduced slowly by the governor control on the panel. The load is reduced until the load comes onto the panel below 100 kW. Once the load is below 100kW, the breaker is pressed and the generator is taken offload. The generator must be allowed to run for five minutes in an idle condition, after which the stop button is pressed on the panel. This completes the manual stopping procedure.

Situations where the generator must be stopped immediately

The generator, being the powerhouse of the ship, requires regular maintenance and overhauling to ensure it operates in an efficient and safe condition. As such, a responsible marine engineer will never wait to conduct maintenance procedures until their machinery is on the cusp of breakdown. Rather, good practice dictates that all necessary precautions are taken to ensure the generator is well maintained and looked after. That said, there is a very thin margin between a fault starting and a fault evolving into a major problem. Determining and diagnosing the presence and cause of faults (minor or major) is a key responsibility of the ship's engineers. Sadly, this is a skill which takes many years of seagoing experience to perfect. To hasten the transition from novice to expert, we have listed the main situations where the generators must be stopped immediately and the standby generator brought on load (as in all situations, individual ships and generator operating procedures will differ. Always seek advice from a senior engineering officer before acting).

1. *Abnormal sounds*. The ship's generator comprises heavy oscillating and moving parts. The attached auxiliaries, such as the turbochargers and pumps, are also high-speed machines which produce loud volumes. Abnormal sounds, no matter how faint or inconsequential, must never be ignored and investigated immediately.
2. *Smoke*. When smoke is coming from or near the generator, it is imperative to stop the generator immediately. There is no need to offload the generator, as the situation

has already likely passed the danger point. Use the emergency stop button provided at the local or remote station. Smoke can be caused by a variety of reasons including friction between moving parts, overheating, and so on. Importantly, never panic. Panic is often our first reaction when we see smoke or fire. Although this is natural, it reduces our ability to think and assess the situation rationally. Try to remain calm and seek assistance if necessary.

3. *Unusual lubricating oil parameters.* If the lubricating oil temperature has increased beyond normal or the oil pressure has dropped below the adequate level, stop the generator immediately and diagnose the issue. This may be as simple as a dirty lube oil cooler or choked filter. If a drop in pressure is noted, usually the first thing that comes to mind is to change to a standby filter. If the standby filter is not primed and put into service in a running condition, bearing damage may occur due to airlock. It is always preferred to stop the machinery and then change to a fully primed standby filter.

4. *Higher differential pressures.* Differential pressure is a term used to assess the condition of the lube oil filter by providing a pressure measurement before and after the filter. The difference between the before and after filter pressures is then displayed by a gauge. If the differential pressure is in the higher range, stop the generator and change to standby filter. On numerous occasions, it has been observed that the generator is allowed to run even when the differential pressure alarm has sounded during manoeuvring. Engineers usually prefer not to risk changing the filter when in a running condition, as doing so may lead to a blackout condition if the filter does not perform correctly. Subsequently, the plan is to change the standby filter once the manoeuvring is complete. However, due to this, it is common for the differential pressure to increase, further resulting in a sudden drop in oil pressure, which in turn trips the generator during the manoeuvring.

5. *Overspeeding.* The generator is a high-speed machine though overspeeding has, in the past, resulted in explosions and fatalities. Overspeeding is due to a problem in the fuel system or, specifically, a malfunction of the governor. If the generator is running above its rated speed and does not trip, the engineers must stop the generator immediately to avoid precipitating a major accident. Once the generator has stopped completely, conduct a thorough crankcase inspection and a renewal of the bottom end bolts. During trial running of the generator after overhauling, governor droop is altered to acquire the required speed as stated in the manufacturer's manual. It may happen that the generator overspeeds due to a wrong setting or a stuck fuel rack. In either case, the importance of conducting thorough inspections and checks cannot be stressed enough.

6. *Cooling water supply.* Cooling water is essential for ensuring the smooth running of all elevated-temperature moving parts. If there is no cooling water supply due to the failure of pumps, the generator should be stopped immediately to avoid overheating damage. Sometimes when there is no cooling water pressure in the line, engineers try to release air from the purging cock provided near the expansion tank line of the generator. This should be avoided, as if there is insufficient water supply (due to the failure of the supply pump), this will lead to further increases in temperature and the stopping of the generator at a later stage. This will inevitably result in the seizure of moving parts. To avoid this, always stop the generator first and then conduct troubleshooting. If the generator has stopped due to starvation of water, the flywheel should be rotated with lubricating oil to avoid further seizure.

7. *Pipe leaks.* If any leaks are found in the fuel, lube oil, or cooling water pipes, this must be rectified only after stopping the generator. This will allow the engineer to rectify the cause of the leak safely. If there is a small fuel oil or water leak from any of the pipe connections, tightening the connection may stop the leak but overtightening may lead to a sudden increase in leakage. If the leakage involves elevated-temperature oil or water, this can lead to severe burns and scalding.
8. *Vibrations and loose parts.* Vibration is one of the main causes of increased wear on moving parts. If loose bolts are found or heavy vibration is detected when the engine is running, stop the generator immediately and find the cause for rectification. It is not a widespread practice to check the tightness of the foundation bolts of the generator and its attached auxiliaries, such as the turbocharger. In fact, many shipping company preventative maintenance systems do not even include the tightening of the foundation bolts as part of their scheduled maintenance. This clearly presents an issue, which is easily addressed by including foundation bolt inspections as part of the standard maintenance routine.
9. *Non-functional alarms and trips.* If at any point an alarm associated with the running generator is found to be inoperable, the generator must be stopped immediately, as there is a possibility that other important alarms and trips are also not working. This can lead to major accidents in the event of a generator failure. Sadly, there is a tendency amongst more experienced engineers to ignore alarms and signals as they assume they are either unimportant or faulty. This is clearly the wrong attitude to have and is certainly taken very seriously by port state control authorities. Always ensure all alarms and safety systems are in good working order.
10. *Water-contaminated oil.* Water leaking into oil decreases the load-carrying capacity of the oil resulting in bearing damage. In these situations, the generator must be stopped, especially where the water content is extremely high. Immediately investigate the cause of the leak and fix accordingly. If necessary, renew or purify the sump oil before bringing the generator back online. There are several well-known cases where generator failures were caused by poor engine room housekeeping. There is a reason generator lube oil tests must be conducted regularly. The effect of even substantial amounts of water contamination may not be seen immediately but will certainly lead corrosion and damage of the crankshaft and bearings in the long term.

The stopping of the generator is not limited to the aforementioned points only. There are many other situations or circumstances that require the generators to be stopped immediately. However, it is the duty of the marine engineer to use their knowledge, experience, and expertise to avoid any kind of breakdown. A good engineer will always think of the worst and hope for the best!

GENERATOR SYNCHRONISATION

Now that we have discussed the power ratings for ships, we can move on to the methods for synchronising an incoming generator or alternator. Generator synchronisation is especially important before paralleling one generator with another. The process of synchronising the generator is done with the help of a synchroscope (during normal operations) or by using the three-bulb method (in the event of an emergency). It is important to remember that prior to paralleling the generators the frequency and voltage of the

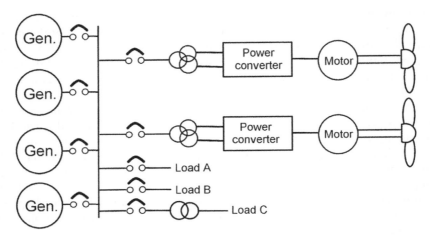

Figure 11.4 Basic components of a diesel-electric propulsion system.

generators are matched. In this section, we will discuss the methods for synchronising and paralleling the generators on a ship. As we said, there are two methods to synchronising a ship's generator: one is the normal method; the other is the emergency method. The synchroscope consists of a small motor with coils positioned on two poles connected across two phases. Let us say it is connected as red and yellow phases of the incoming machine and armature windings supplied from the red and yellow phases of the switchboard busbars. The busbar circuit consists of an inductance and resistance connected in parallel. The inductor circuit has a delaying current effect of 90 degrees relative to the current in resistance. These dual currents are fed into the synchroscope by way of slip rings to the armature windings, which produces a rotating magnetic field. The polarity of the poles will change alternatively in a north-to-south direction with changes occurring in the red and yellow phases of the incoming machine. The rotating field will react with the poles by turning the rotor either in a clockwise or anticlockwise direction. If the rotor is moving clockwise, this means the incoming machine is running faster than the busbar. It is running slower when turning in an anticlockwise direction. It is preferred to adjust the alternator speed to be slightly higher, which will move the pointer on the synchroscope in a clockwise direction. The breaker is closed just before the pointer reaches the twelve o'clock position, at which point the incoming machine is in phase with the busbar (Figure 11.4).

Emergency synchronising lamps (three-bulb method)

This method is used when there is a failure of the synchroscope. In the event of failure, a standby method must be available to synchronise the alternator. This method is the emergency lamp or three-bulb method. Three lamps should be connected between three phases of the busbar. The incoming generator should then be connected. The lamps are connected in this manner because, if they are connected across, the same phase lamps will go on and off together when the incoming machine is out of phase with the switchboard. In this method, the two lamps will illuminate, and one lamp will remain unlit when the incoming machine comes on phase with the busbar. The movement of these illuminated and unlit lamps indicates whether the incoming machine is running faster or slower. For instance, there will be a moment when lamp 'A' will be unlit and lamps 'B' and 'C' will illuminate. Similarly, there will be an instance when lamp 'B' is unlit and lamps 'A' and 'C' are illuminated, and lamp 'C' is unlit and lamps 'A' and 'B' are illuminated. This indicates that the machine is running fast, and the movement of the lamps from unlit to lit suggests a clockwise movement. A clockwise movement indicates a fast running, and an anticlockwise movement indicates a slow-running, incoming generator.

MAINTENANCE AND OVERHAULING OF THE MAIN GENERATOR

A ship simply cannot function without a working generator. It is the lifeline and power production plant of the vessel. The generator is a combination of two separate systems: an alternator and a prime mover, whose capacity depends on the number of machinery or power-consuming items installed on the ship. The alternator is an electro-mechanical device comprising a stator, rotor winding, and an external exciter for supplying excitation voltage. The alternator generates electricity when coupled with the prime mover. The alternator on a ship is often exposed to harsh weather and sea conditions, due to which, its capacity and efficiency tend to reduce. It is particularly important, therefore, to ensure proper maintenance is performed on the alternator part of the generator as per the ship's preventative maintenance schedule, or as and when required. When performing maintenance and overhauling on the alternator, the following points should be observed. Before starting any maintenance work on the alternator, all safety precautions should be taken. The alternator should be shut and locked down. Safety notices should be posted around the engine room and on the bridge advising that the alternator heater is isolated. Clean the alternator ventilation passage and air filter. Check the insulation resistance of the stator and rotor winding. Check the air gap between the stator and the rotor; this should be maintained between 1.5 to 2 mm. The slip rings should be checked for even wear and renewed if required. The carbon brushes should be cleaned and checked for free movement. Check the brush contacting pressure using a spring balance. Check the automatic voltage regulator and wipe off any oil and dust. The lube oil level of the pedestal bearing must be maintained and renewed as per the preventative maintenance schedule. A vacuum cleaner may be used to remove dust accumulations within the inner parts of the alternator. The terminal box cover gasket should be checked for oil and watertightness. All connections in the terminal box must be tightened properly. Check the cable gland for integrity. Forced ventilation around the alternator must be always maintained. Check the heater for proper operation. Finally, check the alternator foundation bolts for tightness. After the maintenance is performed, a zero-load test should be conducted and general conditions, such as noise, temperature, the voltage generated, should be observed and recorded.

DECARBONISATION (D'CARBING)

The decarbonisation or d'carbing, otherwise known as the major overhauling, of the ship's generator is an important and complicated task requiring patience and professional skill. As a marine engineer, it is one of many primary responsibilities to conduct the generator overhauling procedure during routine maintenance or in the event of an emergency. The purpose of d'carbing the generator is not only to clean and remove carbon deposits from the interior parts and spaces involved in combustion but also to check, overhaul, and renew parts involved in the power transmission process, such as the connecting rods, connecting-rod bearings, and main bearings. Thorough knowledge of the generator d'carb procedure is, therefore, critical for marine engineers of all levels employed on ships. Before and during the overhauling process, a variety of tests are performed on various tools and parts of the generator. The following are some of the important tests that are required during the major overhauling of the ship's generator. (1) *Hydraulic jack test*. During overhauling, a variety of hydraulic jack tools are used for opening the generator's cylinder head, bottom end bolts, main bearing bolts, etc. To ensure a smooth d'carb process, these should be assessed before use. (2) *Cylinder head test*. The cylinder heads onboard ships are commonly overhauled and reused. Even the heads supplied from onshore are usually reconditioned. This means it is important to pressure assess the heads for any leaks. Pressure assessing the generator cylinder head is done using water and air. (3) *Bearing cap test*. The serration provided in the bearing housing holds the two caps against each other along with the con-rod bolts. Any damage to bolts will result in damage to the bearing cap. The bearing cap serrations must be checked for cracks by using a die penetrant test kit. (4) *Connecting-rod bolt test*. The bottom cap holds the connecting-rod bearing by way of bottom end bolts. These are subjected to reversal stresses. Crack tests on the connecting-rod bolts are performed during overhauling using a die penetrant test kit. (5) *Connecting-rod bend test*. The connecting rod is subjected to extreme pressures. When overhauling the generator, the connecting rod is checked for straightness by inserting a brass rod into the oil hole of the connecting rod. As the brass rod is slightly larger than the oil bore, any bend in the connecting rod (which cannot be seen with the naked eye) will become visibly obvious, as the brass rod will not pass through the bore. (6) *Fuel injector test*. The fuel injectors are reused after overhauling. With time, the internal parts, which have exceptionally fine clearances, are subject to wear and tear. Any increase in clearance will lead to dripping or other injection problems, eventually resulting in improper pressure injection. To avoid this, the fuel injectors are pressure assessed using an injector testing stand. (7) *Starting air valve test*. Like the fuel injectors, the air starting valves are also overhauled and reused. This means they must be checked for proper operation. The starting air valves are assessed by inserting service air, which will indicate any leaks prior to installing them into the cylinder head. (8) *Relief valve test*. The relief valve of the cylinder head is pressure assessed to check for proper functioning. It is an important part which prevents explosion of the head or damage to the combustion chamber caused by overpressure. Pressure testing is conducted on a bench-mounted test rig consisting of high-pressure air, a pressure control valve, and calibrated gauges. The relief valve is bolted to the accumulator flange, and the air pressure is increased until the valve lifts. The valve is then reset accordingly. (9) *The current test*. This is an important test which is done prior to trying out the generator with fuel after the completion of the d'carb procedure. Once the d'carb is complete, the turning gear is engaged with the indicator cock open. The engine is then turned, with the current continuously monitored. Any fluctuation or increase in the current value indicates some form of obstruction or an issue with the rotating shaft. The last test in our list is (10) the *alarm and trips test*. The alarm and trips of the generator are electrical systems consisting of wirings and contacts. To check their correct operation, tests of the alarms and trips are performed including the lube oil trip, cooling water high-temperature trip, and the overspeed trip. Once the necessary tests and checks have been completed, the actual d'carb procedure can begin.

The d'carbing or major overhauling of a ship's generator is a very tedious task for the marine engineers on board. But, as we mentioned earlier, it is an especially important procedure and forms one of the many key responsibilities of the ship's engineering department. Following a step-by-step procedure, backed up by systematic planning, is the key to completing an efficient and effective generator overhaul. Before d'carbing any of the generators, management- and operational-level engineers must conduct preparation. For management-level engineers, their responsibility will lie in acquiring the necessary permissions and authorisations to isolate the ship's engines and perform the d'carbing procedure whilst alongside. The following preparations and checks should be carried out before any d'carb of the generator is started: (1) ensure that other generators are available for taking up the load of the ship at all times; (2) the d'carb of the generator should be planned so as to not to adversely affect the operation of the ship (if at sea) or impact on other vessel movements (if alongside); (3) ensure all necessary tools and equipment are available onboard and in good condition; (4) check with the engine stores that all spares required for the d'carb are available and in good condition; (5) raise a requisition for all spare parts required for the d'carb operation. Once the five steps have been completed, and the chief engineer has authorised the d'carb to commence, following preparations should be carried out: (1) initiate a toolbox talk with all members of the ship's crew involved in the d'carb operation; (2) isolate the relevant generator from the main switchboard and auto start panel; (3) close all of the systems attached to the generator, such as the seawater system, fuel oil system, air system, and lube oil system; (4) inform the bridge of the impending procedure and situate placards in clearly visible locations on the bridge, outside the engine room and around the generator; (5) review the previous d'carb report and complete the pre-d'carb checklist; (6) if the d'carb includes the crankshaft and bearings, record any crankshaft deflection prior to starting; (7) carry out a detailed risk assessment and ensure all members of the ship's crew involved in the d'carb read and sign the risk assessment.

Once the risk assessment has been completed and signed by all concerned, the d'carb procedure can begin. As with all shipboard operations and activities, safety should be the primary concern, and personal protective equipment should be donned by all individuals. (1) Ensure the tools, hydraulic jack, and lifting devices are stowed in a safe position before commencing the work. (2) All removed parts should be stowed sequentially in a secure location. (3) Cleaning, checking, and measuring should be conducted with any worn-out parts removed and replaced as directed by the manufacturer's instructions. (4) Lifting of all heavy parts by crane or chain block should be supervised by an experienced engineer. (5) Care should be taken not to drop any small parts, such as nuts and bolts or tools, inside the jacket passage or in the crankcase. If this happens, stop the job and remove immediately. (6) All clearances and other critical measurements should be inspected and recorded. (7) All parts should be refitted under the supervision of a senior engineer.

Once the d'carb is complete: (1) the crankshaft deflection should be taken and recorded; (2) the crankcase oil should be removed and the sump cleaned; (3) a crankcase inspection should be carried out to ensure nothing has been left inside; (4) ensure all parts are fitted and tightened as per the manufacturer's instructions; (5) all isolated systems must be returned to operation one by one, starting with the freshwater system. Any leaks should be checked and rectified immediately; (6) fresh oil should be provided, and the lube oil system returned to operation, followed by the fuel oil system; finally, (7) the turning gear should be rotated using a tommy bar to ensure the free movement of crankshaft. Once these actions have been conducted completely, the generators should be slowly brought back on load following the on-load test procedures.

EMERGENCY GENERATOR

It clearly is understood that maintaining continuous power on the ship is one of the most important responsibilities of the engineering department. However, sometimes, accidents are inevitable and due to some unforeseen cause, the ship may suffer a full power failure (or black out). During such a condition, the emergency equipment, such as the lifeboats or navigation lights, must remain in an operational condition as per the regulations set by SOLAS. When a blackout occurs, there must be an alternative source of power which comes on load automatically. This alternative source of power is derived from batteries and the emergency generator. As batteries cannot provide power for extended periods, it is always preferable to use the emergency generator instead. As per SOLAS, the emergency power should come on load within 45 seconds of the power failure. When the power failure takes place, the emergency generator is normally started by a small electric motor which cranks the engine for starting. This motor receives power from the battery which is charged by the emergency switchboard. In the event the emergency generator is unable to start for any reason, an alternative manual method must be available. In accordance with SOLAS, this secondary means of starting should be able to provide an additional three starts within a minimum of 30 minutes.

The most common method for starting the emergency generator is hydraulically, though there are several other options available, including (1) by compressed air, (2) an inertia starts, and (3) through hand cranking. Here, we will discuss the hydraulic method. The hydraulic system for starting the emergency generator works on the principle of hydraulic and pneumatic energy, in which the physical energy is first stored and then supplied or released for starting the engine. The main components of the hydraulic system are the following: (1) The *feed tank and hand pump*. The feed tank is provided with hydraulic oil, which is pumped by hand to the accumulator which helps kickstart the engine. (2) The *hydraulic accumulator*. This is the key component of the system. It is the heart of the system where the energy is stored. It consists of a cylinder in which there is a leakproof sliding piston. Above this piston, the cylinder is pre-charged with nitrogen gas to a pressure of about 200 bar. The oil is pressed against this piston through which pressure of the oil is stored in the accumulator. (3) The *pressure gauge*. This is used to check the pressure in the accumulator. (4) The *relay valve lever*. The operation of this lever releases the energy stored in the accumulator to the starter unit. (5) The *starter unit and engine dog*. The starter unit is attached to the free end of the engine by way of a bracket, and the engine dog is attached to the engine crankshaft by means of a suitable adapter. This starter unit consists of two opposing cylinders with a rack and pinion arrangement. The pinion arrangement has teeth on one end, which drives the dog, having corresponding teeth. Two helical grooves are formed inside the periphery of the pinion which is engaged by spring-loaded balls inside the starter housing. This helps engage and disengage the axial movements. Positive engagement is maintained by the helical tooth from the pinion and racks.

To use the hydraulic starter, we must (1) check whether all the valves for fuel and cooling water are open to the generator. (2) Check the level of the feed tank. Refill if necessary. Ensure the level in the gauge glass is low and the pressure in the accumulator is as per the manufacturer's recommendation. Do not fill the tank, as the oil will return from the starter after starting. (3) Check the pressure in the accumulator. Increase the pressure as required. Pressurise the accumulator as per the manufacturer's recommendations. (4) Operate the relay valve lever. The relay valve lever operates in two stages. First, move the relay valve to an angle of 45 degrees until resistance is felt. In this stage, a small bleed is given to the starter causing a slow rotation, engaging the dog. Second, when the dog is engaged, operate the lever fully. This releases the pressure in the starter causing the engine to start. It is important to avoid any sudden jerks of the relay lever to prevent damage to the gears and the clutching arrangement. (5) When the engine starts turning, release the lever. The lever will return to its normal position. The oil used for starting the engines will also return to the feed tank after starting. (6) Check the pressure in the accumulator. There should be sufficient pressure for an additional two starts. (7) Raise the pressure again for the next emergency.

Emergency generator maintenance

As with all things on board a ship, the emergency generator will work only as well as it is maintained. Considering its high importance, the emergency generator is required to be evaluated regularly in addition to the usual planned maintenance. Keeping the emergency generator in good working condition requires several key tasks, which are listed here.

1. *Change of engine sump oil.* As always, it is important to check the oil level in the sump regularly. Since the emergency generator is kept on auto mode, which ensures the generator starts and comes on load automatically, it is necessary that before starting the engine for operation, the oil level is checked on a regular basis. The condition of the oil needs to be checked for carbon or soot particles and changed accordingly. The running hours for engine oil changes from one manufacturer to another, the engine make, and the grade of oil used. As a rule of thumb, this is usually between 250 and 500 hours. The lifespan of the engine oil must be cut by half when the fuel used in the generator consists of more 0.5% to 1% sulphur.
2. *Air filter.* The combustion air for the engine is passed through an air filter, which may be any of the following types: (a) oil bath air cleaner or (b) dry type air cleaner (i.e., cartridge or dust collector). It is important to clean the air filter at the correct intervals as any delay will lead to clogging, causing less air to be fed into the engine. This will reduce the efficiency of the engine and increase the thermal parameters. When using a dry cartridge, always ensure to replace them at the correct intervals stated by the manufacturer. Again, the typical rule of thumb is one year or after five and seven cleanings.
3. *Check the water separator.* Some emergency generators are provided with a water separator to prevent the mixing of water with fuel. Check the level of water and ensure it is below the level mark. Regularly drain off. This is done to avoid rust and corrosion to the fuel lines, and to avoid incomplete combustion.
4. *Check the electrolyte in the battery.* A battery is used as one of the starting methods for the emergency generator. The electrolyte level in the battery must be checked at regular intervals either by inserting a level stick or by checking the water level in the level tester cap (if provided). Use distilled water to top up a low water level.
5. *Check the alarms and shutdowns.* All the safety devices and alarms fitted in the emergency generator must be checked and evaluated regularly. Generators with 'V' belts have additional alarms which sound in the event of belt failure and or when operated by an idler pulley.
6. *Check the 'V' belt tension.* When a 'V' belt is fitted, inspect for cracks and signs of wear. Renew the belt if exhibiting a cracked appearance. To check the belt tension, press the belt with a thumb in the midway of the pulleys and check the inward

deflection in millimetres. There should be no more than 10–15 mm of deflection depending on the make of the generator.
7. *Clean the oil filter cartridge.* The emergency generator is provided with various oil filters, such as the bypass filter, centrifuge filter, lube oil filter, or fuel feed pump filter. These filters need to be cleaned or renewed as per the manufacturer's instructions.
8. *Check the valve clearance.* The tappet clearance of the inlet and exhaust valve should be checked in accordance with the running hours stated in the maintenance section of the generator's manual. Also, ensure the engine is cold before taking the tappet clearance. Loss of the emergency generators at times when they are needed the most can lead to otherwise avoidable and disastrous situations. incidents. Following a well-planned maintenance system, together with thorough regular checks, will ensure the emergency generators are available if, and when, needed.

In this chapter, we have covered some of the main points relating to the operation and maintenance of marine diesel generators. The importance of the generators really cannot be overstated, therefore good maintenance procedures are needed to keep the generators in working order. In the next chapter, we will look at the ship's electrical distribution system.

Chapter 12

Marine electrical systems

We should now have a basic understanding of the ship's requirement for electrical supply and how the generator provides the supply for the ship's machinery and systems. In this chapter, we will begin to explore the voltages and currents, and the method of electrical supply through the main and emergency switchboards and ancillary installations. The electrical supply on ships is typically three-phase, 60Hz, 440 V. Yet ship designers must react to the industry's demand for bigger and increasingly efficient vessels. But as ship sizes increase, so does the need for more powerful engines and auxiliary machineries. This increase in the size of shipboard machineries and other equipment equals more electrical power demands and thus the provision of higher voltages and power supplies. Any voltage less than 1kV (1,000 V) is referred to as low voltage (LV); voltage above 1kV is referred to as high voltage (HV). Typical marine HV systems operate between 3.3kV and 6.6kV. Many passenger ships, such as the *Queen Mary II*, have HV systems as high as 10kV! To understand how the electrical supply system works on a ship, we can use an example. Let us assume our ship generates eight megawatts of power at 440V, from four diesel generator sets, each of two-megawatt, 0.8 power factors. Each generator feeder cable and circuit breaker must manage a full-load current of:

$$I = 2 \times 106 / (\sqrt{3} \times 440 \times 0.8)$$

Where:

- I = 3,280.4 amps (i.e., approximately 3,300 amps)

Protection devices such as circuit breakers should be rated at approximately 90kA for each feeder cable. Let us now calculate the same if the generated voltage is 6,600V:

$$I = 2 \times 106 / (\sqrt{3} \times 6,600 \times 0.8)$$

Where:

- I = 218.69 amps, or approximately 220 amps

This means the protection devices can be rated as low as 9 kA.
We must also factor in power loss:

$$= I2^* r$$

Where

- I is the current carried by the conductor, and
- R is the resistance of the conductor.

This means the power loss varies as a square root of the current carried by the conductor. If the supply voltage is 440V, then the current carried by the conductor is 0.002P, and if the voltage is raised to 6,600V, the current carried for the same power is

$$(1.515 \times (10^{-4})) \times P$$

This implies that the power loss is reduced by a greater extent if the voltage is stepped up. Subsequently, it is always efficient to transmit power at a higher voltage. Conversely, the power loss can be reduced by decreasing the resistance of the conductor, accordingly:

$$r = \rho \times l/a$$

By increasing the cross-sectional area of the conductor (i.e., the diameter), the resistance of the conductor can be reduced and thus limiting the power loss. But as this involves substantial increases in cost due to the installation of heavy cables and supports, this method is not often implemented if at all. Moreover, a motor (let us assume a bow thruster) may be of a smaller size if it is designed to operate on 6,600V. For the same power, the motor would be of a smaller size if designed to operate on 6,600V compared to 440V. Therefore, ships use HV systems in place of LV systems.

On most vessels over a certain size and deadweight, the electrical supply is rated in kVA. This rating applies to many of the ship's critical machinery, including the generators, transformers, protection devices, and so on. Motors conduct mechanical work, so produce a mechanical output expressed in kW. They also have a fixed power factor. This power factor is always written on the outside of the equipment in kW on a motor nameplate data table. For this reason, we rate motor outputs in kW or BHP (kilowatts/break horsepower) instead of kVA. As far as an electric motor is concerned, its primary function is to convert electrical power into mechanical power, as the load it is connected to is not electrical, but mechanical, and only active power is considered, which must be converted into the mechanical load. Moreover, the motor power factor does not depend on the load as works according to power factor. Using a transformer as an example, the transformer is a static device, which does not perform any mechanical work. Instead, the primary function of the transformer is to step down and step up the voltage ratings. Invariably, when stepping up or stepping down the voltage, it also steps down or steps up the current as an inverse reaction. The transformer is a critical piece of machinery, as it bears two types of losses: (1) copper losses and (2) iron losses or core losses (also known as insulation losses). Copper losses in the transformer (I^2R) depend on the current which is passing through transformer winding. On the other hand, iron losses or core losses depend on the voltage. It can be said that the copper loss depends on the rating current of the load the transformer is supporting; therefore, the power factor can only be determined by the load itself. The power factor of the transformer depends on the nature of the connected load such as resistive load, capacitive load, and inductive load. This means the rating of a transformer can only be expressed as a product of volts and amperes (V/A). *Amp ratings*. The current flowing through the transformer can vary in power factor, from zero PF lead (pure capacitive load) to zero PF lag (pure inductive load) and is decided by the load connected to the secondary. The conductor of the transformer winding is rated for a particular current beyond which it will exceed the temperature at which its insulation is rated irrespective of the load power factor. *Voltage rating*. The maximum voltage which the primary winding can be subjected also has a maximum limit. If the applied voltage to the primary winding exceeds the maximum rated value, this will cause the magnetic saturation of the core leading to distorted output with higher iron losses. Thus, considering both of the aforementioned ratings, it is usual for transformers to be rated in VA. It can further be understood as the product of voltage (V) and current (A). But this does not mean that one can apply a lower voltage and pass a higher current through the transformer contributing to the rated VA value. The VA value is bound individually by the rated voltage and rated current. All electrical equipment in connection with the generation, transmission, and distribution of AC power, such as alternators, transformers, switchgear, and cables, are rated on a kVA basis.

If we examine the following equation:

$$\cos\varphi = kW/kVA.$$

or

$$kVA = kW/\cos\varphi$$

We can see it is evident that the larger the power factor, the smaller the kVA requirement of the machinery. Therefore, at low-power factors, the kVA rating of the equipment must be made higher, which in turn makes the machine larger and more expensive. This means the kVA rating is critically important both at the design stage and during the lifetime operation of the machinery. Transformer sizing is normally conducted according to the following conditions: (1) calculation of peak load. First, we determine the kVA, amperes, or wattage required by the load together with the voltage requirement of the load. This is helpful in determining the secondary voltage (also known as load voltage) or output voltage of the transformer. The load voltage, or secondary voltage, is the voltage required to operate the load (such as lights, motors, and other devices), (2) maintain 10% spare capacity for future loads, and (3) set the load requirement according to the highest rated direct online (DOL) motor.

MAIN SWITCHBOARD

The main switchboard (Figure 12.1) is an intermediate installation in the ship's power distribution circuit connecting the power generators and power consumers. The power generators on ships are auxiliary engines with alternators, and the consumers are different engine room machineries, such as motors or blowers. It is important to isolate any type of fault in the electrical system supplied from the main switchboard; otherwise, it may affect the whole ship's electrical system. If such isolation is not provided, then even a short circuit in a small system can cause a blackout of the whole ship. Therefore, different safety devices are used on board the ship and installed on the main switchboard and electrical distribution panels. This ensures safe and efficient running of machineries and the safety of personnel from electric shock even when one system is at fault.

Figure 12.1 Main switchboard.

The important safety devices fitted on the main switchboard are as follows: (1) Circuit breakers: A circuit breaker is an auto-shutdown device which activates during an abnormality in the electrical circuit. Especially during overloading or short circuit, the circuit breaker opens the supplied circuit from the main switchboard and thus protects the same. Different circuit breakers are strategically installed at various locations. (2) Fuses: Fuses are used for short circuit protection and come in various ratings. If the current passing through the circuit exceeds the safe value, the fuse material melts and isolates the main switchboard from the default system. Normally, fuses are used with 1.5 times full-load current. (3) Over current relay: OCR is used on the local panel and main switchboard for protection from high current. They are installed where a low-power signal is a controller. Normally relays are set equivalent to full-load current with time delay. (4) Dead front panel: It is another safety device provided on the main switchboard, individual panels wherein you cannot open the panel until the power of that panel is switched off. Apart from this, maintenance and operational safety play an important part in the safety of the main switchboard.

BUSBAR

The busbar (Figure 12.2) is a copper plate or bar which is used in the ship's main and emergency switchboards to conduct electricity from the generators or from one electrical terminal to another. Technically, there are no electrical wire connections inside the main

Figure 12.2 Busbar.

and emergency switchboards on ships for connecting the power supply from the generators to these switchboards. Instead, all HV and high current systems are connected by busbars. The busbar's copper plates or bars relate to the help of nut bolts, which transmit the electricity as required. During normal ship operations, the busbar connections are subjected to the harsh maritime environment together with the vibrations generated by the ship and the ship's machinery. These vibrations can cause the nut bolts in the busbar to loosen, which can lead to short circuits. Loose connections inside the switchboard can also lead to sparks that can cause a fire. Moreover, the busbars are meant to carry HV and currents, which tend to heat up the lines due to the flow of energy. For this reason, regular inspections and maintenance of busbars are imperative. If any maintenance is planned for the busbars, the absolute highest standards of safety are required as even the smallest mistake can lead to electrocution and death. Busbar maintenance must therefore be performed only when the complete busbar panel or switchboard is isolated and turned 'OFF'. The best opportunity to conduct busbar maintenance is when the ship is in dry dock. However, as dry docks can be years apart, busbar maintenance may be conducted when the ship is in a full blackout condition, i.e., when the ship's generators are not running and no power is supplied to the main or emergency switchboards. If the main switchboard busbars are to be inspected, or worked on, ensure to keep the emergency generator running. Keep in mind that there will be some portion of the main switchboard which will continue to be fed by the emergency switchboard. To avoid electrocution, always check which parts of the switchboard will remain live and keep away from those areas.

Before conducting maintenance on the busbar, always ensure appropriate safety precautions are implemented. This usually starts with putting a 'lockout' tag on all generators and the emergency generator. Keep the generator system, including the load-dependent, start-stop system, in manual mode. Don rubber gloves even when the switchboard is not in a LIVE condition. Don appropriate personal protective equipment when working on the switchboard. If the ship is in a complete blackout situation, ensure that before cleaning the main and emergency switchboard, the area is well lit with adequate lighting. If the ship is in dry dock, this can usually be arranged via the shore workshop.

To conduct busbar maintenance, (1) open the door to the main and emergency switchboards where the inspection is to be performed; (2) conduct a visual inspection of the copper plate and nut bolts; mark any missing or burnt-out areas; (3) by hand or using a metal or plastic stick (where hand access is not possible), tap the bus plates gently to determine any loose connections; ensure to wear electrical gloves even when the busbar is not LIVE; (4) the busbars are mechanically supported inside the switchboard by means of insulators; these insulators may be manufactured from rubber or ceramic materials; check for any damage to the insulator parts; (5) using only dedicated size spanners or a pre-adjusted torque wrench, tighten the nuts in the busbar connection for the main and emergency switchboards; (6) check the tightness of the wire connections to the circuit breakers; (7) clean the busbar and switchboard area with a vacuum cleaner; (8) if any loose connections or sparks are found, isolate that particular busbar, and the adjacent busbar, before tightening the nut. If any metal pieces or nut bolts are missing or found inside the panel, remove them immediately as they can cause a short circuit and electrical fire.

The ship's electrical officer (or delegated engineer) is required to inspect the busbar periodically for record keeping and as stated in the preventative maintenance system. This is done to avoid any type of accident arising from electrical faults. When conducting electrical inspections, the following safety measures should be followed, as in this instance, the busbar will likely be LIVE: (1) check the load in the running generator by way of the kW metre provided in the main switchboard; (2) open the busbar access door provided at the rear of the main and or emergency switchboards; (3) using an infrared temperature gun, conduct a visual inspection of the switchboard(s) by measure the temperature of the copper plates and the busbar connection. The reading should never be higher or lower than the stated limits depending on the generator load. For example, if the generator load is 50%, and the ambient temperature is 28°C (82.4°F), the busbar temperature must be within 50°C (122°F). If the temperature is too high, then something is clearly abnormal, in which case the cause of the abnormal temperature must be investigated and rectified. Once the inspection and maintenance are completed: (1) close the busbar access doors; (2) remove the 'lockout' tag; (3) restore the main power supply from the generators; (4) inform the senior management team that the inspection/maintenance procedure is complete; (5) reset the main power and confirm whether there are any abnormal sounds in the main and emergency switchboards; (6) monitor the temperature of the busbar with the laser temperature gun; (7) keep the emergency switchboard in auto mode.

GOVERNOR

A governor (Figure 12.3) is a system that is used to maintain the mean speed of an engine, within certain limits, under fluctuating load conditions. It does this by regulating and controlling the amount of fuel supplied to the engine. The governor hence limits the speed

Marine electrical systems 167

Figure 12.3 Governor.

of the engine when it is running in a zero-load condition; i.e., it governs the idle speed, and ensures that the engine speed does not exceed the maximum value as specified by the manufacturer. All marine vessels need a speed control system to control and govern the speed of the propulsion plant used on board, as there can be many variations that impact the engine load. If left unchecked, these may damage the engine and cause loss of life and equipment. The variations in the load on the engine may arise due to several factors including heavy seas, rolling and pitching of the vessel, compromised ship structure, changes in weight of the ship, and many others. Governors are also fitted to the auxiliary diesel engine or generators and the ship's alternators. There are three main types of governors: mechanical, hydraulic, and electro-hydraulic. Mechanical governors consist of weighted balls, or flyweights, which experience a centrifugal force when rotated by the action of the engine crankshaft. This centrifugal force acts as the controlling force and is used to regulate the fuel supplied to the engine via a throttling mechanism connected directly to the injection racks. These weight assemblies are small, and hence the force generated is not sufficient to control the injection pumps of large engines. They can be used where exact speed control is not required. They have a large deadband and small power output. The advantages of mechanical governors include the fact that they are cheap, they can be used when it is not necessary to maintain an exact speed depending on load, and they are simple in construction and have minimal parts. Hydraulic governors, on the other hand, consist of a weighted assembly connected to a control valve, rather than the fuel control racks directly, as is the case with mechanical governors. This valve is responsible for directing hydraulic fluid which controls the fuel racks and thence the power or speed of the engine.

As a greater force can be generated, hydraulic governors are more commonly found on medium- to enormous-sized engines. Today, most ships use hydraulic governors retrofitted with electronic controls. By way of comparison with mechanical governors, hydraulic governors have a higher power output, improved accuracy and precision, and improved efficiency and straightforward maintenance. Electro-hydraulic governors have an actuator with two sections: a mechanical-hydraulic backup and an electric governor. In the event of failure of the electric governor, the unit can be put into manual control on the mechanical-hydraulic backup governor. The mechanical governor is set at a speed which is higher than the rated speed; the electric governor controls the speed and load of the entire system. The system has an electronic control valve that is connected to the armature within an electromagnetic field. An electronic control box sends a signal to the field which positions the armature and causes the control valve to regulate the fuel delivery. The electric control overrides the mechanical-hydraulic mode when the system is set to electronic operation. The main benefits of the electro-hydraulic governor include faster responses to fluctuating load changes, automated control functions are easily built into the governors, and they can be mounted in positions remote from the engine reducing or even eliminating the need for governor drives.

Classification of governors based on their operating principles

1. *Flyweight assembly.* All types of governors are fitted with a flyweight assembly. Two or four flyweights are typically mounted on a rotating ball head that is driven directly by the engine shaft, using a gear drive assembly. The rotation of the ball heads creates a centrifugal force that acts on the flyweights of the assembly and causes them to move outward, away from their axis of rotation. As the speed of rotation is increased and the degree of outward movement of the flyweights also increases, the movement of the flyweights becomes indicative of the engine speed. A spring is installed to counteract the centrifugal force generated on the flyweights, forcing them towards their initial position. This spring is known as the *speeder spring*. The position of the flyweights and their outward movement are transmitted to a spindle (this may be done through a collar), which is free to move in a reciprocating fashion. The movement of this spindle, which forms the control sleeve, actuates a linkage to the fuel pump control and controls the amount of fuel injected. Under normal operational conditions, i.e., constant speed and loads, the control sleeve remains stationary as the force on the flyweights is balanced by the counteracting force exerted by the speeder spring. As the load on the engine is increased, the speed of the engine reduces and the control sleeve moves downward, as the force exerted on it by the speeder spring overcomes the force exerted by the flyweights. The downward movement of the sleeve is linked to the fuel control racks such that there is an increase in fuel delivery and thus the power generated by the engine. The force on the flyweights increases with the engine rpm and once again the system comes back to equilibrium. As the load on the engine is decreased, its speed increases. The flyweights move outward and, in turn, the control sleeve moves up as the centrifugal force overcomes the speeder spring force. The movement of the sleeve actuates the fuel pump reducing fuel delivery; thus, the speed of the engine is reduced, and the system comes into equilibrium.
2. *Hydraulic control.* In this case, the flyweights are linked hydraulically to the fuel control assembly. This system consists of a pilot control valve which is connected to the

governor spindle and a piston. The piston is called the *power piston* and controls the amount of fuel delivered to the engine. It is acted upon by the force of a spring and hydraulic fluid on opposite sides. The amount of oil in the system and, subsequently, the hydraulic pressure on the piston, is regulated by the pilot valve that is controlled by the flyweight assembly. The control valve sleeve is open at the bottom where an oil sump is present in the lower side of the governor housing. A gear pump, which supplies high-pressure hydraulic oil to the system, takes suction from the oil sump. This is driven by the governor driveshaft. A spring-loaded accumulator is present which maintains the required pressure head of oil and allows the drainage of excess oil back to the sump. In the event of constant speed and load operations, the valve is positioned to block the ports in the valve sleeve and hence the passage of oil to the power piston, which remains stationary under the balanced forces. An increase in load decreases the engine speed. In this instance, the flyweights move inwards, causing the governor spindle to move downwards under the action of the force of the speeder spring. This movement lowers the pilot control valve, which directs oil to the underside of the power piston. As the hydraulic pressure on the piston overcomes the spring force acting on it, the piston moves upward, and the fuel supply to the engine is increased, raising its speed. Once the rpm of the engine increases, the control valve falls back to its initial position blocking the delivery of hydraulic fluid to the power piston. Alternatively, as the load on the engine is decreased and its speed increases, the outward movement of the flyweights under the action of the additional centrifugal force causes subsequent upwards movement of the spindle. This leads the pilot control valve to rise as well. This opens the port such that the hydraulic oil in the system flows to the oil sump from under the power piston through a drainage passage. The power piston then moves downwards under the action of the spring force and reduced hydraulic pressure. Subsequently, the amount of fuel supplied to the engine is decreased. This reduces the engine speed, leading to the forces on the flyweights to become balanced once again.
3. *Governor sensitivity.* To increase the sensitivity of the governor and to prevent overcorrection by the system, a compensating mechanism is incorporated in the governor design. In the case of a hydraulic governor, a plunger is present on the power piston shaft and on the drive shaft. These are referred to as the *actuating compensation plunger* and the *receiving compensation plunger*, respectively. The compensating plunger moves in a cylinder which is full of hydraulic fluid. This plunger moves in the same direction as the power piston. The downward movement of the power piston, due to an increase in engine speed, moves the compensating plunger downwards. Because of this movement, the plunger draws oil from a cylinder present below the pilot valve bushing. This creates suction above the receiving compensating plunger, which is a part of the bushing. The bushing moves upwards and closes the port to the power piston. The pilot valve port is opened just long enough to allow the engine speed to return to the set rate, avoiding overcorrection. As the flyweights and pilot valve return to their central position, oil flowing through the needle valve allows the pilot valve bushing to also reach its central position. The bushing and plunger must descend at the same speed to keep the port closed, so the needle valve must be adjusted carefully to allow the correct amount of oil to pass through it. This depends on the engine requirements, as set by the engine manufacturer. In the event of a decrease in engine speed, the actuating compensating plunger moves upwards, and the pressure on the receiving compensating plunger increases. This then moves

upwards with the pilot valve bushing. The port leading to the power cylinder remains closed, and the excess oil is drained out through the needle valve. The bushing is then returned to its central position.

4. *Electronic system.* An electronic governor provides engine speed adjustment from a no-load condition to full-load condition. It consists of a controller, an electromagnetic pickup (MPU) and an actuator (ACT) to conduct the necessary speed control and regulation. The MPU is a micro-generator and has a magnetic field. It consists of a permanent magnet with an external coil winding. The electromagnetic pickup is installed above the flywheel teeth, and depending on its distance from the gear teeth or slot, the magnetic field of the MPU may vary from a maximum to minimum, respectively. Due to the constantly changing internal magnetic field, an AC voltage and frequency are generated in the outer conducting coil. This AC voltage follows the speed of the flywheel. This is the most important aspect of the electronic control system as the governor controller converts the obtained frequency into a DC voltage signal. It then compares this with a set voltage. The results are calculated by a proportional-integral-differential (PID) control, and, finally, the output reaches the actuator which implements the required corrections on the fuel supply to the engine. The electronic controller has different modes of operation to implement various functions. These include (a) detecting the starting of an engine and subsequently directing the fuel supply; (b) suppressing the smoke generated by the engine as its speed increases; (c) adjusting the droop percentage (which is explained further down); (d) remote speed control; (e) idle speed operation, which provides fixed speed control over the entire torque capacity of the engine; and (f) maximum speed control, which is used to eliminate engine over speeding.

Maintenance of the governor

The governor should always be kept clean and free from dirty lubricating oil. Regular flushing of the system with the right lubricating oil should be conducted regularly. The hydraulic fluid and lubricating oil should be of the correct viscosity as mandated by the manufacturer. The system oil levels should be maintained and checked as per the ship's preventative maintenance schedule. The governor should never be tampered with, with any repairs made by a qualified and experienced engineer.

Droop

As the load on the engine increases, the fuel supply to the engine also increases, yet it is allowed to run on a proportionally lower speed. This feature of the governing system is called the *droop*. When more than one prime mover is connected to the same shaft, as is the case when generating electrical power, droop permits a stable division of load between them. The prime mover can be run in a droop speed control mode, wherein its running speed is set as a percentage of the actual speed. As the load on the generator is increased from no load to full load, the actual speed of the engine decreases. To increase the power output in this mode, the prime mover speed reference is increased and results in the flow of fuel to the prime mover increasing. Droop is measured as a percentage, in accordance with the following formula:

Droop % = (No − Load Speed − Full − Load Speed)/No − Load Speed

Speeder spring

The governed speed of the engine is set by changing the tension of the speed-adjusting spring which may also be referred to as the speeder spring. The tension of the spring counteracts the force exerted by the flywheel on the spindle. The pressure of the spring determines the speed of the engine that is necessary for the flyweights to maintain their central position.

Deadband

The deadband of a governor gives the range of speed after which the governor starts operating and making corrective adjustments. Within this range, the governor does not operate at all. The width of the deadband is inversely proportional to the sensitivity of the governor.

Hunting

The continuous fluctuation of the engine speed around the mean required speed is referred to as *hunting*. This occurs when the governor is too sensitive and changes the fuel supply even when there is a slight change in the engine rpm resulting in either too much fuel or too little fuel. This causes the governor sleeve to repeatedly move to its highest position. This cycle continues indefinitely. When this happens, the engine is said to *hunt*.

SAFETY PRINCIPLES OF THE MARINE ELECTRICAL SYSTEM

The safety of marine electrical systems includes safekeeping personnel from electrical shock and causing damage to the machinery due to electrical malfunction. For machinery safety, depending on the size and power rating of the equipment, a relay, circuit breaker or fuse is used to prevent the electrical equipment from overcurrent or overheating. Temperature gauges, rpms motors, direction indicators, amperage metres, and so forth are all different components used locally to monitor the performance of the electrical systems and equipment and understand the general health of the ship's machinery.

Avoiding electrocution

Before we look at the safety appliances and components of the ship's electrical system, it is worth first discussing the risks and hazards of electrocution and how this can be best avoided. When we talk about accidents on a ship, electric shock is the worst of all kinds. Electrical wires and connections are present everywhere on a ship and it is important to recognise them to prevent yourself and others from being critically injured or even killed. Not every incident is the result of negligence, though poor housekeeping and inadequate maintenance have been found to be the main contributors to electrocution on ships. Crew members are most at risk when they first join their ship and after the midpoint of their contract. With the former, this is explained through ignorance of the ship's systems and machineries. For the latter, a respectful appreciation of the dangers can give way to over-confidence. In this section, we will briefly discuss some of the key points to remember when working with, or around, HV electrical systems to avoid the risk of electrocution.

Starting with the first round of the day, check all electrical motors, wiring, and switches for abnormal sounds, variations in temperatures, and loose connections. Ensure that all electrical connections are within the panel box to prevent them from being touched accidentally. In the accommodation block and crew areas, it is important to ensure sockets are not overloaded. Where this is found, carefully dismount the socket and explain the hazards of overloading the socket to the relevant crew member(s). Isolate the system or equipment breaker before starting any work on an electrical system. Use notice boards as much as possible to inform other crew members – and the bridge, if necessary – of any maintenance or overhauling work to avoid accidental starts. Always double-check electrical tools, such as portable drills, for loose or damaged wires before starting any work. Don protective clothing, rubber gloves, rubber knee pads, and safety shoes to reduce the risk of electric shock. Only use insulated handle tools for working on or inspecting electrical systems and machinery. Before starting any electrical work, remove all jewellery, including watches, wristbands, and other conductive items. When working or removing multiple wires, tape off all but the one wire to be worked on. Avoid, wherever possible, working on live systems. Before the work procedure, conduct a toolbox talk and set out all safety hazards and work requirements. Where possible, work as a pair in case some abnormality occurs or injury is sustained. If the system is new, or unknown, never work on it alone; always ask for assistance from a senior officer. Finally, always think about your own and others' safety when conducting electrical work. Never take shortcuts, assume a system is isolated, or assume someone else has done something which you are required to do yourself.

Main switchboard safety devices

Now that we have covered the basic principles of safe electrical working, we can briefly discuss some of the safety devices of the main switchboard. As we know, the main switchboard is an intermediate installation in the ship's power distribution circuit which connects the power generators and the power consumers. The power generators are auxiliary engines with alternators and the power consumers are the various engine room machineries, such as motors or blowers. It is especially important to isolate any type of fault in an electrical system supplied from the main switchboard; otherwise, it will affect the other systems connected to the same power source. If isolation is not provided, then even a short circuit in a small system can cause a blackout of the entire ship (we will cover blackouts shortly). Therefore, different safety devices are installed on the main switchboard and electrical distribution panels. These ensure the safe and efficient running of the ship's machineries and help protect personnel from electric shock even when one system is at fault. The main safety devices fitted to the main switchboard are as follows: (1) *Circuit breakers*. A circuit breaker is an auto-shutdown device which activates during an abnormality in the electrical circuit. During overloading or short circuit, the circuit breaker opens the supplied circuit from the main switchboard. Different circuit breakers are strategically installed at various locations throughout the electrical system. (2) *Fuses*. Fuses are used for short circuit protection and come in various ratings. If the current passing through the circuit exceeds the safe value, the fuse material melts, isolating the main switchboard from the default system. Typically, fuses are rated at 1.5 times the full-load current. (3) *OCR*. The OCR is used on the local panel and main switchboard for protection from high current. They are installed where a low-power signal acts as a controller. Normally, relays are set equivalent to the full-load current with a time delay. (4) *Dead front panel*. The dead front panel is a safety device which prevents access to the switchboard panel unless the power is switched off and the switchboard is isolated.

Air circuit breakers

We touched on circuit breakers (Figure 12.4) earlier in point (1), but given their importance, it is worth explaining their role and function in slightly more detail. The air circuit breaker (ACB) is a safety device designed to overcome defects and safeguard equipment before it breaks down. The main function of ACB is to open and close a three-phase circuit. The ACB opens the circuit automatically when a fault occurs. Faults can be of several types, including under or over voltage, under or over frequency, short circuits, reverse power, earth faults, and so on. The main feature of the ACB is that it dampens or quenches the arcing that occurs during overloading. The ACB has two sets of contacts: a main contact and an auxiliary contact. Each set of contacts consists of a fixed contact and a moving contact. The main contact normally carries most of the load current. All the contacts are made from cadmium-silver alloy, which provides excellent resistance to damage from arcing. When the ACB is closed, a powerful spring is energised which latches the ACB shut. The auxiliary contact makes first and breaks last; i.e., when the ACB is closed, the auxiliary contact closes first followed by the main contact. When the ACB is open, the main contact opens first followed by the auxiliary contact. Thus, the auxiliary contacts are subjected to arcing during the opening of the ACB and can be easily replaced. The main

Figure 12.4 ACBs.

contact closing pressure is kept high so that the temperature rises in the contacts while the carrying current remains within limits. A closing coil operating on DC voltage from a rectifier is provided to close the circuit breaker by operation of a push button. The quenching of the arc is achieved by: (1) using arcing contacts made of resistance alloy and silver tips for the main contacts; arcing contacts close earlier and open later than the main contacts; (2) when opening, the contacts travel a distance at high speed which stretches the arc; this is then transferred to the arcing contact; (3) the cooling and splitting of the arc is achieved by arc chutes which draw the arc through splitters by magnetic action and quickly cool and split the arc until it snaps. The circuit breaker opens when the arc is quenched.

Preferential trips

Like the ACB, the preferential trip is an electrical arrangement which is designed to disconnect non-essential circuits, i.e., non-essential loads, from the main busbar in the event of a partial failure or overload in the main supply. The non-essential circuits or loads on ships include air conditioning, exhaust and ventilation fans, and galley equipment, which can be disconnected momentarily and reconnected after fault diagnosis. The main advantage of the preferential trip is that it helps prevent the operation of the main circuit breaker trip causing a loss of power on essential services and thus prevents blackout and overloading of the generator. The preferential trip circuit consists of an electromagnetic coil and a dashpot arrangement which provides a delay to disconnect affected non-essential circuits. Along with this, there is also an alarm system which functions as soon as an overload is detected, and the trips start operating. There are some mechanical linkages in the circuit which instantaneously operate the circuit and complete the circuit for the preferential trip. The dashpot arrangement consists of a piston with a small orifice into which is placed a small cylinder assembly. This piston moves up against fluid silicon. The time delay is governed by the orifice in the piston. The current passes through the electromagnetic coil and the linkages are kept from contacting using a spring arrangement. As soon as the current value increases the limit, the electromagnetic coil pulls the linkage up against the spring force, instigating the instantaneous circuit and the alarm system. The lower linkage completes the circuit for the preferential trip circuit. The current passes through the coil in the preferential trip circuit which pulls the piston in the dashpot arrangement. The movement of this piston is governed by the diameter of the orifice and the time delay. The preferential trip operates at 5-, 10-, and 15-second intervals, and the load is removed accordingly. If the overload persists, then an audible and visual alarm is activated. Without doubt, the preferential trip is one of the most important safety devices on a ship's electrical system. It helps remove excessive load from the main busbar, thus preventing blackout situations which can render the vessel dead in the water. This is obviously an extremely dangerous condition, especially in busy or congested channels, and in heavy seas.

Blackout conditions

Blackout is one condition every marine engineer is all too familiar with and afraid of. Indeed, a blackout at sea is one of the most terrifying situations that any ship can find itself in. From the bridge to the engine room, the galley to the mess, everyone on board is affected by a blackout. In this section, we will discuss first what a blackout is and then what actions need to be taken in response to a blackout occurring. A blackout condition is a scenario on a ship where the main propulsion plant and associated machineries, such as the boiler, purifier, and other auxiliaries, stop operating due to a failure of the generator and alternator. With modern technologies and increasing automation, ships are provided with systems to avoid blackout conditions such as autoloading sharing systems and auto-standby systems. These work by running the generator sets in parallel or on standby so that one comes on load automatically should the running diesel generator fail. Should a blackout occur, it is imperative to remain calm and focused and not panic. The emergency generator will restore power in short order. The first action is to inform the officer of the watch (OOW) about the condition. Request additional support – this might include extra lookouts on the bridge and the bridge wings and or in the engine room (if needed). If not already present, inform the chief engineer. If the main propulsion plant is running, bring the fuel lever to the zero position. Close the feed of the running purifier to avoid overflow and wastage of fuel. If the auxiliary boiler was running, shut the main steam stop valve to maintain the steam pressure. Investigate the cause of the issue and, if possible, rectify. Before recommencing the generator set, start the pre-lubrication priming pump (if the supply is provided by the emergency generator); if not, use the manual priming handle. Start the generator and bring it on load. Immediately start the main engine lube oil pump and main engine jacket water pump. First, reset the breakers for the essential machinery then reset the breakers for the preferential tripping sequence (i.e., non-essential machinery). Once this procedure is complete, the power should return, and the vessel can continue safely. Blackouts are a common occurrence on ships, which means there is no reason to panic. That said, they require both skill and patience to resolve, especially so when the vessel is underway or engaged in manoeuvring. The best way to tackle situations such as these is to be calm and composed and to know the engine room and its machinery inside out.

In this chapter, we have covered some of the basic elements of shipboard electrical systems. As this is quite a wide and complex area, it is not possible to cover every detail, but hopefully, we should understand the criticality of the electrical system however shipboard power is measured, and personal safety when conducting electrical work. In the next chapter, we will begin to examine the role and function of air compressors, which are a vital component of the engine room infrastructure.

Chapter 13

Electrical distribution systems and redundancy

Power management and conservation are integral parts of the marine engineering operations on board ships. Today, ship engineers are strongly encouraged to practice best power-saving practices whilst performing their various onboard duties. Power management consists of two main aspects: (1) automatic power management systems, which use automation to conserve power, and (2) using best practices and management guidelines to reduce power consumption. Most modern ships are built with provisions for periodically unattended machinery spaces (PUMS). On such vessels, an automatic power management system manages the power supply and use of equipment and machinery. For example, not only does the power management system do away with manual synchronisation of the generators, but it also efficiently regulates the number of generators on the busbar according to the changing load. Some of the major functions performed by the power management system are (1) cutting in and out of the generators according to increases and decreases of load, (2) gradually loading and unloading of generator alternator sets to minimise thermal and frictional stresses, and (3) performing load sharing operations amongst the generators symmetrically or asymmetrically (depending on auto/manually set parameters). The diesel generators are the primary components of the power management system. All generators have a manufacturer's specific minimum and maximum load criteria, and optimum load criteria. When the generators are synchronised with the ship's power management system, the engineers have the option of changing the minimum and maximum point beyond which the generator cannot be loaded. This prevents various stresses from developing on the physical components of the generator. The loading and unloading of power from the alternator are driven by time lag functions, which often means that a sudden spike in load cannot be compensated by the power management system. To counter this, a hardwired preferential trip is used as a redundant backup to prevent sudden blackout. Some ships are also fitted with a shaft motor. This compensates for a sudden drop in load but also minimises shaft torque on engines with a long propulsion shaft. Moreover, some advanced vessels are fitted with a combined shaft motor/generator set, which is entirely regulated by the power management system. When generator sets are run in parallel, including the shaft generators, diesel generators, and/or steam-driven turbine generators, the power management system completely regulates the load on each component. In the case of generators with equal load capacity, the load on the busbar is distributed symmetrically on the alternators.

For efficient fuel consumption, it is always desirable to run the minimum number of generators, each at a load that is optimum. For instance, one generator running at 30% load may be more fuel efficient than two running at 15% and, conversely, one generator running at 70% may consume more fuel than two running at 35% load each. Thus, performance evaluation of generators according to their maximum and optimum rated capacity must be conducted regularly. At the beginning of each passage, the marine engineers must discuss the power management plan and consider numerous factors such as the number of reefers on board (if relevant), the use of stabilisers during the voyage, maintenance to be carried out on any generator during the voyage and determine which and how many generators are to run. Let us briefly look at some of the factors which help in reducing power consumption on board.

- *Reefers.* Many container ships are required to carry specially designed refrigerated containers called reefers. Unlike conventional containers, which are large empty metal boxes, reefer containers are refrigerated. They are used for the transportation of fresh produce such as fruit and vegetables or temperature-sensitive cargo, including medicines and some types of chemicals. Because reefer containers require a constant source of power for their refrigeration systems to function, they are plugged directly into the ship's main power supply. This obviously causes a major power drain, which needs to be compensated accordingly. Specialist refrigeration ships, called reefers, only carry refrigerated cargo. This cargo may be loaded directly into the ship's hull or in reefer containers. In either case, the stowage plans must be checked so that reefers requiring ventilation are carried on open decks. When placed in cargo holds, the efficient usage of the reefer cooling water system is a more effective method than using heavy inlet and exhaust fans to facilitate cargo-hold ventilation. Hence, it is imperative the freshwater cooling system for reefers, which includes freshwater and seawater pumps, expansion tanks and pipelines, is kept in good working condition. *Ballast pumps.* Most ballast pumps are heavy-duty pumps which consume lots of power. Ballast plans should be formulated with the aim of using the ballast pumps only when required. Filling of tanks, where practical, must be conducted by gravity. Similarly, ejectors should only be used during the final stripping of tanks and not continuously when deballasting. *Fuel transfer pumps.* Using service steam to heat the fuel in the storage tanks is an effective power reduction method. Fuel to be transferred must be kept at the temperature stated in the fuel specification manual. Low fuel temperatures can result in the pumps frequently tripping, as well as prolonged running of the pumps to transfer the same amount of fuel. *Air compressors.* Any air leaks in the starter, service, or working air must be repaired as soon as detected to prevent the continuous running of the compressors. Doing so also prevents frequent compressor loading and unloading. The running hours of the compressors must be managed closely with planned maintenance on the compressors conducted according to the manufacturer's instructions. *Freshwater.* Most ships use hydrophore tanks to pump freshwater onboard for domestic and service purposes. These tanks must be topped up frequently with air to minimise the running of the hydrophore pumps to achieve the set pressure in the tank.
- *Central cooling water system.* Care must be taken when establishing the number of seawaters, elevated temperature, and low-temperature pumps which are running. Sometimes, additional pumps may start, resulting in higher power consumption. *Engine room ventilation.* Ventilation fans are large power consumers. Engine room pressure and temperature must be carefully evaluated to run only the required number of fans. Where fan motors are dual speed or of a variable frequency type, the selection of lower speeds, where practical, will go a long way in reducing the ship's overall power consumption. *Lights.* A simple, yet ignored factor, is switching

off lights which not in use. Cargo-hold lights, steering gear room lights, and deck lights should be switched on only when in use. These methods should demonstrate how important passage planning is to minimise the ship's consumption of power. Considering the number of ship operations conducted, close coordination amongst the deck and engineering departments is critical to achieve optimum efficiency.

One key area where all members of the ship's staff can contribute to the efficiency of the ship is in managing accommodation power consumption. The accommodation of a ship is the living space where the cabins for the crew are located, together with the galley and mess, provision stores and refrigeration rooms, recreational rooms, gymnasium, meeting rooms, lockers, etc. The accommodation is supported by a water supply, sewage system, air-conditioning system, and fire safety systems, amongst many others. These systems are quite different when compared to the conventional systems found on shore. In this section, we will look at some of the important systems provided in the ship's accommodation and how the ship's staff can contribute to conserving power.

Domestic freshwater system. Freshwater may be taken from shore or produced onboard. With the latter, the freshwater generator produces water for use in the domestic freshwater system. The system works by distilling water. The water passes through a mineraliser (hardening filter) to reduce acidity and increase the mineral content of the water before being directed to one of the freshwater storage tanks. If the distilled water is required by the boilers, the mineraliser stage is bypassed. A silver ion steriliser is supplied to destroy bacteria and deposit silver ions in the stored water. This provides effective sterilisation whilst the water is stored. Normally, one domestic freshwater tank is used, with the second being filled or ready for use, on standby.

Steriliser. The onboard steriliser may be of two types: a silver ion steriliser (as discussed earlier) or an ultraviolet steriliser. The silver ion steriliser is used to treat water already in the freshwater storage tank. The hydrophore unit draws water from one of the freshwater storage tanks and passes it through the steriliser unit. The water then passes between electrodes where silver ions are introduced. This produces a concentrated sterilising solution. This solution is then pumped into the selected freshwater storage tank as required. The concentration is such that when diluted, the residual level of the silver ions in the tank remains toxic to bacteria. If the water remains in the tank for an extended time, it may become necessary to retreat the water to restore the required ion balance. Samples of the water in the storage tanks, the supply system, and the steriliser should be taken and analysed at regular intervals as recommended by the manufacturer. The ultraviolet steriliser is a device that conducts the sterilisation of water instantaneously by way of ultraviolet rays. Accordingly, it does not have a remaining effect, unlike that of chlorine sterilisation. It consists of a quartz tube, a germicidal lamp, and a water filter. The germicidal lamp emits ultraviolet rays which kill any waterborne germs. Avoid looking directly at the germicidal lamp with the naked eye, as ultraviolet rays are extremely harmful.

Mineraliser. The mineraliser or rehardening filter is designed to treat distilled water from the freshwater generator, therein producing water that is more suitable for human consumption. The mineraliser consists of dolomite stones. As the water passes through the mineraliser, acidic components are neutralised by a reaction with calcium and magnesium salts in the dolomite, resulting in an improved pH. This should be between 7.5 and 10 though the pH level can be easily adjusted by increasing or reducing the quantity of dolomite in the filter. A part of the dolomite is dissolved in the water, supplying the necessary mineral salts and hardness. Domestic freshwater for widespread use is stored in two freshwater tanks: ports and starboard. There are two freshwater hydrophore pumps (one running and the other on standby), which draw from the freshwater tanks and deliver to the hydrophore pressure tank. This is provided with an air cushion which is topped up from the service air system. The pressure in the hydrophore tank controls the starting and stopping of the hydrophore pumps. As water is consumed, the tank pressure drops, which automatically starts the selected pump and refills the tank. When the pressure increases to a predetermined value, the pump automatically stops. One pump is normally in use, with the second pump in shutdown or kept ready for use.

Marine freshwater hydrophore. From the hydrophore tank, the water flows into three systems: (1) The *domestic cold water system and accommodation services*. This system covers the water supply for drinking water fountains and the accommodation for use in cabins, pantries, and the galley. (2) The *domestic hot water system*. This system supplies continuous hot water to the accommodation for domestic purposes. Water is circulated continuously by the hot water circulating pump, passing through a calorifier, where it is heated by steam or electricity to the correct temperature. Topping-up of the system is from the hydrophore tank. This arrangement of constant water circulation ensures that hot water is available at an outlet immediately, which reduces the flow of cold water until hot water is available. (3) The *engine room and deck service system*. This system supplies water to the deck freshwater hose connections, engine room freshwater hose connections, and many other places in the engine room where freshwater is required; for example, the filling connection for the expansion tank, the filling connection for chemical dosing tanks, the generator turbocharger water-washing connection, the stern tube seal unit, bilge oily water separator, and so forth.

Domestic refrigeration system. There are three refrigerated chambers where all food provisions are stored. These are the vegetable room, the meat room, and the fish room. The refrigeration plant is automatic in operation and consists of two reciprocating type compressors, two condensers with an evaporator coil, and a fan unit in each of the three refrigerated chambers. Cooling for the meat, fish, and vegetable rooms is provided by a direct expansion R134a system. Liquid R134a refrigerant is passed to the evaporator coil for the compartment, and the expansion valve regulates the amount of liquid flowing to the evaporator in accordance with the gas outlet temperature; if the temperature rises, more refrigerant is passed into the evaporator. The liquid expands to the gas stage in the evaporator coil by extracting heat from the air in the refrigerated chamber. The air in the cold chamber is circulated over the evaporator coils by an electrically driven fan. The supply of refrigerant to the expansion valve is regulated by means of a temperature-controlled solenoid valve in the supply line. The refrigerated room evaporator is equipped with a timer-controlled electric defrosting element. The frequency of defrosting is determined by means of a timed defrosting relay built into the starter panel. Under normal conditions, one compressor/condenser unit is in operation, with the other ready for manual start-up, with all valves shut until required. The system is not designed for parallel operation of the compressor units, and the valve on the compressor unit, which is out of service, must be fully closed. The compressor draws R134a vapour from the cold chamber evaporators and pumps it under pressure to the condenser where it is cooled by water circulating from the central cooling freshwater system. The gas is condensed under pressure into a liquid. The compressors are protected by high-pressure, low-pressure, low lubricating oil pressure, and condenser cooling water failure cut-out switches. The liquid refrigerant passes through a filter/dryer to the cold room evaporators. Thermostats in each chamber enable temperature controllers to operate the solenoid valves independently to reduce the frequency of compressor starts and running time.

Marine refrigeration system. The evaporators accept the refrigerant as a super-cooled vapour from the expansion valves. The opening of the expansion valve is regulated by the refrigerant gas temperature at the outlet from the evaporator. This vapour extracts heat as it passes through the evaporator but is still colder than the liquid stage. The cold vapour then returns to the compressor, passing through the heat exchanger where it cools the liquid refrigerant further. The solenoid valves at the air coolers (evaporator units) are opened and closed by the room thermostats, allowing refrigerant gas to flow to the evaporator when open. With the solenoid valves closed, no gas flows to the evaporators and so no gas flows back to the compressor suction. The low-pressure switch will stop the operating compressor. Any leaks of refrigerant gas from the system will result in the system becoming undercharged. The symptoms of an undercharged system include low suction and discharge pressures, with the system eventually becoming ineffective. Bubbles will appear in the liquid-gas flow sight glass. When required, additional refrigerant can be added through the charging line, after first venting the connection between the refrigerant bottle and the charging connection. This prevents any air or moisture in the connection pipe from entering the system. The added refrigerant is dried before entering the system. Any trace of moisture in the refrigerant system will lead to problems with the thermostatic expansion valve icing up, causing blockage. The meat and fish room operating temperatures are $-20°C$ ($-4°F$), and the vegetable room operating temperature is $+4°C$ ($32.9°F$). The temperatures in the chambers are regulated by thermostats which activate the associated solenoid valve supplying gas to the air cooler/evaporator.

Accommodation air-conditioning system. Cooled air is supplied to the accommodation by an air handling unit (AHU) located in the air-conditioning unit room. The AHU consists of an electrically driven fan drawing air through the following sections from the inlet to the outlet:
- 1 × air filter
- 1 × steam preheating unit
- 1 × enthalpy exchanger
- 1 × reheat section
- 2 × air cooler evaporator coils
- 1 × humidifier section
- 1 × water eliminator section
- 1 × fan section
- 1 × discharge section

Humidification of the air is arranged by an automatic control fitted at the outlet section of the AHU. The air is supplied through the distribution trunking to the accommodation. Cooling is provided by a direct expansion R134a system. The plant is automatic and consists of two compressor/condenser units supplying the evaporators contained in the accommodation AHU. The expansion valves for the coils are fed with liquid refrigerant from the air-conditioning compressor; the refrigerant having been compressed in the compressor, then cooled in the condenser, where it is condensed to a liquid. The liquid R134a is then fed, via the filter dryer, to the evaporator coils where it expands under the control of the expansion valves, before being returned to the compressor as a gas. The phase change (liquid to gas) takes place in the evaporator coils where it extracts heat from the air passing over the outside of the coils. The compressors are fitted with an internal oil pressure–activated unloading mechanism which provides automatic starting and variable capacity control at 100%, 75%, 50%, and 25% of full capacity by unloading groups of cylinders. This variable capacity control allows the compressor to remain running even when the load is light and thus avoids the need for frequent stopping and starting. The compressor is protected by high and low-pressure cut-out switches, a low lubricating oil pressure trip, a cooling water pressure trip, and high-pressure and oil supply pressure differential trip. A crankcase heater is provided for use when the compressor is not running. Any leakage of refrigerant gas from the system will result in the system becoming undercharged. Indications of the system being undercharged include low suction and low discharge pressure, resulting in the system becoming ineffective. A side effect of the low refrigerant gas charge is an apparent low oil level in the sump. A low charge level will result in excess oil being trapped in the circulating refrigerant gas, causing the level in the sump to drop. When the system is charged to full capacity, this excess oil is separated out and returned to the sump. During operation, the level as shown in the condenser level gauge will drop. If the system does become undercharged, the system pipework should be checked for leaks. The only reason for an undercharge condition after operating previously with a full charge is that refrigerant is leaking from the system. When required, additional gas can be added through the charging line, after first venting the connection between the gas bottle and the charging connection. The added refrigerant is dried before entering the system. Any trace of moisture in the refrigerant will lead to problems with the thermostatic expansion valve icing up, causing blockage. Cooling water for the condenser is supplied from the low-temperature central freshwater cooling system. Air is circulated through ducting to outlets in the cabins and public rooms. The airflow through the outlets can be controlled at the individual outlets.

Sewage treatment plant. Sewage (black and grey water) from the accommodation is drawn by gravity or by vacuum through the pipe system to the ejector on the sewage collection tank. The vacuum in the system is maintained by circulating fluid from the collection tank through the ejector. The sewage in the collection tank is discharged to the sewage treatment plant located in the engine room. The sewage treatment plant is a biological unit which works on the *aerobic-activated sludge principle*. The plant treats black and grey water and is fully automatic in operation. Air is supplied to the sewage treatment unit by an independent aeration blower. This sewage treatment plant consists of a tank and four main compartments: (1) A *bioreactor with matrix* (aeration compartment). The sewage in this compartment comes from the lavatory pans, urinals, and sickbay. The incoming effluent passes through a screen to prevent the passage of inorganic solids into the bioreactor compartment, where it mixes with the activated sludge already present. Passage through the vacuum system breaks down the raw sewage into small particles which mix easily and encourages bacterial action. The matrix unit in the compartment ensures movement of the effluent and rapid biological breakdown of the raw sewage by bacteria. Air is supplied by means of a blower and distributed evenly through the tank by aerators. The gases produced during the bacterial action are vented to the atmosphere via the funnel. Oxygen intake is essential for the aerobic activity of the bacteria to occur. These organisms require oxygen for digesting the raw sewage, and it also assists by agitating and mixing the incoming sewage with water, sewage sludge, and the bacteria already present in the compartment. (2) *Settling or clarification compartment.* Effluent from the bioreactor compartment flows to the sedimentation tank compartment where the sludge is separated by gravity. The sludge is then returned to the bioreactor compartment screen section by means of an airlift supplied with compressed air from the aeration blower. The effluent then passes into the bottom of the filter tank. (3) *Filter compartment.* The activated carbon filter in this chamber breaks down any remaining micro-organisms and filters out any solid material. Air is supplied by means of the blower and distributed evenly through the tank by the aerators at the bottom. The filter requires backflushing with steam every month. The clean effluent flows from the filter compartment into the sterilisation compartment. (4) *Clean water sterilisation or discharge compartment.* This compartment is provided with float-operated switches which activate the discharge pump when the elevated level is reached and stops the pump when the compartment is empty. Sterilisation of the treated effluent is chlorination with sodium hypochlorite by means of a chemical injection pump or by manually adding chlorine tablets. The sewage treatment plant works automatically once it is set, though periodic attention is required, and the unit must be monitored for correct operation. The treatment plant discharge pump may be set to discharge overboard, into the double-bottom sewage collecting tank or to the port and starboard deck connections for discharge onshore. The sewage collecting tank can be pumped using the same method.

Garbage disposal system. Annex V of MARPOL 73/78, i.e., the Regulations for the Prevention of Pollution by Garbage from Ships, controls the way waste material is treated onboard ships. The regulations require the vessel to have a garbage management plan (GMP) in place. The plan should outline the procedures for the handling, segregation, storage, and subsequent disposal of the vessel's garbage. The plan must be clearly displayed in locations used for the handling of waste and name the person on board responsible for plan management. Although it is permissible to discharge a wide variety of garbage at sea, preference should be given to disposal utilising shore facilities where available. Only food waste is permitted to be disposed of inside mandated special areas and not less than 12 NM (13.8 mi, 22.2 km) offshore. These mandated special areas are the Antarctic, Baltic Sea, Black Sea, Gulf of Aden, Mediterranean Sea, North-West European Waters, Persian Gulf, Red Sea, and the Wider Caribbean Area. Every ship must have a standard GMP which outlines the responsibilities of crew members and the location of all the garbage bins and collection areas.

Firefighting in the accommodation. At various places, including inside each cabin, smoke detectors are present to detect smoke from fire. Each deck consists of portable fire extinguishers of a dry chemical powder (DCP) type; usually two in number, one port side, and one on the starboard side. Fire hydrants and fire hoses are present outside on the deck wings on each deck. The galley has a separate fire extinguishing system of its own.

Galley CO_2 fire extinguishing system. The galley exhaust duct has a local CO_2 system consisting of a single CO_2 cylinder, which is positioned within a small compartment adjacent to the galley. This provides an extinguishing capability in the event of a galley fire or fire in the galley exhaust duct. In the event of fire, switch off the galley fans and close the uptake fire damper in the galley deckhead. The emergency stop switch for the galley fans should be located outside the galley, and on the bridge, and in the fire control station. Evacuate all personnel from the galley and then release the CO_2 cylinder by opening the cylinder storage door. Fully open the cylinder outlet valve. Exit the galley and close the galley door. Some galleys are also equipped with deep fat fryer wet chemical extinguishing media. Fat fryers are particularly difficult to protect due to the amount of stored heat that is contained in a large quantity of cooking oil. The deep fat fryer appliance in the galley is protected by a fixed fire suppressant system. The protection system comprises a single stainless-steel storage cylinder containing an extinguishing agent. The cabinet is in the galley and is activated by pulling a release handle located close to the system cabinet. When activated, the chemical extinguishing agent is discharged into the fire extinguishing pipework. The discharge from the cylinder is led via piping to fixed spray nozzles. The extinguishing chemical has an expected storage life of 12 years. The extinguishing wet chemical used is R-102 Ansulex Low pH Liquid Chemical.

In this chapter, we have looked at the primary electrical distribution systems and redundancies on ships. In the next chapter, we will briefly discuss the role and function of the air compressor.

Chapter 14

Air compressor

The air compressor is one of the many machineries on board a ship that serves multiple functions. The main purpose of air compressor (see Figure 14.1), as the name suggests, is to compress air or, indeed, any fluid to reduce its volume. Some of the main compressors used on ships are the main air compressor, the deck air compressor, the air-conditioning compressor, and refrigeration compressors. In this chapter, we will learn specifically about air compressors. The air compressor is a device with vast applications in every industry and household requirement. In the maritime industry, the air compressor is a critical piece of equipment. It can be employed in any number of processes ranging from cleaning filters to starting the main as well as auxiliary engines. Air compressors produce pressurised air by decreasing the volume of air and in turn increasing its pressure. To explain this

Figure 14.1 Typical marine air compressor.

in a technical sense, an air compressor is a mechanical device through which electrical or mechanical energy is transformed into pressure energy in the form of pressurised air. The air compressor works on the principles of thermodynamics. According to the *ideal gas equation*, a gas – without any temperature difference – when subject to an increase in pressure, reduces its volume accordingly. The air compressor works on the exact same principle in the sense that it produces compressed air by reducing the volume of air. This reduction in volume results in an increase in air pressure without causing a difference in temperature. Air compressors on ships can be classified into two diverse types according to their function: the main air compressor and the service air compressor. The *main air compressor* is a high-pressure compressor which operates on a minimum pressure value of 30 bar and is used by the main engine. *The service air compressor* compresses air to a low pressure of only 7 bar and is used in the service and control airlines. Alternatively, air compressors may be categorised according to their design and working principles. Of these categories, there are four: (1) the centrifugal compressor, (2) the rotary vane compressor, (3) the rotary screw compressor, and (4) the reciprocating air compressor. On ships, the reciprocating air compressor is the most widely used. A reciprocating air compressor consists of a piston, connecting rod, crankshaft, wrist pin, suction valve, and discharge valves. The piston is connected to the low and high sides of the suction and discharge line. The crankshaft rotates which in turn rotates the piston. The downward-moving piston reduces the pressure in the main cylinder. This pressure difference opens the suction valve. The piston is taken down by the rotating crankshaft and the low-pressure air fills the cylinder. The piston reciprocates upward. This upward movement starts increasing the pressure, closing the suction valve. When the air is pressurised to its specific value, the discharge valve is opened, and the pressurised air starts moving through the discharge line and is stored in the air bottle. This pressurised air in the air bottle can be used to run the main and auxiliary engines.

The third way of categorising air compressors is by their use. Normally, air compressors on board ships are employed as the (1) main air compressor, (2) topping-up compressor, (3) deck air compressor, and (4) emergency air compressor. *Main air compressor*. This is used to supply highly pressurised air to start the main and auxiliary engines. The air compressor produces and then stores pressurised air in the air bottle. There are different capacity main air compressors available, but the capacity must be adequate for starting the main engine. The minimum air pressure for starting the main engine is 30 bar. A pressure valve is provided which reduces the pressure and supplies controlled air from the air bottle. The control air filter controls the input as well as output air into and out of the air bottle. *Topping-up compressor*. This type of compressor is used to counteract the effects of any air leaks in the system by 'topping up' the volume of pressurised air as required. *Deck air compressor*. The deck air compressor is used on deck and as a service air compressor. It may have a separate service air bottle for this purpose. These are lower capacity pressure compressors, as the pressure required for service air is in between the range of 6 to 8 bar. *Emergency air compressor*. An emergency air compressor is used for starting the auxiliary engine in an emergency or when the main air compressor has failed to fill the main air receiver. This type of compressor may be motor driven or engine driven. If motor driven, it should be supplied from an emergency source of power. The air compressor will work efficiently if designed and installed to proper specifications. It is important that all appropriate crew members are trained and proficient in the use of the emergency air compressor. Air compressor efficiency can be further increased using the following techniques and installations. *Pressure-bar*. The pressure-bar or pressure gauge should be installed on compressors to ensure that the air pressure and discharge air are at the specified pressure. Without this device, if air is pressurised below the required value, it cannot support the system to which it is employed. *Safety devices*. These are the devices used to reduce the loss of energy from the air compressor and increase efficiency. Safety devices automatically shut down the input and output air when adequate compressing is reached. This saves the device from developing overpressure.

There are various components which are common to all types of air compressors. These include the following: (1) The *electricity or power source*. This is the key component of any type of compressor and is essential for the running operation of the compressor. The power source or electric motor is used to run the compressor efficiently and at a constant speed without incurring fluctuations. (2) *Cooling water*. Cooling water is used to cool the compressor in between the various stages of operation. (3) *Lube oil*. Lube oil is necessary to keep the mobile parts of the compressor lubricated. This lubrication reduces the friction in the moving parts of the compressor and thus extends the working life of the components. (4) *Air*. This is the key component without which the air compressor cannot function. The air around us is low pressure and thus serves as the primary input to the compressor. (5) *Suction valve*. The suction valve is provided with a suction filter that sucks in the input air. This input air is then compressed in the main compartment of the compressor. (6) The *discharge valve*. This valve allows the output air to be discharged into a storage tank or air bottle. We have already lightly touched on the main uses of compressed air on board a ship, but there is no harm done in examining these uses a little further. On board a ship, compressed air is used for several especially important purposes. As we know, the air compressor is used to provide the starting air to the main engine. In addition to the main engine, other systems also require compressed air. These systems include the many control valves, throttle controls, and other monitoring systems which work using pressurised air. As most of the compressed air is supplied as high-pressure 30 bar air for starting the main engine, when this air is needed for low-pressure functions, the high-pressure air must be reduced to lower working pressures by way of pressure reduction valves. This air is typically reduced to a pressure of 7–8 bar. Some of the main uses of this low-pressure air are starting the auxiliary engines and the emergency generators, charging the freshwater and drinking water hydrophores, blowing the foghorn, providing spring air for the exhaust valves of the main engine, dry washing the main engine turbochargers, sewage treatment plants for aerobic sewage breakdowns, soot blowing of the boilers, pneumatic pumps for oil transfers, and many more applications such as service air for cleaning and painting operations, chipping, and the operation of pneumatic tools such as grinders and chisels. One more important branch of this 7–8 bar compressed air is used as control air. The control air is a filtered branch of the service air which is made free of any moisture and oil carry-over. This controlled air is used for pneumatic controllers and is important for the operation of shipboard machinery. The system through which this compressed air is distributed is called the air line. There are three types of air line on a ship: the main high-pressure airline, the service airline, and the control airline. Each air line services a separate and specific function according to the air pressure it carries and the machinery it feeds.

Most modern vessels are employed with multi-stage reciprocating air compressors which have intercoolers and aftercoolers with auto-draining and unloading arrangements. The capacity of the main air compressor is expressed in terms of free air delivery (FAD) and is stated in cubic metres per hour (cu ms/hr). FAD can be defined as the volume of air discharged by the compressor, in any one given hour, that would occupy one cubic metre if expanded to atmospheric pressure and cooled to atmospheric temperature. The discharge of the main air compressor is led into a main air bottle or reservoir which stores this pressurised air at a maximum of 30–32 bar. A ship may have two or three main air compressors depending on many factors such as the air compressor capacity, the volume of air needed for starting the main engine, and the demand for air for other services on a particular ship. The marine boilers and exhaust gas economisers on certain ships are soot blown with air. These ships may be employed with higher capacity air compressors at the design stage depending on the requirements set by the shipowner. As per SOLAS requirements, a ship's main air compressor should be able to fill its air reservoirs from zero to maximum pressure (30 bar) within one hour.

Most ships are equipped with a set of two air bottles. These are sometimes referred to as air reservoirs. They may be of a vertical or horizontal design. Air bottles are hydraulically pressure tested to 1.5 times their working pressure. As per SOLAS regulations, the total capacity of the air bottle must be sufficient to provide at least 12 consecutive main engine starts for a reversible engine and at least 6 consecutive starts for a non-reversible engine, without needing replenishment. There must be two identical main air receivers and one emergency air bottle for every vessel. Each air bottle should be equipped with the following mountings: (1) *Fusible plug*. The fusible plug consists of a bismuth (50%), tin (30%), and lead (20%) composition. The purpose of the fusible plug is to release the compressed air in the event of an abnormally high compressed air temperature. The fusible plug has a melting point of 104.4°C (220°F) and is fitted to the bottom of the air bottle. Where a relief valve is not directly fitted to the air bottle, the fusible plug is usually found on the ship's side. (2) *Atmospheric relief valve*. The atmospheric relief valve is provided as an overpressure protection device and serves as a backup to the fusible plug. In the event of an engine room fire when CO_2 flooding is required, this valve must be opened before evacuating the engine room. As an additional safety feature, the air receiver relief valve opening may be located either outside the engine room through the ship's funnel or inside the engine room itself. In the case of the latter, CO_2 bottle calculations for fighting engine room fires must be conducted accordingly. This demand for extra CO_2 is taken into consideration during the ship's design stage. (3) *Spring-loaded safety valve*. The spring-loaded safety valve is used for setting the pressure at 32 bar (for a 30-bar working pressure) with an equal to or greater than 10% rise in accumulation of pressure. (4) *Compensation ring*. When a hole is cut or machined into a pressure vessel, higher stresses will apply to the material around the hole. To reduce this, compensation rings are fitted. This is a flange on which a valve or fitting is mounted. A compensation ring provides structural integrity for the air pressure vessel. Additional fixtures typically include a manual drain valve or automatic drain valve, an assortment of pressure gauges, an access door, the main starting air valve, auxiliary starting air valve, fitting valve, and service air or whistle air valve.

Large cylindrical air bottles usually have one longitudinal welding seam. The longitudinal and circumferential seam is machine welded with full penetration welds. The welding details are governed by the air pressure to be stored in conjunction with classification society regulations. All welded air receivers must be stress relieved or annealed at a temperature of about 600°C (1,112°F). The welding must be radiographed for integrity analysis. Air receivers are subject to statutory survey and inspection and must be periodically hydraulic tested at 1.5 times the working pressure every ten years. Specific testing requirements may apply for smaller or unique air bottle designs. Outside the official testing regimen, air receivers must be inspected as per the ship's preventative maintenance schedule. The main purpose of the inspection is to check for signs of corrosion, cracks, and fissures. Moisture in the air receiver can give rise to corrosion and despite the proper operation of compressor cooler drains, it is common for copious amounts of condensate to collect, particularly in humid conditions. It is always good practice to check the air reservoir drains regularly to assess the quantity of condensate. In extreme conditions, the drains may need to be emptied as many times as twice or thrice daily to remove accumulated condensate emulsion. In most cases, the worst corrosion can be found near the air receiver drain. If the structural integrity of the air receiver remains following a thorough visual inspection, thickness measurements can be taken using an ultra-sonic thickness gauge. If the thickness of the air receiver is compromised, it is necessary to reduce the air pressure to be contained in that specific air bottle. This, after certain calculations, can be done by changing the cut-in and cut-off settings of the air compressor when that receiver is in use and the relief valve settings require readjustment. Furthermore, it is worth remembering the air receiver can be isolated completely and kept on standby to be filled manually with caution, whenever required. If the air bottle is too small to enter, then internal inspections can be conducted using a probe-mounted camera. The internal surface of the air bottle may be coated in a graphite suspension in water, linseed oil, Copal varnish, or an epoxy coating. Whichever type of coating material is used, it must have anti-corrosive, anti-toxic, and anti-oxidising properties.

The control air system consists of a branched airline which passes through a pressure-reducing valve. Pneumatic control equipment is especially sensitive to the contaminants which may be present in compressed air. Viscous oil and water emulsions can also cause moving parts in the control equipment and control valves to stick. Rubber parts such as diaphragms, spools, and washers are particularly susceptible to oil damage. Water can cause rust build-up which may result in moving parts sticking or suffering damage through the abrasive action of rust particles. Metallic wear and other small particles can cause damage by abrasion. Moreover, any solids mixed with oil and water emulsions can conspire to block small orifices. This means it is crucial to thoroughly clean and dry the control air before use. When the source of the control and instrument air is the main air compressors and the main air reservoir itself, then special provisions are necessary to ensure that the air quality is superior. The pressure-reducing valve which brings the main air pressure to the 7 or 8 bar required by the control air system can be affected by emulsion carry-over and often requires frequent cleaning to prevent control air contamination. Automatic drain traps may be fitted to the control air system, but these often require manual drainage by the engine crew daily. Copious amounts of free moisture and oil emulsion carry-over in the air can be removed by special control air membrane filters installed in the control air line. A typical control air filter arrangement consists of an oil and moisture collecting filter followed by a membrane air dryer filter. The treatment of air through these membrane filters results in the air being filtered and dried, which removes virtually all traces of oil, moisture, and air impurities. A simple line air filter can be produced using a small plastic float and an auto-drain arrangement. The filter may also be drained manually if the vessel enters a particularly humid environment, where frequent draining will be required. The filter dryer unit consists of a primary filter, secondary filter, and membrane hollow fibre elements.

The control air enters the dryer chamber through the line filter located in the lower part of the dryer unit. In the dryer unit, the primary filter removes coarse rust particles, dust, and other large impurities. The secondary filter acts as a coalescer, separating water droplets and oil mists up to $0.3\mu m$. A differential pressure gauge indicates the condition of the primary and secondary filters. A higher differential pressure indicates a dirty membrane filter. These membrane elements must be renewed in accordance with the ship's preventative maintenance schedule. The high-pressure air piping from the air compressor to the receiver should be as smooth as possible and absent of unnecessary bends to allow the air to flow freely to the receiver without restriction. Bends in the piping can create backpressure in the line; this is especially likely to occur when accumulated moisture or oil emulsion is left in the line. The emergency air compressor is a small independent air compressor which can be driven by an independent prime mover, such as the engine, or a separate power supply from the emergency switchboard. It is used to fill the emergency air bottle. This reserve air is usually only needed to kick-start the auxiliary engine (generator) of a dead ship. The control air onboard is also used in the emergency shut-off, quick-closing valve system. This system operates on 7-bar pressure and is typically used for fire and funnel dampers. As per SOLAS regulations, the emergency air bottle, which feeds the quick-acting valves, must be kept in a continuous state of readiness. In the event of an uncontrollable engine room fire, the quick-closing valves are operated. These valves use controlled air to shut off the outlet valves for the fuel oil and lube oil tanks, as well as the engine room funnel and blower dampers. On gas tankers, the control air supply is also used by the emergency shutdown system (which is not discussed in this book).

MAIN COMPONENTS OF THE AIR COMPRESSOR SYSTEM

The main parts of the air compressor are as follows: (1) The *cylinder liner*. This is made from graded cast iron and is accompanied by a water jacket to absorb the heat produced during the compression process. It is designed to provide a streamlined passage for the pressurised air resulting in minimum pressure drop. (2) The *piston*. For non-lubricating type compressors, a lightweight aluminium alloy piston is used for lubrication. Graded, cast iron pistons are used, in conjunction with piston rings, for sealing and scraping off excess oil. (3) *Piston rod*. In high-capacity compressors, which are usually substantial in size, a piston is attached to a piston rod made from alloy steel. These are fitted with anti-friction packing rings to minimise the risk of compressed air leaks. (4) *Connecting rod*. The connecting rod helps to minimise thrust to the bearing surface. It is usually made from forged alloy steel. (5) *Big end bearing and main bearing*. These are designed to provide rigidity to the running rotational mechanism. They are made from copper lead alloy and will enjoy a long operational life if proper lube oil and lubrication are provided. (6) *Crankshaft*. The crankshaft uses counterweights for dynamic balancing during high-speed rotations to avoid twisting from torsional forces. The connecting rod, big end bearing, and main bearing are connected to the crankshaft via a crank pin and journal pin. These must be regularly polished to ensure the bearings have a long working life. (7) *Frame* and *crankcase*. Normally, these have a rectangular shape and accommodate all the moving parts of the air compressor. The frame and crankcase are typically made from rigid cast iron. The main bearing housing is fitted on a bore in the crankcase and is made to the highest precision to avoid misalignment. (8) *Oil pump*. A lubricating oil pump is fitted to supply lube oil to all the bearings, which can be chain or gear driven, through the crankshaft. The pressure of the oil can be regulated by way of a regulating screw provided in the pump. A filter in the inlet of the pump is also attached to supply clean and particle-free oil to the bearings. (9) *Water pump*. Some compressors may have a water cooling pump attached which is driven by the crankshaft via a chain or gear. Some systems do not use pumps, as they have water supplied from the main or auxiliary cooling system. (10) *Suction and discharge valve*. These are multi-plate valves made from stainless steel and are used to suck and discharge air from one stage to another, and into the air bottle. (11) *Suction filter*. This is an air filter formed from copper or soft steel with a paper material to absorb oil. The filter has a wire mesh which prevents metal or dust particles from entering the compression chamber. (12) *Intercoolers*. Intercoolers are normally fitted between two stages of the compression process to cool down the air temperature and to increase the volumetric efficiency of the compressor. Some compressors have integrated copper tubes for cooling, and some have exterior intercooler copper tube assemblies. (13) *Driving motor*. An electric motor is attached to the compressor, which is then connected to the flywheel. These are the main components of the marine air compressor. Actual parts may vary according to the design of the air compressor and or the specific requirements of the system or ship.

SAFETY FEATURES OF THE COMPRESSOR

Every air compressor on a ship is fitted with several safety features to avoid the development of dangerous conditions. Some of the main safety features of the air compressor system include the *relief valve*. A relief valve is fitted after every stage to release excess pressure. *Bursting disc*. A bursting disc is a copper disc which bursts when the internal pressure exceeds the pre-determined safe value. *Fusible plug*. Generally located on the discharge side of the compressor, the fusible plug fuses if the air temperature is higher than the operational temperature. The fusible plug is designed to melt at elevated temperatures. *Lube oil low-pressure alarm and trip*. If the lube oil pressure falls lower than normal, the alarm is activated followed by a cut-out trip signal to avoid damage to the bearings and crankshaft. *Water-elevated temperature trip*. If the intercoolers are choked or the flow of water is reduced below safe limits, then the air compressor will overheat. To avoid this, a high-water temperature trip is activated which automatically cut-offs the compressor. *Water no-flow trip*. If the attached pump is not working or the flow of water inside the intercooler is insufficient to cool the compressor, then the moving parts inside the compressor will seize due to overheating. A no-flow trip is provided which continuously monitors the flow of water and trips the compressor when the water level falls below the safe limit. *Motor overload trip*. If the current taken by the motor during running or starting is abnormally high, then there is the possibility of causing damage to the motor. An overload trip is thus fitted to avoid this situation from developing.

AIR COMPRESSOR MAINTENANCE

General maintenance regimes. The air compressor requires proper planned routine maintenance for safe and efficient operation and to avoid breaking down. Routines for maintenance obviously depend on the manufacturer's recommendations as provided in their manual. That said, the following are generic maintenance checks that should be conducted in accordance with the running hours stated:

@ 250 running hours:
- Clean the air filter.
- Check the unloader operation.
- If a belt is provided for driving the cooling water pump, check its tightness.

@ 500 running hours:
- Change the lube oil and clean the oil sump.
- Clean the lube oil filter.
- Check and renew the suction and discharge valves.

@ 1,000 running hours:
- Conduct a crankcase, main, and big end bearing inspection.
- Overhaul the relief valves.

@ 4,000 running hours:
- Overhaul the piston and big end bearing and conduct piston ring renewal.
- Clean the intercoolers.
- Perform motor overhauling.

Remember, these are generic assumptions, and actual running hour requirements may differ from one manufacturer to another.

Bumping clearance. Bumping clearance is the top clearance in the main air compressor between the top of the piston and the cylinder head when the piston is at top dead centre. This clearance is determined by the cylinder head gasket which should be of the correct size. It is strongly advised to use genuine parts only. The bumping clearance is important as it relates to the volumetric efficiency of the air compressor. When the bumping clearance of the compressor is large, then the clearance volume is big enough to accommodate a large volume of air when the piston is at the end of its stroke. When the piston moves downwards, then a large suction stroke becomes ineffective, as the suction will happen only when the cylinder pressure falls below atmospheric pressure. This means the piston stroke is wasted by expanding a large amount of air that has accumulated in the clearance volume due to a large bumping clearance. (1) *Measuring the bumping clearance.* The bumping clearance is fixed by the thickness of the cylinder head gasket as installed by the manufacturer, but it can be measured by a lead ball placed between the top of the piston, assembling the compressor, and the manufacturer-recommended cylinder head gasket. As the piston falls back to top dead centre, the ball will compress. This can then be measured by a micrometre. Most bumping clearances fall within a range of between 1.2 mm and 1.8 mm. The bumping clearance should be calculated every 3,000 running hours or depending on the performance of the compressor. (2) *Adjusting the bumper clearance.* To adjust the bumping clearance, we must change the thickness of the cylinder head gasket by adding or removing shims from the foot of the connecting rod and the bottom end bearing of the compressor. (3) *Checking basic screwed connections.* All the unions and screwed connections must be checked for tightness and must be retightened, if necessary. This includes all cooler and air lines, unions on pipes and hose lines, cylinder heads, cylinders, electric motors, measuring and switching devices, bearings, and any other accessories. This must be conducted at a minimum of every 250 running hours.

Replacing the air filter cartridge. The air filter must be checked at frequent intervals, and certainly not less than every 250 running hours or whenever the air discharge temperature becomes too high. The air filter element must be replaced with a new filter. It must be noted that the air compressors are installed in strategic locations where there is sufficient air supply to ensure the suction is never devoid of air. this means we must take extra care to ensure the old cartridge is not blown with air and re-used, as this may damage the air filter. Dirty air filters affect the volumetric efficiency of the compressor as does the temperature of the inlet air to the compressor. The air filter cartridge should be renewed no less than every 2,000 running hours.

Changing the oil. It is important that the oil is changed according to the maintenance frequency, as the lube oil is not only used for the lubrication of the running parts of the compressor but also to remove the heat generated during the operation of the compressor. The compressor must be run for some time before the oil is changed, as this ensures that all the sludge and particles are in suspension. The oil should be completely drained by removing the drain plug and flushed with fresh oil before the oil is replenished. The crankcase should be thoroughly cleaned at the same time with a lint-free cloth. It is critical that no two grades of oil are used and mixed in the system. Oil changes must be conducted at least every 1,000 running hours.

Cleaning oil strainers. The oil strainer is typically a mesh plate which is supposed to be cleaned at defined maintenance intervals. The filter should also be cleaned when the oil is changed. It should be cleaned with a solvent and blown with air. This should take place every 1,000 running hours.

Checking the valves. This is one of the most important jobs in the maintenance of the compressor. Extreme care must be taken when performing this critical task. When removing and installing a valve, extreme care must be taken that no valve parts are damaged. This applies to the sealing surfaces of the valve. It is worth noting that diverse types of compressors have diverse types of valves and removal methods. In any case, the valves must be removed and checked for diesel oil or water leaks. The valves must never be held in a vice. All the parts must be meticulously cleaned after opening the valve and must be checked for signs of carbonisation. Overcarbonisation indicates that an oil top-up is more than required. Failure to do so can lead to the sticking of the valves, eventually resulting in their breaking. Sticking valves also reduce the volumetric efficiency of the compressor as the quantity of air discharged is comparatively less. The valve plate should be checked by pushing the valve plate from the side of the valve seat with a screwdriver. The valve plate will move a distance which is equivalent to a valve lift. The springs and valve plates should be changed if required. If necessary, the plate can lap if no spares are available on board. The frequency for the disassembling and cleaning of the valve is usually given at 3,000 running hours, 3,000 running hours for the replacement of the valve plate, and 3,000 running hours for checking for spring fatigue.

Installing the valves. All valves must be refit with new gaskets and rings. Only use original spare parts as installing non-original gaskets may lead to compressor leaks and damage.

Replacing the valves. It is recommended by most manufacturers that the valves are not repaired but replaced, as they may fail due to fatigue. Replacing the valves is done in the same way as mentioned in the checking of the valves.

Checking the intercoolers. The intercoolers are arranged after every stage of the main air compressor. They reduce the work done in compressing the air by successfully cooling it down between each stage. The intercooler is made from copper tubes bent into a 'U' formation. The air passes through the tubes and water circulates around them. Intercoolers may be either a single tube type or straight tube type. The intercoolers are provided with purge spots to collect and drain water or oil which may ingress to the air coolers. The intercoolers are opened as per the maintenance frequency and are designed to be cleaned with a solvent. When checking the intercoolers, it is equally as important to check the integrity of the tubes. The intercooler is provided with a bursting disc to relieve abnormal pressure in the event a tube bursts. The single tube type coolers are the most difficult to clean and the rate of wear is higher than the straight tube types. Straight tube types, on the other hand, are susceptible to plugging when they leak. The intercoolers should be checked at least every 3,000 running hours.

Cylinder liner, cooling jacket, and piston. Ambient air naturally carries a lot of moisture, which can ingress into the air compressor and condensate on the cylinder liner. If this happens, the condensate emulsion is likely to wash away the lubricating oil film from the liner, thus causing liner wear. This can also result in scoring to the liner. Proper drainage of the compressor is necessary to prevent excessive liner and piston ring wear. The cylinder liners can become lubricated by the oil splashed from the crankcase. They may also receive lubrication from lubricating oil quills like the main engine. In these cases, the oil is delivered by an engine-driven lube oil pump. The piston rings can wear out causing blow past. This will eventually reduce the volumetric efficiency of the compressor. The cylinder liners, along with the cooling jacket, should be inspected and cleaned every 3,000 running hours. The pistons, piston rings, and oil scraper rings should be checked every 6,000 running hours.

Lubricating oils. Different manufacturers have different requirements for the use of lubricating oils. Most manufacturers do not permit the use of synthetic lubricating oils with three-stage, air-cooled compressors. This is because the good hydrolytic properties of synthetic oils cause moisture to condense in the crankcase increasing the risk of corrosion and drive damage. Because of their design, three-stage, air-cooled compressors have low final compression temperatures, rendering the high-temperature stability of synthetic oils useless.

AIR COMPRESSOR STARTING AND STOPPING PROCEDURES

Certain steps and systematic procedures need to be followed to start or stop the air compressor. In this section, we will briefly cover the generic procedures for starting and stopping an air compressor and establish what checks are needed to be made before starting the air compressor and during its operation. Before starting the air compressor, the following steps must be followed. (1) Check the lube oil in the crankcase sump by means of a dipstick or sight glass. (2) All the valves of the compressor discharge must be in an open condition. (3) If any manual valve is present in the unloader line, this must always be kept open. (4) All alarms and trips – for example, the lube oil low pressure, water-elevated temperature, overload trip – must be checked for correct operation. (5) All valves in the cooling water line must be in an open position. (6) The cocks for all the pressure gauges must be in the open position. (7) The air intake filter must be clean. (8) Finally, if the compressor has not been started for a long time, it should be turned on manually with a tommy-bar to check for the free movement of its parts.

Unloading the compressor. Unloading is a normal procedure during the starting and stopping of the compressor. It is conducted for the following reasons. (1) When starting a compressor motor, as the load on the motor is exceedingly high, the starting current is also high. To avoid further loading of the compressor an unloader arrangement is provided which is normally a pneumatic or solenoid control, and which releases the pressure during the starting of the compressor. Once the current reduces to the running value, the unloader closes automatically. Normally, a timer function is used for the opening and closing of the unloader. (2) Air contains moisture, and during the compression process, some amount of moisture gets released. Liquid in any form is incompressible, and if some amount of oily water mixture is present inside the cylinder, it will damage the compressor. To overcome this problem an unloader is used. During the starting procedure, the unloader comes into action and releases all the moisture accumulated inside the cylinder. (3) Intermediate operation of the unloader may be selected so that during the process of compression any moisture or oil accumulation cannot take place. (4) During the stopping procedure, the compressor unloader is operated to ensure, for the next starting procedure, the cylinder is moisture-free.

Checks during the operation of the compressor. During the operation of the air compressor, the following checks should be carried out regularly: (1) check to ensure all the pressure gauges are showing the correct readings for the lube oil pressure, water pressure, etc.; (2) check for any abnormal sounds such as knocking or banging; (3) check for any lube oil or water leaks; (4) if cylinder lubrication is provided, check the supply via the sight glass; (5) check if the discharge pressure for all units is within normal and expected parameters; (6) check the air temperature after the final stage is under the limit; (7) check the flow of cooling water via the sight glass. If attached, the cooling water pump should be checked for free rotation; and (8) check the relief valve of all units for leakages. With some compressors, provision is given to check the relief valve with a hand lever. If provided, check all units accordingly.

AIR COMPRESSOR TROUBLESHOOTING

The air compressors on ships require special attention and care to ensure their smooth operation. It is only through routine maintenance and inspections that any faults or impending issues can be diagnosed and remedied. The compressor, however, is a peculiar piece of equipment which tends to suffer some problems irrespective of the maintenance regime. In this section, we will go through some of the key issues that can arise during the operation of the air compressor and suggest ways to troubleshoot these problems.

1. The following causes are usually indicative of low-pressure lube oil in the air compressor: (a) faulty pressure gauge, (b) cock to the pressure gauge is in a closed position, (c) low oil level in the sump, (d) leaks in the supply pipe, (e) suction filter is choked, (f) oil grade in the crankcase is incompatible, (g) attached lube oil gear pump is faulty, (h) worn-out bearings and excessive clearance.
2. Abnormal noises during operation can be indicative of (a) loose foundation bolts; (b) worn-out bearings, excessive clearance; (c) imbalanced crankshaft resulting in high-end play; (d) valve plate broken or faulty; (e) relief valve lifting below the setting pressure; (f) bumping clearance is insufficient; (g) piston is worn out, or the piston ring is broken.
3. Vibration in the machinery usually indicates: (a) the foundation bolts are loose (b) discharge pressures are high, or discharge valve plates are faulty; (c) the liner and piston are worn out; (d) insufficient bumping clearance; (e) the cooling water temperature is too high.

4. The cooling water temperature can go high because of the following reasons: (a) inlet or outlet valve for the cooling water is closed; (b) the intercooler is choked; (c) the cooling water in the expansion tank is low; (d) the pipe passage has narrowed due to scale formation; (e) the water-pump belt or gear drive is broken; (f) the first stage discharge pressure is too high.
5. In the event the first stage discharge pressure is high, check for signs of (a) a faulty pressure gauge; (b) a choked intercooler air passage; (c) the second stage suction valve not closing properly, allowing air to escape from the second stage to the first stage; (d) the discharge valve of the first stage malfunctioning, and remaining in the closed position; and (e) the spring of the discharge valve is malfunctioning.
6. In the event the first stage discharge pressure is low, this may be caused by (a) a faulty pressure gauge; (b) the suction filter is choked; (c) the unloader of the first stage is leaking; (d) the first stage suction valve is not closing properly, resulting in compressed air leakage; (e) the first stage suction valve is not opening fully, leading to less intake of air; (f) the discharge valve is faulty and remains permanently open; (g) the relief valve after the first stage is leaking; (h) the piston ring of the first stage is badly worn out, allowing air to pass.
7. In the event of high discharge pressure in the second stage, the reasons may be (a) faulty pressure gauge; (b) the discharge valve to the air bottle is shut; (c) the second stage discharge valve plate is worn out, or the spring is worn out; (d) the valve is stuck in the closed position; (e) the aftercooler air passage is choked; (f) the air bottle is overpressurised.
8. When the second stage discharge pressure is low, this may be caused by (a) a faulty pressure gauge; (b) the suction valve for the second stage is malfunctioning and in the open position; (c) the suction valve for the second stage is not opening fully and is taking in less intake air; (d) the discharge valve is faulty and remains open during operation; (e) the piston rings of the second stage are worn out, leaking compressed air; and (f) the relief valve of the second stage is leaking.
9. If the relief valve of the first stage is lifting, this may be indicative of (a) the spring of the relief valve is malfunctioning, thus lifting at lower than expected pressure; (b) the discharge valve of the first stage is not opening; (c) the intercooler air passage is blocked; (d) the suction valve of the second stage is in a stuck position; (e) there is water inside the compression chamber due to a crack in the jacket and water is leaking inside.
10. If the relief valve of the second stage is lifting, this may be indicative of (a) the relief valve is malfunctioning, and lifting at a lower than setting pressure; (b) the main discharge valve to the air bottle is closed; (c) the discharge valve plates and spring are worn out, leaving the valve in a closed position; (d) the aftercooler air passage is blocked; (e) there is water inside the compression chamber due to a crack jacket.

In this chapter, we have covered a lot of ground related to the ship's air compressor. It cannot be stressed enough that the air compressor and its associated systems are critical to the safe and efficient operation of the ship. This requires good housekeeping, regular maintenance, and investment in genuine parts and stores. In the next part of this book, we will begin to examine the heating, ventilation, and air-conditioning (HVAC) systems found on board ships. First, we will look at the role and function of the marine boiler.

Part III

Heating, ventilation, and air conditioning

Part II

Chapter 15

Marine boiler

On today's ships, marine boilers are used for auxiliary purposes and on vessels that run off marine diesel engines or diesel-electric engines. For ships that operate steam turbines (for example, naval frigates and destroyers) the boiler is an integral part of the main propulsion system. In this chapter, however, we will focus primarily on auxiliary boilers, i.e., the types of boilers used for running the auxiliary systems on board merchant ships. Marine boilers are designed and built according to the specific needs required by the vessel and its operational demands. When designing a new type of boiler, naval architects must consider a variety of factors. Firstly, they need to rate the boiler. This means they need to estimate the steam output required from the boiler for the ship being built. For this, there are three key requirements: (1) Requirement 1 – the steam consumption requirement for compensating for heat loss in the tanks, (2) Requirement 2 – the steam consumption requirement for raising the temperature of the fuel oil in the tanks, and (3) Requirement 3 – the steam consumption requirement for all other services.

REQUIREMENT 1 – COMPENSATING FOR HEAT LOSS

Most ships run by diesel engines have fuel oil tanks that are used to store high-sulphur fuel oil. As the viscosity of high-sulphur fuel oil is extremely high, the high-sulphur fuel oil is as dense as tar; this means it is almost impossible for high-sulphur fuel oil to flow freely. To transfer the high-sulphur fuel oil to the settling tanks and then the high-sulphur fuel oil service tank, this viscosity needs to be enhanced to a level that is more manageable. For this, the high-sulphur fuel oil storage tanks are equipped with heating coils that store the high-sulphur fuel oil at a steady temperature. The heating fluid used in the heating coils is steam produced by the auxiliary boiler. When designing the boiler system, the architect must first identify where each high-sulphur fuel oil storage tank will be located as per the general arrangement drawing for the ship. Then the architect needs to account for the surrounding space. This means noting the adjacent fixtures to each tank bulkhead. Depending on the surrounding area of each tank bulkhead (for example, the engine room, void space, ballast water tanks, and sludge tanks) the anticipated ambient temperature is analysed. From this point, the architect can determine the amount of steam flow rate required to maintain the required temperature of each tank. This is calculated accordingly using the following formula:

$$Q1 = U A (T2 - T1)$$

Where:

- Q = heat loss from the bulkhead (W).
- U = overall heat transfer co-efficient (W/m2 in °C/°F).
- A = area of the tank bulkhead under consideration (m2).
- T2 = temperature of the tank to be maintained (°C/°F).
- T1 = temperature of the adjacent medium of the bulkhead considered (°C/°F).
- Qt = sum of heat loss from all six bulkheads of the tank.
- Q1 = sum of heat loss from all tanks.

As we now know the heat transfer rate, we can calculate the mass flow rate of steam using the following formula:

ms = Q1/Δh

Where:

- ms = mass flow rate of steam (kg/s).
- Q1 = calculated heat transfer (kW).
- Δh = enthalpy drop of the steam (kJ/kg).

REQUIREMENT 2 – RAISING THE FUEL OIL TEMPERATURE

Not only is steam required to compensate for heat loss from the fuel oil tanks, but steam is also needed to heat the fuel oil to the required temperature before it can be pumped into the engine. For this, the time (t) in hours, required to heat up the oil in each type of tank, is calculated using the following formula:

ΔT/t

Where:

- Storage tank – 0.2°C (33°F) / hour rise in temperature.
- Service and settling tank – 4°C (39°F) / hour rise in temperature.
- All other tanks – 1°C (33.8°F) / hour rise in temperature.

This calculation involves three steps:
   The calculation of heat (Q in watts) required to heat the contents of each tank.
   The summation of the individual heat requirements to obtain the total heat transfer required to raise the temperature of fuel oil in tanks (Q2).
   Using the obtained heat requirement to find the required mass flow rate of steam for this purpose.
   A worked example of this calculation is shown next, where the amount of heat required to raise the temperature of the fuel oil tanks may be expressed as:

Q2 = m Cp dT/t

Where:

- Q2 = means the heat transfer rate (kW).
- m = mass of fuel oil in the tank (kg).
- Cp = specific heat capacity of the fuel oil (kJ/kg °C/°F).
- dT = change in temperature of the fuel oil (°C/°F).
- t = total time over which the heating process occurs (hours).

As we know the heat transfer rate, the mass flow rate of steam can be calculated using the following formula:

ms = Q2/Δh

Where:

- ms = mass flow rate of steam (kg/hr).
- Q2 = calculated heat required to raise the temperature (kW).
- Δh = enthalpy drop of the steam (kJ/kg).

REQUIREMENT 3 – ALL OTHER SERVICES

Steam is also used by ships to cater for other heating requirements, such as a heat exchange medium in high-sulphur fuel oil purifiers, light diesel oil (LDO) purifiers, and lube oil purifiers; as a heating medium in the booster modules; to preheat the main engine jacket cooling water; and as a heating medium in calorifiers (calorifiers are high-pressure storage units of heated water, which is used in gantry and toilet utilities). The heat requirements for all such services are calculated individually and then added together. For simplicity, we can refer to the obtained heat requirement as Q3. Once the heat requirements for the three purposes (mentioned earlier) are obtained, they are added to obtain the total heat rate and total steam mass flow rate required for the boiler. This is expressed as:

Total Heat Rate Required (Q) = Q1 + Q2 + Q3 (kW)

Total Mass Flow Rate Required is calculated from the relation:

mS = Q/Δh (kg/hr)

Where:

- Δh = enthalpy drop of the steam (kJ/kg).

BOILER RATINGS

Now, there are two rating systems used to obtain a suitable boiler. The first is the *from and at rating (FAR)* and the second is the *kilowatt rating (kW rating)*. In Figure 15.1, the vertical axis corresponds to the steam output as a percentage of the 'from' and 'at'

Figure 15.1 Boiler control position.

ratings, at different pressures. So, for example, at 15 bar, assuming the feed water temperature is 68°C (155°F), we can calculate the percentage FAR from the graph which is 90%. This means if the boiler has a rated steam output of 2,000 kg/hr, the actual steam output of the boiler will be 90% of the rated output, which is 1,800 kg/hr. Now, when the architect chooses a boiler, they need to specify the rated steam output to the boiler manufacturer. The boiler manufacturer then provides the boiler 'from' and 'at' rating graphs for the proposed boiler. The above calculations are conducted for various boiler pressures and feed water temperatures, to check that the actual steam output is more than the steam flow rate (mS) obtained in the initial design calculations that we have previously discussed.

While some boiler manufacturers prefer 'from' and 'at' ratings, other manufacturers prefer to use a different system called the kilowatt rating system. In effect, this is just a unique way of expressing the same data. To obtain the actual steam flow rate from the kW rating of a boiler, the following relation is used:

$$\text{Steam output}\left(\frac{\text{kg}}{\text{h}}\right) = \text{boiler rating}(\text{kW}) \times \frac{3600\,\text{s/h}}{\text{energy to be added}\,(\text{kJ/kg})}$$

In the previous expression, the energy to be added refers to the amount of energy added to the boiler by the feedwater (which in turn depends on the feedwater temperature). The architect should ensure the *steam output obtained is more than the steam flow rate (m^s)* obtained in the initial design calculations that we discussed previously. The previous expression checks are conducted at various working pressures of the boiler and at different ranges of feedwater temperature, depending on the steam requirement at various sailing conditions. It must be assured that the chosen boiler meets the vessel's requirements in all conditions and at different load combinations. When deciding which type of boiler to install on a ship, the architect must also consider the functionality of the boiler and any space constraints. For most auxiliary boilers, a shell and tube type is typically chosen, where the boiler drum holds the water reserve, and fire tubes run along the length of the drum. The hot gases produced by the burner are carried in the fire tubes as they provide greater surface area for transferring heat energy to the water medium. In most cases, where there are no space constraints, auxiliary boilers are horizontally oriented, as this prevents the type of pressure fluctuations which are more commonly experienced with vertically oriented boilers. However, for exhaust gas economisers or exhaust gas boilers (these are boilers that do not have a furnace; rather, it is the exhaust gases from the engine that pass through the fire tubes to heat the water in the boiler drum) vertical configurations are preferable, as these provide less back pressure on the exhaust gas system. Exhaust gas boilers are only ever used when the vessel is underway. When in port, the auxiliary boiler must be used instead.

PROCEDURES FOR STARTING AND STOPPING THE BOILER

Starting the boiler. The boiler is one of several key pieces of machinery that keeps the vessel operating. It is an extremely dangerous piece of equipment that generates steam at extremely high pressure, and it is for this reason that proper care should be taken when operating the boiler. In this section, we will briefly discuss a step-by-step procedure for starting and stopping a generic marine boiler. It should be noted that the following steps may not apply to all types of boilers and each boiler may require additional steps to be followed as per the system design. That said, the basic procedure remains unaltered. First, we must ensure the vent valve on the boiler is open and there is no pressure in the boiler. Then, we must check that the steam stop valve is closed. Check that the valves for the fuel are open and allow fuel to circulate through the system until it reaches the temperature required as per the manufacturer's recommendations. Open the feedwater valves to the boiler and fill the inside of the boiler drum with water to just above the low water level. This is done, as it is not possible to start the boiler when the water level is below the low water level due to the safety features that prevents the boiler from starting. It also prevents the water inside the boiler from expanding and overpressurising the boiler. Start the boiler in automatic mode. The burner fan will start the purging cycle, which will remove any gases present in the furnace, forcing them out through the funnel. After the pre-set purge is complete, the pilot burner will ignite. The pilot burner consists of two electrodes through which a large current is passed via the transformer. This produces a spark between the electrodes. The pilot burner is supplied with diesel oil. When the oil passes over, the transformer ignites. The main burner, which is supplied with heavy oil, catches fire with the help of the pilot burner. Check the combustion chamber from the sight glass to ensure the burner has lit and the flame is satisfactory. Maintain a close watch on the water level as the pressure increases. Open the feedwater when the level of water inside the gauge glass is stable. Close the vent valve after steam starts to form. Open the steam stop valve. Once the working steam pressure is reached, blow down the gauge glass and float chambers to check for any alarms.

Stopping the boiler. If the boiler is going to be stopped for a long duration, for example, during maintenance or survey, change the fuel type to distillate fuel. If a separate heating arrangement for heavy oil is present, then there is no need to change over to distillate fuel and the current oil can be kept in circulation mode. To stop the boiler, put the boiler into automatic cycle. Close the steam stop valves. Close the boiler feed-water valves. When the boiler pressure has reduced to just over atmospheric pressure, keep the vent valve open to prevent a vacuum forming inside the boiler.

BOILER MISFIRES AND MALFUNCTIONS

Failure of the boiler to start is a common phenomenon on most ships. There can be several reasons why the boiler might fail to start. In this section, we will learn about some of the most common reasons why boilers fail.

1. *Fuel inlet valve to the burner is in the closed position.* The fuel line for the boiler's burner consists of several valves located in the fuel tank, pump suction, discharge valve, or before the boiler burner. Any of these may be in the closed position resulting in fuel starvation and boiler failure.
2. *Line filter at the inlet of the fuel line for the burner is choked.* If the system runs on heavy oil, then there is a high chance of the filters in the line becoming choked. To avoid this from happening, the boiler system is normally designed for changeover from diesel to heavy oil during starting and heavy oil to diesel when stopping. This helps keep the filters and the fuel line clean.
3. *Boiler fuel supply pump is not running.* There are two reasons for the fuel pump to stop working. Normally when the pumps are in pairs, the changeover auto system is kept in the manual position, and if the operating pump trips, the standby pump will not start automatically. Another reason may be the tripping of the pump due to a short circuit in the system.
4. *Solenoid valve in the fuel supply line is malfunctioning.* Today most systems have adopted advance automation, but there remains the possibility for the solenoid in the fuel supply line to malfunction, preventing it from opening.
5. *Flame eye is malfunctioning.* A flame eye is a photocell-operated flame sensor fitted directly on the refractory to detect whether the burner is firing or not. If the flame eye unit is malfunctioning, it will give off a trip signal even before the burner starts firing up.
6. *Air or steam ratio setting is incorrect.* For proper and efficient combustion, the air fuel ratio is especially important. If the supply of air is excessive, then this will result in the production of smoke. If the air volume exceeds more than the normal level, the flame will drown causing flame failure.
7. *Forced draft fan flaps malfunctioning.* For removing excess gases trapped inside the combustion chamber, forced draft fans (FDF) are used for pre-purging and post-purging operations. These are connected to a timer which shuts off the fan flaps. If the flaps malfunction, then continuous forced air will enter the chamber, preventing the burner from producing a sufficient flame.
8. *Any contactor switch inside the control panel is malfunctioning.* The boiler control panel consists of several contactors and PLC cards. Just one contactor malfunctioning may result in the boiler failing to start.

9. *Trip not reset.* If any previous trips – such as the low water level, flame failure, emergency stop – have tripped and not been reset, then the boiler will not start.
10. *Main burner atomiser is clogged.* Main burners consist of an atomiser for the efficient burning of fuel. If the atomiser is clogged by sludge and fuel deposits, then the burner may not be able to produce a sufficiently productive flame.
11. *Pilot burner nozzle is choked.* The pilot burner nozzle is exceedingly small and can be easily blocked by carbon deposits and sludge, resulting in flame failure. Some pilot burners have a small filter which can be easily clogged following continuous operation. If the nozzle or filter is blocked by an accumulation of carbon deposits, then the flame with fail to ignite.
12. *Electrodes are not generating a spark.* The initial spark used for generating the flame is produced by an electrode. If the electrode is covered in carbon deposits, then the spark will not generate, causing the boiler to fail.

MARINE BOILER FAILURES

Every engine room machinery system requires a specific procedure for starting and stopping. The boiler, being one of the most important systems on board ships, requires exceptional care and attention during operation and maintenance. Failure to do so can – and has – led to catastrophic accidents. There are several key operating mistakes which are the leading cause of boiler failure.

1. *Starting the boiler without pre-purging the furnace.* All boilers come with an automation system for starting and stopping. This comprises a programme for pre-purging and post-purging the furnace before the burner is fired. Never ignore or isolate this safety feature. If the boiler must run manually, it must be pre-purged by means of the FDF for at least two minutes. Ignoring this step can lead to blowback and even boiler pressure explosion.
2. *Ignoring furnace blowback.* Several accidents in the past have involved furnace blowbacks, some of which have led to fatalities. Even so, it is common for marine engineers to overlook the danger of furnace blowback even when the first attempt ends with flame failure and blowback. One of the main reasons for this is not pre-purging the furnace (see point no. 1).
3. *Bypassing the safety alarm.* There is a widespread practice among marine engineers to bypass some of the safety or automation sequences in a bid to shorten the operating procedures of the boiler. Such systems are installed to ensure the safety of the ship and her crew and should never be tampered with or ignored.
4. *Not checking the boiler refractory.* The boiler refractory provides improved heat-exchanging efficiency and closes/seals any gaps to restrict fire, heat, and ash inside the boiler. Regular inspection of the boiler refractory is important, as a damaged refractory will expose the boiler shell to flame and heat, leading to bulging or even cracking of the boiler shell.
5. *Dirty gauge glass.* The gauge glass, as installed on the boiler, is the only physical means of checking the water level within the boiler. Every marine engineer should know the procedure for blowing the gauge glass. Neglecting this task can lead to an erroneous water level indication causing severe damage to the boiler and tubes.

6. *Pilot burner check*. It is common for marine engineers to remove the pilot burner to check the electrode spark. This practice can lead to electrical shock and even fire if the surroundings are not thoroughly cleaned or if the pilot burner is kept on an oily floor plate wrapped with rags. The best way to check pilot burner operation is to fit it in place and watch the firing via the boiler inspection manhole located opposite the burner.
7. *Poor clean-up after burner maintenance*. When any maintenance work is conducted on the burner assembly (pilot or main burner) and the surrounding area is not cleaned before evaluating the boiler, there is a high possibility of fire or explosion as blowbacks are common during starting-up after maintenance. Furthermore, any oil spilled inside the furnace and over the burner assembly can exacerbate the situation, leading to a blowback into a full explosion.
8. *Cold condition thermal shock*. Never fire a boiler continuously when starting from a cold condition. This is to avoid thermal shock. In a cold condition, the boiler should be started by following an intermediate 'firing' pattern, for example, two minutes of firing followed by a 'break' of ten minutes. Moreover, once the boiler begins to warm up, the 'break' time should gradually decrease with the 'firing' time gradually increasing.
9. *Wrong operation of the exhaust gas boiler circulating pump*. The exhaust gas boiler normally comes with a water-circulating pump. It is important to start this pump well ahead of time – at least two hours – before starting the main engine. It should be stopped no less than 12 hours (the time may reduce depending upon the capacity of the boiler and the geographical condition) after stopping the main engine to avoid thermal shock and exhaust gas boiler fire.
10. *Cleaning of exhaust gas boiler tubes*. The exhaust gas boiler tubes are arranged in the passage of exhaust gases, which heats the water in the tubes. If these smoke tubes are not cleaned regularly, they can lead to soot deposits developing over the tubes. During low-load operation or improper combustion, oil can mix with the soot to form a combustible sludge. When ignited, the sludge can burst into flames causing a major soot fire. This can easily turn into a hydrogen or iron fire.

BOILER FEEDWATER CONTAMINATION

The quality of boiler feedwater plays a key role in determining marine boiler efficiency. Contamination of boiler feedwater can lead to several issues including corrosion and scale formation. To eradicate these problems, it is important to understand the principles of good boiler water management and how contaminants can enter the boiler feedwater. Inside the boiler system, the boiler feedwater passes through a series of pipelines, tanks, and other ancillary systems. The feedwater is always chemically treated to reduce the effects of harmful minerals and gases. The main problem starts when a substantial volume of this boiler feedwater is lost through leaks and processes such as boiler blowdown, and soot blowing. This means that make-up water is needed to compensate for the loss of feedwater. It is this make-up water that brings most impurities into the boiler system. *Contaminated make-up feedwater*. New feedwater is introduced to the boiler system in one of two ways:

1. From the freshwater tanks, whose water is meant for drinking
2. From the seawater distillation plant or freshwater generators

Boiler feedwater is taken from the freshwater generator. Seawater contains a large volume of salts and other dissolved minerals. Alternatively, the potable water generated in the freshwater generator often contains small droplets of saltwater. Salt droplets also result from saltwater leaks in the distillate condenser. Feedwater therefore contains several types of dissolved minerals and salts. The dissolved gases in the seawater – which are either absorbed from the air or because of decayed plant and animal carcasses – are also carried over within the vapour of the distilled water. Many of these gases and impurities are harmful, and if left untreated, they will eventually lead to boiler problems. *Contaminated boiler feedwater*. The problems caused by contaminated boiler feedwater can be broadly classified into two main types:

1. Corrosion
2. Scale formation

Technically, the aforementioned problems are interlinked. Both result in a loss of boiler efficiency and can cause boiler tube failures and an inability to produce steam. *Corrosion*. One of the most common causes of boiler corrosion is the effect of dissolved oxygen in the make-up and feedwater. Corrosion leads to a failure of machinery from within the boiler and reduces overall boiler efficiency as the extent of the corrosion worsens over time. *Scale Formation*. Scale formation, or mineral deposits in the boiler, results from hardness contamination of the feedwater. The minerals in water that make feedwater 'hard' are Calcium (Ca^{++}) and Magnesium (Mg^{++}). These minerals form a scale over the surface of the piping, water heaters, and everything else they contact. Hardness contamination of feedwater may also result from either deficient softener systems or leaking raw water condensates. This kind of scale or mineral deposit acts as an insulator which lowers the heat transfer rate. The insulating effect of deposits also causes the boiler metal temperature to rise and if left untreated can lead to tube failure through overheating. Copious amounts of such deposits throughout the boiler can reduce heat transfer sufficiently to decrease overall boiler efficiency.

BOILER BLOWDOWN PROCEDURES

As we now know, the boiler is one of the most important systems on board ships. An efficiently working marine boiler requires timely maintenance and exceptional care when starting and stopping. Routine clean-ups are extremely useful in increasing the working life of the boiler. In this section, we will discuss one of the most important procedures in marine boiler maintenance: the boiler blowdown. To be effective, boiler blowdowns need to be performed regularly. The water which is circulated inside the boiler tubes and drum contains Total Dissolved Solids (TDS) along with other dissolved and undissolved solids. During the steam-making process, i.e., when the boiler is in operation, the water is heated and converted into steam. However, these dissolved solids do not evaporate and are separated from the water or steam. They tend to settle at the bottom of the boiler shell due to their weight. This layer of semi and undissolved minerals prevents the transfer of heat amid the gases and water, which will eventually lead to an overheating of the boiler tubes or shell. Different dissolved and undissolved solids lead to scaling, corrosion, and erosion. These solid impurities are also carried over with the steam into the steam system, leading to build-ups of deposits inside the heat exchanger surface where the steam is the primary heating medium. To minimise these problems, a boiler blowdown is done, which helps to remove the carbon deposits and other impurities.

Boiler blowdown. The boiler blowdown is conducted to remove carbon deposits and other impurities from the boiler. The blowdown is done to remove scum and bottom deposits; remove the precipitates formed because of chemical additions to the boiler water; remove solid particles, dirt, foam, or oil from the boiler water. This is done via the scum valve through a procedure known as 'scumming': reducing the density of water by reducing the water level and removing excess water in the event of an emergency. Inside the boiler, the blowdown arrangement is provided at two distinct levels: at the bottom level and at the water surface level. The latter is known as a 'scum blowdown'. When the bottom valve is used, the procedure is known as a boiler blowdown, and when the scum valve is used, the process is known as 'scumming'. The boiler water blowdown can be done in two ways depending upon the type, design, level of automation used, the capacity of the boiler, and the characteristics of the boiler feedwater system.

Intermittent or manual blowdown. When the blowdown is conducted manually by the boiler operator at regular intervals in accordance with the preventative maintenance schedule, it is referred to as a 'manual blowdown'. This type of blowdown is handy for removing sludge formations or suspended solids in the boiler. It is also particularly useful when there is oil contamination within the boiler water caused by leaks in the heat exchanger. By using manual scumming, the oil present on the water surface can be easily removed. The major drawback to manual blowdown is heat loss caused by hot water running out of the water drum. This is caused by the valve being opened slightly, allowing a small quantity of water to go into the blowdown. Even so, the result is a significant loss of heat and pressure.

Continuous blowdown. Many modern boilers are provided with some form of blowdown automation. This allows the continuous blowdown of the boiler water, which helps keep dissolved and suspended solids under boiler operating limits. This system is known as continuous blowdown. In this system, the automation monitors the blowdown continuously and in turn checks the quality of feedwater and the quality of water inside the boiler shell for dissolved and undissolved impurities. Accordingly, it automatically opens the blowdown valves if the boiler water TDS exceeds the permissible operating limit. As the blowdown valves are precisely controlled, the water discharged from the blowdown removes the maximum of dissolved impurities with minimum heat and water loss from the boiler water, maintaining boiler efficiency. Most boilers with continuous blowdown automation are fitted with some form of heat recovery system; i.e., the hot water from the boiler blowdown is first sent to a heat exchanger unit which uses the heat of the water (for example, to preheat the feedwater by installing a heat exchanger or heat recovery equipment) before it is discharged overboard. The choice of blowdown system, i.e., whether manual, continuous, or automatic, will depend on numerous factors, and the blowdown valves will be fitted with suitable accessories as per the system. To calculate the percentage of blowdown, use the following formula:

Quantity blowdown water/Quantity feedwater $\times 100 = \%$ blowdown

Procedure for scumming and bottom blowdown

For bottom blowdown, the blowdown valve located at the bottom of the boiler is opened. To perform scumming, instead of opening the bottom blowdown, the scum valve is opened. The procedures for conducting the blowdown procedure are as follows:

1. Open the overboard or ship side valve first.
2. Open the blowdown valve, this valve is a non-return valve.
3. The blowdown valve adjacent to the boiler should be opened fully to prevent cutting off the valve seat.
4. The rate of blowdown is controlled by the valve.
5. After blowdown, close the valve in reverse order.
6. A hot drainpipe, even when all valves are closed, indicates a leaking blowdown valve.

If the boiler is blown down for inspection, the firing needs to be stopped first to allow the boiler to cool off. Open the boiler vent plug to allow natural cooling at atmospheric pressure. Ensure the overboard valve (non-return) is functioning properly so that no seawater can enter the boiler pipeline; otherwise, it will create a vacuum due to sudden steam cooling, resulting in burst pipes. Once the boiler blowdown is complete, open the belly plug to remove the remaining content in the engine room bilges. Ensure the scum blowdown is completed before the bottom blowdown; otherwise, the scum settled on the water's surface – or any oil content in the boiler water – will get agitated, contaminating the boiler water.

Advantages and disadvantages of boiler blowdown

Regular blowdowns of the boiler water help to keep the total dissolved solid impurities under the rated limits. The process helps in preventing corrosion, as it removes the impurities which accelerate the corrosion process. It also helps in preventing scaling of the boiler tubes and internal surfaces, and it prevents the carryover of impurities and contaminants with the steam. Lastly, it prevents scaling of the internal parts of the heat exchanger where the pure steam is used as the heating media. If the blowdown procedure is not done correctly within the preventative maintenance schedule, the blowdown of boiler water tends to increase heat as well as pressure losses. The heat and pressure losses from the boiler water blowdown together reduce boiler efficiency. If the blowdown arrangement is manual, additional work hours will be needed to conduct the operation safely.

If there is a visible oil sheen in the boiler gauge glass or hot well inspection glass, do not perform a scum blowdown, as this will contaminate the water causing oil pollution. The cause of the oil leakage inside the boiler must be stopped and all efforts made to clear the oil from the hot well by filling the reservoir with freshwater, displacing the oily water. Always ensure the boiler operator knows the Vessel General Permit (VGP) areas and complies with Chapter 12 of the VGP. It is never permissible to discharge any wastewater from a boiler blowdown in restricted areas except for safety reasons.

The vessel must not discharge any boiler water from boiler blowdowns in port waters. This is because the water consists of various chemicals or other additives which are added to reduce impurities and prevent scale formations. These chemicals are often harmful to aquatic life and the marine environment. To avoid contaminating coastal waters, the boiler blowdown must be conducted as far from inshore as possible. Before conducting the blowdown, the master and duty officer on the bridge must be informed, and a record made in the ship's log. The boiler blowdown operation must be recorded in the engine room logbook, which must include the time the blowdown started and the end time. If the boiler blowdown or hot well water is transferred to the bilges, this must be recorded in the oil record book and engine room logbook. If necessary, the boiler blowdown may be performed in inshore or harbour waters only in the following conditions:

- If the ship is entering drydock
- For any safety reasons

There are two primary methods for reducing the number of boiler blowdowns needed to keep the boiler in an efficient state of operation. *Chemical treatment.* The main aim of doing a boiler blowdown is to reduce the dissolved impurities in the boiler water, which leads to scale formation. Scale formation directly leads to heat transfer within the internal surface of the boiler leading to a reduction in boiler efficiency. If the boiler water can be evaluated regularly and treated accordingly using chemical additives in the hot well, the feedwater will have fewer impurities, making it good for use. *Boiler water blowdown reduction.* With an increase in boiler blowdown, the water and fuel consumption of the boiler water will increase. The best practice is to remove the manual blowdown system and to install an automatic boiler water measurement and blowdown system. This system will effectively monitor the impurities in the boiler and open the discharge blowdown valve accordingly.

CLEANING THE GAUGE GLASS

Gauges are used in many places around the ship. Gauge glass is a type of level indicator which shows the amount of fluid in a tank or any other storage vessel on a ship. The gauge glass for a boiler has two different compartments: the top side and the bottom side, which are connected to two different sections of the boiler. The top side of the gauge glass is connected to the steam side of the boiler and the bottom side is connected to the water side of the boiler. The pressure on both sides equalise which means the water level within the boiler can be sighted. To maintain the gauge glass in optimal working condition, it is necessary to conduct some simple maintenance tasks. First, check the nut and tighten if necessary. Check the bolts on the boiler flanges; if loose, tighten them. Check if the union nuts are loose; if so, tighten them. *Gauge glass blowdown procedure.* The gauge glass should be blown before lighting up the boiler, after stopping the boiler, and regularly if the level in the gauge glass is suspected erroneous. *Cleaning the water side of the gauge glass.* To clean the water side of the gauge glass, close the V and W valves. Open the cock and determine if there is water coming out of the drain valve, indicating the drain line is clear. Close the drain valve and keep the cock open. Monitor to see whether the water level rises in the gauge glass; this indicates the line to the gauge glass is clear. Repeat each step two or three times to remove any sludge and deposits.

In this chapter, we have covered some of the key operational, maintenance, and safety factors associated with the ship's marine boiler. In the next chapter, we will turn our attention to the heating, ventilation, and air-conditioning systems (HVAC) on board ships.

Chapter 16

Central cooling system

The machinery systems fitted onboard ships are designed to work with maximum efficiency and run for extended periods without interruption. The most usual form of energy loss from machinery is heat. This loss of heat energy must be reduced or carried away by some form of cooling media, such as a central cooling water system, to avoid the machinery malfunctioning or breaking down. There are two cooling systems used onboard for this purpose. The first is the *seawater cooling system*, where seawater is directly used in the machinery systems as a cooling media for the heat exchangers, and the second is the *freshwater* or *central cooling system* (see Figures 16.1 and 16.2 for examples). Freshwater is used in a closed circuit to cool down the engine room machinery. The freshwater returning from the heat exchanger after cooling the machinery is further cooled by seawater in a seawater cooler. As discussed earlier, in the central cooling system, all major shipboard machinery is cooled using circulated freshwater. This system comprises three different circuits: (1) *Seawater circuit*. Here, seawater is used as a cooling media in large

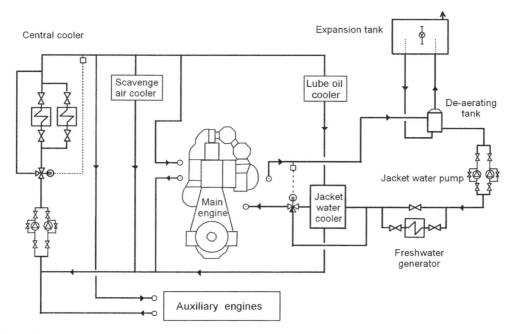

Figure 16.1 Typical seawater cooling circuit.

Figure 16.2 Typical marine heat exchanger.

seawater-cooled heat exchangers to cool the freshwater of the closed circuit. These are the central coolers of the system and are normally installed in duplex. (2) *Low-temperature circuit*. The low-temperature circuit is used for low-temperature zone machinery, as this circuit is directly connected to the main seawater central cooler; hence, its temperature is lower than that of the elevated temperature circuit. The low-temperature circuit comprises all auxiliary systems. The total quantity of low-temperature freshwater in the system is maintained in balance with the elevated temperature freshwater cooling system by an expansion tank which is common to both systems. The expansion tank used for these circuits is filled and made up from the hydrophore system or from the distilled water tank using the freshwater refilling pump. (3) The *high-temperature circuit*. The elevated temperature circuit in the central cooling system consists of the jacket water system of the main engine where the temperature is quite high. The elevated water temperature is maintained by low-temperature freshwater, and the system normally consists of the jacket water system of the main engine, freshwater generator, the diesel generator during standby conditions, and the lube oil filter for the stuffing box drain tank. The elevated temperature cooling water system is circulated by electrical cooling water pumps, one in service and one on standby. During standby, the generator is kept warm by the circulating system from the generator in service. When the main engine is stopped, it is kept warm by elevated temperature cooling water from the generator. If this is insufficient, the water may be heated by a steam freshwater heater. The loss in the closed circuit of the central cooling freshwater system is continuously compensated by the expansion tank which also absorbs the increase in pressure due to thermal expansion. The heat absorbed by the elevated temperature circuit is transferred to the low-temperature circuit by the temperature control valve junction. The outlet temperature of the main engine cooling water is kept constant at 85°C–95°C (185°F–203°F) by means of temperature control valves. These work by mixing water from the two central cooling systems, i.e., the low-temperature system and the high-temperature system.

The advantages of the central cooling system include the following: (1) *Minimal maintenance costs*. As the system runs with freshwater, cleaning, maintenance, and component replacement are minimal. (2) *Low corrosion*. Since the seawater system is only in the central part, the corrosion of pipes and valves is also minimal. (3) The *high speed of fluid provides better heat exchange*. Higher speeds are possible in the freshwater system, which results in reduced piping and lower installation costs. (4) *Use of less expensive materials*. Since the corrosion factor decreases, less expensive materials can be used for valves and pipelines. (5) *Constant temperature level is maintained*. Since the temperature control is irrespective of seawater temperature, a stable temperature is maintained which helps in reducing machinery wear. (6) *Reduced wear on engine parts*. The cylinder liner is maintained at a constant temperature by the jacket which helps reduce cold-condition corrosion. (7) The *system is ideal for unmanned engine rooms*. The greater reliability and temperature controllability of the system makes it ideal for unmanned machinery spaces. On the other hand, the disadvantages of the central cooling system are the high installation cost and the limitations involved in achieving low temperatures.

HEAT EXCHANGERS

Several types of heat exchangers are used on board ships. The type of heat exchanger used for a particular purpose depends on the application and requirement. All systems on board ships depend on a heat exchanger where the fluid is either cooled or heated. The types of exchangers are defined by their construction and are either *shell and tube-type heat exchangers*, or *plate-type heat exchangers*.

Shell and tube-type heat exchanger

This is the most popular design with a shell accompanying several tubes. The flow of liquid to be cooled is carried through these tubes, whereas the secondary liquid flows over the tube inside the shell. Shell and tube-type heat exchangers are extremely economical to install and easy to clean; however, the frequency of maintenance is higher than all other types. With this heat exchanger, the complete shell is fitted with a tube stack. This is colloquially known as the shell. There are two end plates which are sealed on both sides of the shell and a provision is made at one end to cater for expansion. The cooling liquid passes through the tubes which are sealed on either end into the tube plate. The tubes are secured in the tube plate by bell mouthing and expansion. The shell is enclosed with water chambers which surround the tube plates completely. The coolers may be either single pass or double pass. Gaskets are fitted between the tube plates and the shell, as well as between the tube plate and the end cover to cater to the leakages from the cooler. The other side of the tube plate, which is not fixed but free to move, has seals on either side of a safety expansion ring. The main engine cooling freshwater cooler and the main engine lube oil cooler are conventionally circulated with seawater, which passes through the tubes of the cooler. The shell, on the contrary, is in contact with the liquid being cooled, lube oil or distilled freshwater. To avoid damage, corrosion-inhibiting chemicals are added directly to the expansion tanks to keep a thick protective layer inside the pipelines. Drew Marine offers LIQUIDEWT® and Unitor supplies ROCOR NB®, which is used for corrosion inhibition. The shell is usually made from cast iron or steel. It is recommended that the coolers are installed vertically to ensure automatic venting of air from the system, as the airlock causes excessive overheating, which reduces the effective cooling surface area of the liquid being cooled. Baffles are fitted on the tube bundle, which leads the liquid to be cooled up and down, thus increasing the effective surface area of cooling. They also support the tubes, providing strength and rigidity to the bundle. Aluminium brass alloy is used for the construction of the tubes, which are 76% copper, 22% zinc, and 2% aluminium. Sacrificial anodes are used in the seawater side for corrosion prevention. These work by preventing the material engulfed with seawater from becoming the electrolyte. The tubing may fail prematurely due to contaminated coastal water or through excessive turbulence caused by seawater flow rates. To prevent this, the velocity of the seawater should be kept below 2.5 mps where aluminium brass alloy is the piping material. It is worth noting that a little turbulence is required to reduce silting and sediment settlement in the tubes.

> *Operation of the shell and tube-type cooler.* To operate the shell and tube-type cooler, a leak test of the piping should be conducted in advance. The cooling liquid and the liquid to be cooled should be circulated, flushed, and checked for leaks. It is advised to run clean cooling fluid in the tubes during the initial phase of circulation as debris can erode the protective layer in the tubes. The seawater inlet and outlet valves must be kept fully open to allow the liquid to be cooled to bypass, if needed, by a three-way temperature control valve. Vents are provided on either side of the medium, the cooling liquid and the liquid to be cooled. The vents should be opened first after the initial circulation of fluids or after maintenance to purge any trapped air. Drain plugs are mounted in the coolers at the lowest points to drain the cooler completely during maintenance. Single-pass, vertically mounted coolers, ensure automatic venting. Horizontally mounted coolers, such as the inlet cooling water branch, should be faced downwards, and the outlet water should be faced upwards to allow automatic venting of air.
>
> *Maintenance of the shell and tube-type cooler.* During maintenance, the fluids should be isolated completely, after which the cooler may be opened. The heat transfer surfaces should be cleaned properly. Seawater fouls the cooling surfaces through plant and animal growth, which is indicated by an increase in temperature difference between the cooling liquid and the liquid being cooled. Furthermore, the pressure changes will indicate that excessive corrosion is occurring, which can lead to leaks in the tubes.

Plate-type heat exchanger

The plate-type exchanger consists of thin corrugated plates joined in a parallel fashion, creating a cavity for fluid to flow inside it. Alternate sides of the plate carry two different fluids, between which the heat transfer is performed. Installation of this type of heat exchanger is much more expensive than shell and tube-type installations; however, the maintenance cost is much lower. The efficiency of the plate type is also higher than the shell and tube-type when compared size for size. The former can also withstand higher pressure ratings. The first and last plates are called the innermost and the outermost plates. They are held together by the frames on either side and are further set in place by tie bolts.

Four branch pipes on the pressure plate, which are aligned with ports in the plates, allow the two fluids to pass freely. Seals around the ports are arranged such that one fluid flows in alternate passages between the plates and usually in opposite directions. The plate corrugations promote turbulence in the flow of both fluids and so encourage efficient heat transfer. Turbulence, as opposed to smooth flow, causes more liquid to pass between the plates. It also breaks up the boundary layer of liquid which tends to adhere to the metal and functions as a heat barrier when the flow is slow. The corrugations make the plates stiff so permitting the use of thinner materials. They additionally increase the plate area as well. Both factors contribute to this type of heat exchanger's efficiency. As excessive turbulence can cause the erosion of the plate material, moderate flow rates must be maintained. Titanium plates, though expensive, are used, as they offer the best resistance to corrosion and erosion. The rubber seals between the plates are bonded to the plate surfaces by special adhesives and must be removed by acetone. The rubber seals are not suitable for extremely elevated temperatures, as they lose their elasticity, harden, and become brittle. Once brittle, the rubber seals are very easily broken when the cooler is opened for cleaning and inspection. The rubber joints are squeezed and then tightened by clamping bolts. This provides the best seal. The length of the tie bolts should be measured correctly before opening the plate-type cooler for maintenance and inspection as over-tightening of the plates can bend the tie bolts, rendering them useless. Alternatively, undertightening can cause the plates to leak. Subsequently, it is important to ensure the tie bolts are tightened with the correct torque as provided in the manufacturer's instructions. In summary, plate-type coolers are awfully expensive compared to shell and tube-type coolers because of the use of titanium plates, but their low maintenance and minimal operational cost offer better efficiency and, ultimately, are more cost-effective in the long term.

Repair of plate-type cooler. When plate-type coolers are opened for inspection, they should be checked for holes which could leak. Furthermore, the joints should be checked for their sealing surfaces and pasting integrity. These types of coolers are fitted with a filter on the seawater side, which should be opened regularly to debris. If the coolers are to be put out of service for extended periods, then the seawater side should be drained completely, flushed with freshwater, and then dried completely.

In addition to the shell and tube-type and plate-type heat exchangers, there are several other variations available, which are briefly discussed in the following sections.

Plate-fin heat exchanger

The plate and fin type heat exchangers are not dissimilar to plate-type exchangers. The difference is in the fixture of fins, which helps improve the efficiency of the exchanger. There are three types of fins: offset fins, which are fixed perpendicular to the direction of flow; straight fins, which are parallel to the direction of flow; and wavy fins, which have a curved form. The efficiency of this type of heat exchanger is higher than standard plate-type units but the installation and maintenance costs are also higher.

Dynamic scraped surface heat exchanger

With this type of heat exchanger, a continuous scraping of the surface extends the lifespan of the unit and helps improve heat transfer efficiency. The scraping is achieved by a blade unit operated by a motor-driven shaft with a timer moving inside the frame. As the scraper moves along the surface of the unit, it removes fouling such as organic material and sediment. This heat exchanger is normally reserved for the heat transfer of highly viscous fluids. Maintenance costs are comparatively lower than the previous two, primarily because of the auto-cleaning process.

Phase change heat exchanger

As the name suggests, the phase change heat exchanger is used to change the phase of a medium from solid to liquid or liquid to gas by using the principle of heat transfer. This type of exchanger is normally operated in a freeze cycle and melt cycle. The unit is constructed like a shell and tube-type exchanger but consists of at least two divider walls which provide an upper and lower annular space which facilitates flow passage. It also consists of fins in both passageways for efficient heat transfer.

Spiral heat exchanger

This type of heat exchanger consists of concentric-shaped flow passages which help in creating a turbulent flow of a fluid. This in turn increases heat transfer efficiency. Initial installation costs are higher, but so is the efficiency of the unit compared to most other designs. As the spiral heat exchanger is compact in size, it is more easily installed within smaller spaces, making it an attractive option for ship designers. Of all the heat exchangers, the spiral type has the lowest maintenance costs which are mainly due to its small size. The flow of fluid in the spiral type is a rotary current flow which itself possesses self-cleaning properties, which facilitates the semi-automatic removal of fouling inside the spiral body.

Direct contact heat exchanger

With direct contact heat exchangers, there is no separating wall within the unit. Both mediums are in direct contact during the heat transfer process. Direct contact type heat exchangers can be further categorised in one of three ways: (1) gas-liquid, (2) immiscible liquid-liquid, and (3) solid-liquid or solid-gas.

Charge air cooler

The charge air coolers, commonly known as air coolers, are fitted after the turbochargers to decrease the temperature of the air before it enters the engine cylinder. The charge air coolers are provided with fins to increase the heat transfer surface as the air itself has poor heat transfer properties. Solid drawn tubes are passed through copper fin plates and are bonded by soldering for maximum heat transfer effect. The ends of the tubes are fixed to the tube plates by expansion and soldering. The air is cooled to its dew point temperature so that the moisture content in the air is removed by precipitation. This is because any water content will cause sulphur corrosion if it enters the engine cylinder. The condensation goes to the air cooler tank, if fitted; otherwise, it is directed to the bilge well.

Inter- and aftercoolers (for air compressors)

These types of coolers are remarkably like the charge air coolers as their function is to reduce the temperature of air to its dew point. The purpose being to remove any moisture which might enter the compressor cylinder. These types of coolers are fitted with pockets and drain valves for the removal of moisture and oil. The coolers have special U-shaped tube coolers made of copper. The shape ensures they occupy less space. These coolers may be referred to as coil coolers. Inter- and aftercoolers are awfully expensive to install and difficult to clean. Furthermore, they have poor heat transfer efficiencies due to the necessarily large tube diameter.

Hopefully, between this chapter and the chapters on power management and the air compressor, we have developed a good understanding of the role and function of the central cooling system, and why they are integral systems for the ship. In the next chapter, we examine the refrigeration and air-conditioning systems onboard. Again, because of the nature of these systems, there will be some overlap, but this should serve to reinforce their working principles and functions.

Chapter 17

Refrigeration and air conditioning

The refrigeration plants on merchant vessels (see Figure 17.1) are essential for ships involved in transporting refrigerated cargo and of course for keeping perishable provisions for the crew. On reefer ships, the temperature of perishable or temperature-sensitive cargo such as food, chemicals, and liquefied gases is controlled by the refrigeration plant. The main purpose of the ship's refrigeration plant is to prevent the growth of micro-organisms, oxidation, fermentation, and the drying out of cargo. All refrigeration units consist of the same types of components, which are typically as follows: (1) *Compressor*. Reciprocating single or two-stage compressors are normally used in refrigeration systems for compressing and supplying the refrigerant. (2) *Condenser*. Shell and tube-type condensers are used to cool down the refrigerant to the required temperature. (3) *Receiver*. The cooled refrigerant is supplied to the receiver, which is also used to drain out the refrigerant from the system during maintenance. (4) *Drier*. The drier consists of silica gel, which is used to remove any

Figure 17.1 Typical marine refrigeration plant.

moisture from the refrigerant. (5) *Solenoids*. Different solenoid valves are used to control the flow of refrigerant into the hold or compartment. A master solenoid is provided in the main line, and slave solenoids are located throughout the individual cargo holds and compartments. (6) *Expansion valve*. An expansion valve regulates the refrigerants to maintain the correct hold or compartment temperature. (7) *Evaporator unit*. The evaporator unit acts as a heat exchanger to cool down the cargo hold or compartment by transferring heat to the refrigerant. (8) *Control unit*. The control unit consists of different safety and operating circuits which ensure the safe operation of the refrigeration plant.

The compressor, acting as a circulation pump for the refrigerant, has two safety cut-outs. The first is the low-pressure cut-out and the second is the high-pressure cut-out. When the pressure on the suction side drops below the set value, the control unit stops the compressor. When the pressure on the discharge side rises, the compressor trips. The low-pressure cut-out is controlled automatically, i.e., when the suction pressure drops, the compressor stops and when the suction pressure rises again, the control system restarts the compressor. The high-pressure cut-out is provided with a manual reset. The hot compressed liquid is passed to a receiver through a condenser to cool down. The receiver may be used to collect the refrigerant when any major repair work must be conducted. The master solenoid is fitted after the receiver, which is controlled by the control unit. In the event of a sudden stoppage of the compressor, the master solenoid also closes, preventing the flooding of the evaporator with refrigerant. The cargo holds or compartment slave solenoid and thermostatic valve regulate the flow of the refrigerant into the hold or compartment to maintain the temperature of the room. For this, an expansion valve is controlled by a diaphragm movement. This works according to the pressure variations which are managed by a bulb sensor filled with expandable fluid, fitted at the evaporator outlet. The thermostatic expansion valve supplies the correct amount of refrigerant to the evaporators where the refrigerant takes up the heat. This is then boiled off into vapours, resulting in a temperature drop.

A major concern of every marine engineer working onboard a ship is the quality of the refrigerants used and its purity. The moisture content in the refrigerant is of immense importance as too much moisture can cause serious operational issues. Water can enter the system through sub-quality refrigerants during top-up. Excessive moisture in the refrigeration system can quickly lead to freezing up, corrosion, and the formation of sludge. We will briefly expand on why these issues are best avoided. (1) *Freeze ups*. Freeze ups occur when moisture picked up by the refrigerant starts to freeze, building ice crystals that block the refrigerant passage in narrow passageways, for example, in the expansion valve. This effect is called intermittent cooling, as the compressor stops intermittently due to the blockage in the expansion valve. It starts again when the ice crystals have melted allowing the refrigerant to freely pass. This is a periodic process of constant freezing and melting, which causes high-frequency compressor stops and starts. If left unchecked, this can cause the compressor to malfunction, leading to the spoiling of cargo and crew provisions. (2) *Corrosion*. Moisture is a major cause of corrosion. However, moisture in combination with an HCFC refrigerant[1] containing chlorine (for example, R-22 or R-409A) creates a more severe type of corrosion, as the chlorine hydrolyses with the water to form hydrochloric acid (HC^l), which is aggressive to most metals. Heat adds significantly to the problem by accelerating the acid-forming process. For HFC refrigerants[2] (such as R-404A or R-407C), polyolester oils are very hygroscopic and may decompose at elevated temperatures forming hydrofluoric acid. This is again overly aggressive to metal. (3) *Sludge formation*. Acid inside a system can emulsify with compressor oil to form an aggressive oily sludge that reduces the lubrication properties of the lube oil. This can lead to serious compressor damage. Sludge can also cause a variety of other problems in the system, such as blockages in the strainer, expansion valve, and other small-diameter passages.

It is worth noting that the aforementioned problems do not normally happen overnight but instead build up over time if inferior quality or incorrect refrigerants are used. A refrigeration system breakdown is costly and time-consuming, especially when it results in a loss of refrigerant charge. Although the system will be repaired and returned to operation, the repair does not eliminate the cause of the problem. It is therefore vital that refrigerants comply with appropriate quality standards with regard to purity and composition. As the prices for more environmentally friendly alternatives are higher than traditional refrigerants, such as R-22, R-401A, or R-409A, it is common for cheaper refrigerants to be used. Whilst this may save money in the short term, in the longer term, it will inevitably lead to costly repairs. To ensure the safe and efficient operation of the refrigerant system, it is strongly recommended to use approved refrigerants such as the *Unicool* refrigerants supplied by Wilhelmsen Ships Service, as these comply with the ARI 700-2006 standard, which defines and benchmarks the purity and composition of industrial grade refrigerants.

As we have said, the primary purpose of the refrigeration plant is to keep cargo and food provisions at low temperatures to prevent them from spoiling. Like all systems onboard ships, the refrigeration plant requires constant maintenance to keep working. The main task for marine engineers is to maintain the refrigerant level in the system. This is called *charging*. Although the volume of refrigerant will gradually reduce over time with use, the main cause for a reduction in refrigerant is piping leaks. If the refrigerant level is left to reduce below the safe limit, several issues can evolve, which will affect the efficient working of the refrigeration plant. These issues include short cycling of the compressor, low suction pressure, difficulty maintaining the desired temperature of the hold or compartment, and a reduction in plant efficiency. Should any of these problems arise, it is usually necessary to recharge the refrigerant. On most systems, there are two methods for charging the refrigerant: gas charging and liquid charging. Today, gas charging is by far the most preferred method as it is both safer and less environmentally damaging. *Gas charging the refrigeration plant*. For gas charging, a special T-piece valve block with a mounted pressure gauge is provided to combine the three connectors. These connectors are the (1) vacuum pump, (2) charging cylinder, and (3) charging point. To gas charge the system, complete the following steps: (1) Connect the gas bottle or charging cylinder, vacuum pump, and charging point to the valve block. (2) Connect the discharge of the vacuum pump to an empty recovery bottle. (3) Open the valve between the vacuum pump and the charging bottle located on the valve block. Do not open the main valve of the charging cylinder. This will remove any air trapped inside the pipe. Once a vacuum is achieved, close the charge cylinder valve. (4) Open the valve of the charging point pipe in the valve block and run the vacuum pump until a vacuum is reached. This will remove any trapped air in the pipe. Shut the valve in the valve block. (5) Keep the system idle for five minutes to ensure there is no pressure drop. A drop in pressure will indicate there is a leak in the system. (6) Open the charging bottle pipe valve and the charging point pipe valve, which are located on the valve block. This will set the line for charging. Ensure the vacuum pump valve is shut. (7) Open the main valves in the charging cylinder and the charging point. (8) Do not overfill the system. Make sure the receiver has at least 5% free space for expansion. During the charging process, always ensure that no refrigerant leaks out into the environment, as even gas-charged refrigerants are still hazardous to the environment, albeit less so than liquid-type refrigerants.

As the reefer system is the backbone of ships carrying refrigerated cargo, any malfunction of any of the components of the system can lead to degradation and wastage of perishable and cold storage cargoes. This includes the provisions for the ship. It is therefore important to maintain and run the refrigeration plant properly to avoid any breakdowns. In this section, we will cover some of the safety features of the refrigeration plant which help ensure breakdowns do not happen. The refrigeration system has various safety devices installed which include alarms, cut-offs, and trips. The main safety devices we are concerned with are the *low-pressure cut-off*. This is a compressor safety device which cuts off the compressor in the event of a pressure drop in the suction line. The pressure of the suction line is continuously sensored by the control unit. When the pressure falls below the set value, which means the hold or compartment is appropriately cooled, the low-pressure cut-out will auto-trip the compressor. When the pressure rises, indicating there is flow of refrigerant in the line due to an increase in room temperature, the low-pressure switch will restart the compressor. *High-pressure cut-out*. As the name suggests, the high-pressure cut-out activates and trips the compressor when the discharge side pressure increases above the set limit value. The high-pressure cut-out is not auto-reset and must be reset manually. The justification behind this is to force the engineers to personally address the cause of the fault, which is leading to a rise in pressure; otherwise, this may lead to an overloading of the compressor and a breakdown. *Oil differential cut-out*. This safety device is again used by the compressor as it is the only machinery in the circuit which has rotational parts which require continuous lubrication. In the event of low supply or no supply of lube oil to the bearing, the differential pressure will increase and activate a trip signal. This safeguards the bearing and crankshaft from potentially damaging friction. *Relief valves*. Relief valves are fitted to the discharge side of the compressor. They work by lifting the compressor in the event of overpressure. One relief valve is also fitted to the condenser refrigerant line to avoid damage to the condenser should high-pressure develop in the discharge line. *Solenoid valves*. A master solenoid valve is fitted to the common or main line after the condenser discharge. This closes when the compressor stops or trips to avoid an overflow of refrigerant into the evaporator. All holds and compartments are fitted with individual slave solenoid valves which control the flow of refrigerant to that hold or compartment. *Oil heater*. Last of all, the oil heater is provided for the compressor crankcase oil and prevents the compressor from becoming excessively cold, which will affect the quality of the lube oil.

Refrigeration systems, by their nature, have unique hazards that require care and attention. No injury, no matter how minor, should go unattended. Always obtain first aid or medical attention immediately. When working in or around the reefer, always wear safety gloves and glasses. This is especially important when charging the refrigerant. Always keep hands, tools, and clothing, clear of the evaporator and condenser fan. No work should be performed on any unit until the circuit breakers and start-stop switches are turned off and the power supply isolated. Never, under any circumstances, bypass any electrical safety device. When performing arc welding on a unit or condenser, disconnect the wire harness connection from the module in the control box. Never remove the wire harness from the module unless grounded to a unit frame with a static-safe wrist strap. In the event of an electrical fire, open the circuit switch and extinguish with CO_2 extinguisher or other electrical firefighting appliance. All ships' staff are responsible for their own and each other's safety. This means recognising and reducing the hazards associated with handling refrigerant gases such as phosgene gas hazards (due to elevated temperature) or asphyxiation hazards in non-ventilated spaces. When charging the refrigerant, always ensure to oversee the compressed gas bottle appropriately and stow it in a secure location and position when not in use.

As we mentioned earlier, some ships specialise in the transport of refrigerated cargo. Indeed, some ships only carry refrigerated cargo. These types of vessels are known as reefer ships. Many reefer ships have fully refrigerated cargo holds which are used to transport temperature-sensitive cargoes such as fresh produce. The most famous reefer ships are the 'banana boats' operated by companies such as Geest and Hamburg Süd. Not every vessel that is required to carry temperature-sensitive cargo is a reefer ship, which means specially adapted containers must be used. These containers consist of an enclosed refrigeration unit attached to the exterior of the container. Because these containers are self-contained, i.e., they do not have an internal power source, they must be connected to the ship's main power supply. In this section, we will briefly examine the design and function of reefer containers and the role of the ship's engineers in ensuring they are kept in good working order. When reefer containers are loaded onto ships, the power supply for the refrigeration unit is provided by the power generated from the ship's diesel generator. If the vessel's generator capacity is not sufficient to support the additional power consumption required by the refrigerated containers, mobile power packs are used instead. There are a wide variety of refrigerated shipping containers available on the market today including the closed reefer, the modified or controlled atmosphere (MA/CA) reefer, and the automatic fresh air management (AFAM) container. The *closed reefer* is the most conventional type of refrigerated container and consists of a single ISO container with an integral front wall and all-electric automatic cooling and heating unit. The *MA/CA reefer container* is an insulated shipping container which maintains a constant atmosphere by replacing consumed oxygen using an air exchange system. This is designed to maintain an ideal atmosphere in equilibrium with the cargo's deterioration rate. The automatic fresh air management containers or AFAM container uses advanced technology to regulate the air combination by automatically adjusting the scale of fresh air exchange. It works in a comparable manner to the MA/CA container by controlling the composition of oxygen, carbon dioxide, and other gases. The controls of the AFAM refrigerated container can be adjusted to extend the shelf life of the cargo carried.

Some important points to note about container refrigeration are the container refrigeration unit is always fitted to the front of the container and serves as the container's front wall. Some units use dual voltage and are designed to operate at 190/230, or 380/460 V, AC, three-phase, 50–60 Hz power. The operating control power is provided by a single-phase transformer, which steps down the AC supply power to 24 V, one-phase control power.

Without dwelling too much on the subject, it is worth explaining the main sections of the reefer unit, as the engineering department is frequently called upon to resolve issues with the refrigeration system. As with all refrigeration systems, the reefer unit has a compressor section. This consists of a compressor (with a high-pressure switch) and a power cable storage compartment. A power transformer may be installed where the ship's power supply is different from that of the container. It also contains modulating and suction solenoid valves for controlling the quantity of gas flow. Safety fittings in the section include a moisture liquid indicator, pressure relief valve, and filter drier. Safety of the system is further enhanced by electronic monitoring provided by compressor, suction, and discharge sensors; supply air temperature sensor; supply recorder sensor; and ambient temperature sensor. The second main part is the condenser section. The condenser section contains the condenser fan and motor, an air-cooled condenser coil, and condenser saturation sensor.

For air-cooled condensers, air is normally pulled from the bottom and discharged horizontally through the centre of the unit. Some units may have a water-cooled condenser receiver, though these are uncommon as they are expensive. The third part of the unit is the evaporator section. This section contains a temperature sensing bulb, return recorder bulb sensor, and a thermostatic expansion valve for regulating the flow of refrigerant, and maintenance of inside temperature. The assembly consists of an evaporator coil and heater, drain pan and heater, defrost, and heat transmission switches. The evaporator fan circulates air throughout the container by pulling air into the top of the refrigeration unit and directs air through the evaporator coil where it is either heated or cooled and then discharged out through the bottom of the refrigeration unit into the container. The fourth and final component is the fresh air make-up vent. The purpose of this vent is to provide ventilation for commodities that require fresh air circulation. This vent must be closed when transporting frozen foods. Air exchange depends on static pressure differential which will vary depending on how the container is loaded.

The most widespread problem associated with shipboard refrigeration systems is leaking refrigerant. To diagnose a leak in the refrigeration system, there are several simple tests that can be conducted. The first is the *soap water test*. This is done on low-pressure lines by spraying soapy water on the pipes. If the soapy water meets a leak, bubbles will form indicating the location of the fault. The second is the *halide lamp test*, which is used for all pressure lines. The third is the *electronic leak detector*. This is a portable unit which is used to conduct leak detections. It should not be used in noisy locations. The fourth test is the *permanent or fixed-type leak detection unit*. Some reefer containers are installed with fixed-type leak detection units which emit an audio-visual alarm should a leak occur. With reference to reefer container units, all the pipes in the refrigerated container unit are insulated. Before testing the piping for leaks, make a small hole in the insulation and check for leaks near the hole (using soapy water, the halide lamp, or an electronic detector). Once the source of the leak is confirmed, remove the insulation from the complete pipe section to isolate the exact location of the leak.

The ship's air-conditioning system helps to lower the ambient temperature of a compartment, cabin, or other location used for human occupation (such as the mess, ship's offices, meeting rooms, the bridge). The air-conditioning system is fed off the ship's central refrigeration system. Each compartment has its own load demand, which means the air-conditioning plant's output must be controlled to achieve the required cooling. This procedure is referred to as capacity control. The capacity control of a refrigeration plant can be defined as a system which monitors and controls the output of the plant as per the load on demand. As the load (temperature) of one compartment is reached, there is no more need for refrigerant to cool that space. Hence, the solenoid valve supplying the refrigerant to that compartment will shut. The most common methods for controlling compartment temperature are variable speed motors (normally found in small air-conditioning units), controlling the on-off cycle of the compressor, and using the cylinder unloading method for keeping the suction valve in the open position. This is accomplished by introducing a capacity controller valve in the compressor which is operated by lube oil pressure (hydraulic type) or by using a solenoid-operated control valve. Where the first two methods are used, it is common to experience the following malfunctions. *Motor overheating*. An excessively high switching frequency will cause excessive motor heating. Bearing damage. During the start-up phase, the oil pressure is low and bearing lubrication is not optimal. This can lead to a reduction in the shelf life of the bearings and connected components. Oil return in intermittent mode. More oil enters the refrigerant cycle during start-up than during continuous operation. Frequent switching prevents adequate lubrication through poor oil return.

The main components of the capacity control valve system are the (1) compressor lube oil pump supply, (2) the capacity control valve, (3) the capacity control regulating valve, and (4) the unloader assembly. The compressor lube oil pump supplies oil to the bearings, and one connection is provided to the capacity control valve. The capacity control valve is provided with high-pressure oil from the lube oil supply pump from the compressor. This valve had several grooves bored into its periphery and is connected to the unloader mechanism at various locations in the units. A spring piston is provided which controls the spreading of high-pressure oil supply into the bore chamber. The spring piston is pressed by the oil supplied through an orifice. This pushes the piston and aligns the unloader holes, providing high-pressure oil to the unloader unit. The unloader assembly comprises an unloader piston held by a spring. The unloader piston is connected to a rotating cam ring, which has lifting pins attached to the suction valve. The lifting pins always act on the suction valve, i.e., unloading the unit in a stop condition. When the bores on the control valve align with the unloader bores, oil passes through, pressing the unloader piston. This then rotates the cam releasing the unloader pins from the suction valve. The capacity control regulating valve is responsible for controlling the pressure (i.e., the opening and closing of capacity control valve ports with the unloader ports). One end is connected to the crankcase and another end to the capacity control valve. As the pressure in the crankcase drops due to a reduction in load, the oil in the capacity control valve is drained into the crankcase causing the closure of the unloader ports, the lifting of the suction valve, and the cutting out of the cylinder unit. This likewise means that all cylinders are unloaded at start-up, which releases unnecessary load on the motor during the starting-up procedure.

Where the capacity control valve is solenoid operated, solenoid valves are used in conjunction with the servo valve to operate the opening and closing of the suction valve. It is fitted to the top of the cylinder near the suction valve. In an energised position, the solenoid closes the access between the two cylinders or stages in the compressor by keeping the suction valve open and bypassing the hot discharge gas directly to the suction line. This reduces the pressure of the unit to zero bar, reducing the capacity of the compressor by half. With the solenoid valve de-energised, the gas ports in the valve plate and cylinder head are open. The only disadvantage of this type is that the spring in the solenoid valve may malfunction. Moreover, its operation is easily affected by high variations in ambient temperatures.

In this chapter, we have discussed the functions and working principles of the ship's refrigeration and air-conditioning systems. In the next part of this book, we will turn our attention to the water management system, first examining ballast water management, then oily water management, and last of all, wastewater management.

NOTES

1. HCFCs are compounds containing carbon, hydrogen, chlorine and fluorine.
2. HCFC refers to hydrochlorofluorocarbon, while HFC refers to hydrofluorocarbon.

Part IV

Water management systems

Chapter 18

Ballast water management

Although ballasting and de-ballasting is an operation managed by the ship's deck department, it is a critical system that is overseen by the ship's engineering staff. In this chapter, we will begin by explaining the role and function of ballast water management, and the implementation of the ballast management plan, before moving on to the technical aspects of the ballast water system. Ballasting or de-ballasting is a process by which seawater (ballast) is pumped in and pumped out of a ship when the ship is at port or underway. Ballast or ballast water is carried by a vessel in its ballast tanks to ensure its trim, stability, and structural integrity. Ballast tanks are constructed in ships with a piping system and high-capacity ballast pump which performs the operation. In ancient times, ships would carry solid ballast for stability, as cargo was often minimal, or there was no cargo to be carried at all. However, as time, passed difficulties were encountered during the loading and discharging of solid cargo. The process of transferring solid cargo was also time-consuming, and for this reason, solid ballast was replaced by water ballast. As seawater was a readily available and inexhaustible supply, it was quickly adopted as ballast media. Today, ballasting, or de-ballasting is required when ships enter channels, such as the Panama and Suez Canals, during loading or unloading of cargo, and when the ship is coming alongside. When no cargo is carried by the ship, the ship becomes lighter in weight, which can affect its stability. For this reason, ballast water must be brought on board and stored in dedicated tanks to help stabilise the vessel. However, when the ship is filled with cargo, the stability of the ship is maintained by the weight of the cargo itself; hence, there is no requirement for ballast water. Moreover, if the ship is fully loaded with cargo on one side, then ballast must be pumped on board to balance the ship; otherwise, the ship may roll over and capsize. Subsequently, the ballast system on ships is of critical importance.

To reduce the harmful effects of ballasting on the marine environment, the International Maritime Organisation (IMO) adopted the *International Convention for the Control and Management of Ship's Ballast Water and Sediments, 2004* (usually shortened to the *Ballast Water Management (BWM) Convention*) to control and manage ships ballast and sediments on 13 February 2004. The convention aims to stop the inadvertent spread of aquatic microorganisms[1] transferred from one marine area to another through the ballasting operations of ships. In response, Port State Authorities around the world implemented their own requirements for ballasting and de-ballasting operation for ships sailing in their territorial waters. To simplify the requirement for the control of ballast water, the convention demands that a "BWM plan" is established and implemented on all ships engaged in international trade. The BWM plan consists of several mandates and requirements, including (1) international rules and regulations for different Port State Controls around the world, (2) locations of ports providing shore-based discharge facilities for sediments and ballast water, (3) roles and responsibilities of the ship's staff engaged in ballasting operations, (4) operational procedures together with approved methods for ballasting, (5) locations of different coastal waters approved for ballast exchange, and (6) example sampling points and treatment methods. In terms of the actual ballast management plan itself, which is provided in the convention as a pro forma template, the following sections must be completed before, during, and after the ballasting operation: (1) date of the ballasting operation, (2) record of the ship's ballast tank used in the operation, (3) temperature of the ballast water, (4) salinity of the ballast water (i.e., the salt content in parts per million (ppm), (5) position of the vessel ship (i.e., latitude and longitude), and the (6) volume of ballast water involved in the operation. Once the data has been collected and recorded, this must be signed by a responsible officer (normally the chief officer) although the master is in overall charge of the operation and is responsible for countersigning the ballast management plan once complete. In addition, ships are required to record the date when the ballast tank was last cleaned. If there is an accidental discharge of ballast exchange, this too must be recorded and signed. The ballast management plan must be accurate and up to date, as any Port State Control has the authority to demand sight of the ballast management plan. Any discrepancies or erroneous information can lead to severe penalties.

On 8 September 2017, the *BWM Convention* became effective and applies to all new and existing ships designed to carry ballast water and are of 400 gross tonnes and above. To demonstrate compliance with the requirements of the convention, every ship must carry a valid International Ballast Water Certificate, a BWM plan, and a ballast water record book. The convention includes two regulations that define BWM standards: (1) regulation D-1, which addresses the ballast water exchange standard, and (2) regulation D-2, which details the ballast water performance standard towards the treatment of ballast water using Type Approved BWM Systems. The accountable authorities, which includes scientists, shipowners, and ship operators, as well as Flag states, determined that the method of ballast water exchange must provide an effective means of preventing the unintended transfer of harmful marine organisms. The convention prohibits all ballast water exchange anywhere at sea except in specific circumstances, where certain requirements must be complied with to perform a ballast water exchange at sea. These include the following: (1) the vessel must be at least 200 NM (230 mi, 370 km) from the nearest land and in water with a minimum depth of 200 m (656 ft), and (2) when a ship cannot meet these criteria due to reasons such as short passage duration or enclosed waters, the exchange must be carried out as far from the nearest land as possible, with a minimum distance of at least 50 NM (57 mi, 92 km) from the nearest land and in a water depth of at least 200 m (656 ft). A Port State, in consultation with adjacent or other states, may designate areas where ballast water exchange can be performed where are no locations which meet the above requirements. In any case, the IMO must be consulted where a Port State wishes to implement national BWM regulations.

The quantity, distribution, and circulation of ballast water are determined by the master of the vessel and are based on operational and environmental conditions. The vessel's master and the designated BWM officer are responsible to the authorities for the implementation of the BWM plan.

BALLAST WATER EXCHANGE METHODS

In accordance with regulation D1: Exchange, ballast water exchange is based on the principle that organisms and pathogens contained in ballast water taken on board from coastal waters will not survive when discharged into deep oceans or open seas, as these

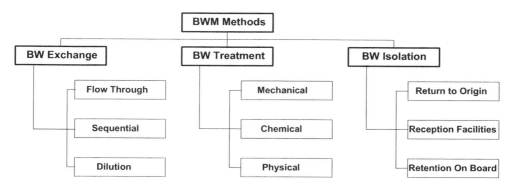

Figure 18.1 BWM methods.

waters have different temperatures, salinity, and chemical compositions. There are three methods for conducting regulation D1 ballast water operations (see Figure 18.1). The first is the *sequential method*. In this process, the ballast water tank is first emptied and then refilled with replacement ballast water to achieve at least 95% volumetric exchange. The ballast water in each tank should be discharged until suction of the pumps is lost. The stripping pumps, or eductors, should be used where possible, to avoid a situation where organisms are left in the bottom of the tank. The tank is then refilled with new water. The emptying of the tanks can be done individually or in pairs. The second is the *flow-through method*. The flow-through method is a process by which replacement ballast water is pumped into a ballast tank intended for the carriage of ballast water, allowing water to flow-through overflow or other arrangements to achieve at least 95% volumetric exchange of ballast water. Pumping through three times the volume of each ballast water tank is considered sufficient to meet the requirements of regulation D1. The third method is the *dilution method*. The dilution method is a process by which replacement ballast water is filled through the top of the ballast tank intended for the carriage of ballast water with simultaneous discharge from the bottom at the same flow rate. This achieves a constant level in the tank throughout the ballast exchange operation. At least three times the tank volume must be pumped through the tank. Commonly two ballast pumps are used simultaneously, whereby one is acting as the filling pump and the other as a suction pump. As it is essential to keep the filling level in the tanks constant, exact control of the pumped volume of both pumps must be maintained.

Regulation D2: Performance defines the performance standard for the ballast water treatment system. The D2 standard specifies the maximum volume of viable organisms allowed to be discharged, including specified indicator microbes harmful to human health. Ships conducting BWM in accordance with this regulation are required to discharge:

1. Less than 10 viable organisms per m^3 > 50μ in minimum dimension,
2. Less than 10 viable organisms per ml < 50μ and >10μ in minimum dimension, and
3. Less than the following concentrations of indicator microbes:
    - Toxicogenic Vibrio Cholera less than 1 colony-forming unit (cfu) per 100 ml, or less than 1 cfu per 1 g zooplankton samples;
    - Escherichia coli less than 250 cfu per 100 ml; and
    - Intestinal Enterococci less than 100 cfu per 100 ml.

Compliance with the performance standard (D2) is only achievable by use of a ballast water treatment system. In general, treatment systems that comply with the standard D2 must be approved by the ship's classification society.

BWM PLAN AND IMPLEMENTATION

The BWM plan was developed to provide guidance and assistance to the ship's staff to support effective and compliant operation of the ballast water exchange system. The implementation and management of the plan enable the vessel to prepare for the steps and proceedings required when conducting ballast water exchange. Since ballast water exchange operations are more hazardous than normal port operations, vigilant and calculated planning is necessary to ensure the ship's safety, to remain in compliance with national and international regulations, and to safeguard the marine environment. It is the responsibility of the vessel's master and the BWM officer, usually the chief officer, to develop detailed measures and processes related to the ballast water exchange. All members of the ship's staff involved in the procedure must be trained and familiar with the safety aspects of ballast water exchange and the BWM plan, and in particular the methods of exchange used on board their vessel. The chief engineer is usually also involved in the ballasting operation as a liaison between the bridge and the engine control room. The procedure for ballasting requires (1) the approved vessel's loading conditions to be used during ballast water exchange; (2) the vessel's ballast pumping and piping arrangements, positions of associated air and sounding pipes, positions of all compartment and tank suction and pipelines connecting them to the vessel's ballast pumps, and, in the case of use of the flow-through method of ballast water exchange, the openings used for the release of water from the top of the tank together with overboard discharge arrangements; (3) the means of confirming that the sounding pipes are clear and that the air pipes and non-return valves are in good working order; (4) the distances offshore required to undertake the various ballast water exchange operations (this also includes the time required to complete individual tank discharges and filling); and (5) the methods to be used for the ballast water exchange at sea, for example, flow-through and dilution, and the need to constantly monitor the ballast water exchange operation.

The contents of the BWM plan are summarised here. (1) *Ship-specific particulars*. This includes the vessel's class, IMO number, Flag state, gross registered tonnage (GRT), main dimensions, total ballast water capacity, number and capacity of ballast pumps, total number of segregated ballast tanks, ballast tank capacities, centres of gravity, maximum free surface, units to be used for ballast measurement, and details of the appointed BWM officer. (2) *Plans and diagrams*. This includes the tank arrangement and capacities, a piping diagram of the ballast system and a layout diagram of the ballast control system, and a list or set of diagrams showing sampling and access points in the pipelines and tanks. These sampling points are to be provided to enable Port State Control or other approved authorities to confirm that a ballast water exchange has been conducted prior to the discharge of ballast water. (3) *Safety considerations*. There are various circumstances that may affect the ship's safety during the ballast water exchange process, including stability considerations such as the minimum required metacentric height (GM), strength considerations (avoiding overpressurisation of tanks), limits of longitudinal and torsional strengths, and the training of officers and crew. (4) *Preferred method or methods of ballast water handling*. (5) *Duties of the appointed BWM officer*. (6) *Sediment management*. Residual sediment taken into the ballast water tanks can contain an assortment of microorganisms, including resting stages. When tanks are later filled with ballast water, the accumulated sediment and associated biota may be resuspended and discharged at subsequent ports of call. Ballast water tanks and their internal structures should therefore be designed to minimise the accumulation of sediments and allow for easy cleaning and maintenance, as required by the BWM Convention. The volume of settled-down sediments should be continuously monitored and regularly removed in accordance with the BWM plan. Any release of sediments during the cleaning of ballast tanks is prohibited and should be avoided as far as is possible.

The pre-planning of ballast water exchange should include the following data collection (which is usually done by or with the support of the marine engineering department):

1. Establishing which tanks are subject to BWM
2. Establishing which exchange method to use
3. Calculation of each intermediate stage with the loading instruments on board (for example, the sequential method)
4. Calculation of the estimated time span
5. Establishing in which areas ballast water exchange is possible

Because of the possibility that a partial exchange may encourage the regrowth of aquatic organisms, ballast water exchange should only be commenced in any tank where there is sufficient time to complete the exchange for that tank in full. If a tank couple will be operated simultaneously with both ballast water pumps running, the tank levels must be controlled continuously and, if necessary, one line reduced. In any case, it should be ensured that a single tank is filled or discharged by only one pump to avoid unacceptable high-pressure developing. Throughout the ballast water exchange process, the master must take into consideration the vessel's position including traffic density, weather forecasts and sea conditions, the vessel's stability and loading conditions, condition and performance of the vessel's machinery and manoeuvrability. When it is necessary to load ballast water on board, a few of the following key points must be kept in mind to minimise the uptake of potentially harmful aquatic organisms and pathogens, or indeed sediments containing such pathogens: (1) areas identified by the Port State in connection with warnings provided by ports concerning ballast uptake and any other port contingency arrangements in the event of emergency situations; (2) in darkness when organisms may rise up in the water column; (3) in very shallow water; (4) where the ship's propellers may stir up sediment; (5) in areas with large phytoplankton blooms (i.e., algal blooms such as red tides); (6) nearby sewage discharge points; (7) where a tidal stream is known to be more turbid; (8) where tidal flushing is known to be poor; (9) in areas close to aquaculture (such as salmon farms, etc.); and (10) where dredging is or has recently been carried out. In addition to these points, it is always strongly recommended to avoid ballasting in areas with naturally elevated levels of suspended sediments, such as river mouths, and delta areas, or in locations that have been affected significantly by soil erosion caused by inland drainage.

Working in conjunction with the ship's engineers, the duties of the BWM officer are to (1) ensure the safety of the vessel and crew; (2) ensure that BWM procedures are followed and recorded; (3) be familiar with the requirements of the Port State authorities with respect to ballast water and sediment management; (4) where ballast exchange is required, ensure the steps of the ballast exchange sequence are followed in the relevant order; (5) ensure adequate personnel and equipment are available for the execution of the planned BWM operations; (6) ensure all required BWM records are maintained and up to date, including the ballast water record book; (7) where required, prepare the appropriate national or port ballast water declaration forms prior to arrival; (8) assist the Port State Control or quarantine officers for any samplings that may need to be taken; (9) oversee crew familiarisation and training of BWM requirements and applicable shipboard systems and procedures; and (10) perform any other duties, as specified by the ship owner or operator. The chief engineer is usually requested to, or will delegate the responsibility for, ensure the manholes of the specific tanks are opened prior to commencement of the flow-through method or instead, for tanks with no direct access to open deck, the vent heads are removed. These need to be re-secured after completion of the operation. It is necessary that the BWM officer keeps the master advised on the progress of the BWM operations and any envisaged deviations from the agreed plan.

RECORD KEEPING

Each procedure concerning ballast water exchange must be fully recorded in the ballast water record book, which forms an integral part of the BWM plan. These records are a legal document and may be used as evidence should the vessel be prosecuted for breaching the convention or any local, national, or international laws. The records that must be kept include (1) the location (latitude and longitude) of where the ballast water exchange took place; (2) a detailed position and description of the watertight and weathertight closures (for example, manholes, the opening of vents and air pipes) which may have been opened during the ballast exchange (and since re-secured); (3) descriptions of the procedures required to conduct ballast water exchange and the estimated volume of ballast water, including the following:

- When ballast water was taken onboard
- Whenever ballast water is circulated or treated for BWM purposes
- When ballast water is discharged out to sea
- When ballast water is discharged to a reception facility
- Accidental or other exceptional uptake or discharge of ballast water
- Additional operational procedures and general remarks

There are codes provided at the front of the ballast record book. All entries are to be made in reference to these codes. Figure 18.2 provides an illustration of the ballast sequence planned for a medium-sized container vessel. Importantly, authorised Port State Control officers may inspect the ballast water record book on board the vessel at any time the vessel is within territorial waters. These officials may choose to make a copy of the entries in the record book and require the vessel's master to certify that the copy is a true replica. Any copy certified as such may be permissible in any legal proceedings as evidence of the facts stated in the BWM plan. Hopefully, it is quite clear from the earlier explanation that our aim is to prevent marine pollution caused by ballast water from one location being discharged in another location. It should be noted that the ballast water exchange standards covered under the provisions of D1 are temporary, and eventually, all ships will be required to comply with the performance standards set out in D2. This means all vessels will have to carry some form of ballast treatment plant. It should be understood, therefore,

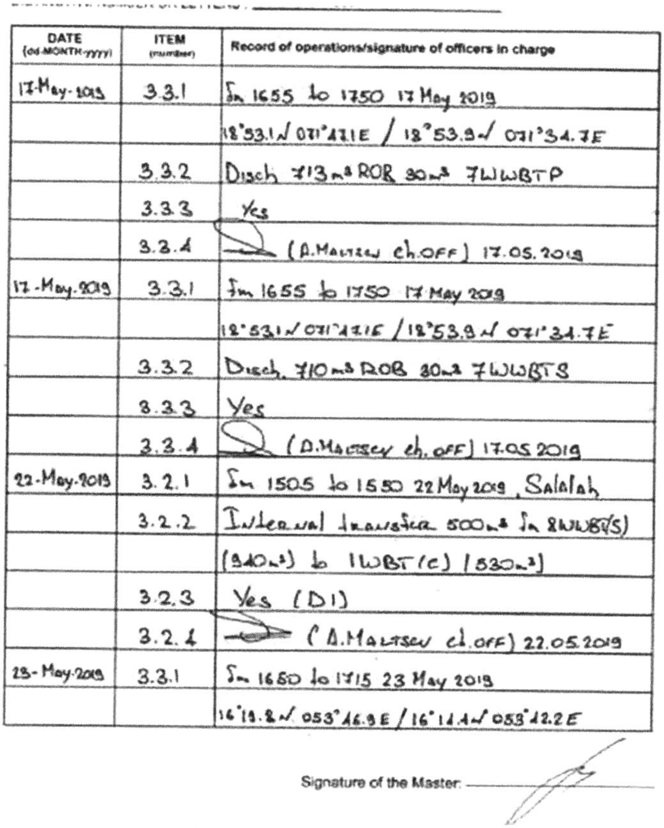

DATE (dd-MONTH-yyyy)	ITEM (number)	Record of operations/signature of officers in charge
17-May-2019	3.3.1	Fm 1655 to 1750 17 May 2019
		18°53.1N 071°27.1E / 18°53.9N 071°34.7E
	3.3.2	Disch 713 m³ ROB 30 m³ 7LWBTP
	3.3.3	Yes
	3.3.4	(A.Maltsev ch.OFF) 17.05.2019
17-May-2019	3.3.1	Fm 1655 to 1750 17 May 2019
		18°53.1N 071°27.1E / 18°53.9N 071°34.7E
	3.3.2	Disch 710 m³ ROB 30 m³ 7LWBTS
	3.3.3	Yes
	3.3.4	(A.Maltsev ch.OFF) 17.05.2019
22-May-2019	3.2.1	Fm 1505 to 1550 22 May 2019, Salalah
	3.2.2	Internal transfer 500 m³ In 8WWBT(S) (310 m³) to 1WBT(C) (530 m³)
	3.2.3	Yes (D1)
	3.2.4	(A.Maltsev ch.OFF) 22.05.2019
23-May-2019	3.3.1	Fm 1650 to 1715 23 May 2019
		16°13.2N 053°16.9E / 16°11.1N 053°12.2E

Signature of the Master: _____

Figure 18.2 Typical ballast water sequence log.

that vessels must comply with either regulation D1 or regulation D2. The scheduling deadline for compliance with regulation D2 is:

- All new ships (i.e., those ships built on or after 8 September 2017) must comply with regulation D2 performance standards.
- All existing ships (i.e., ships built before 8 September 2017) are required to meet the regulation D2 standards at the first IOPP renewal survey after 8 September 2019.
- All vessels must comply with regulation D2 standards before 8 September 2024.

In summary, since the *International Convention for the Control and Management of Ships' Ballast Water and Sediments, 2004,* entered into force on 8 September 2017, ships engaged in international traffic are required to manage their ballast water and sediment operations to a certain standard and must carry a BWM plan, a ballast water record book, and an International BWM Certificate. BWM systems must be approved by the ship's class whilst accounting for IMO Guidelines. The carriage of ballast water is indispensable to maintaining acceptable load and trim conditions. When conducting ballast water exchange, the engineering staff are required to maintain a diligent watch to respond to power failures, ballast pump or pipe failures, and or structural failures that could impact on the safety and integrity of the ship.

BALLAST TANKS ON SHIPS

Now that we have covered, in some degree of detail, the regulatory aspects of ballasting, we should be in an advantageous position to recognise the importance of keeping the ballast system in good working order. Whereas the actual performance of ballasting rests with the deck department, the critical job of maintaining the ballast systems lies with the engineering department. In this section, we will begin to explore the design and structure of a ship's ballast system. During the design and construction stages of a new ship, the ballast tanks are introduced at various locations for maintaining the stability of the ship during passage. The concept of ballast is not new and has been implemented since ancient times. Before the introduction of pumps and ballast tanks, ships loaded dry ballast such as sacks of sand and rocks, and iron blocks, as well as using barrels of food and potable water. This method helped to a certain extent to maintain the stability of the ship and its seaworthiness. Modern ships carry liquid ballast, which includes freshwater, seawater, or brackish water in various ballast tanks. As ships have expanded in size, and the cargo carried by the vessels varies from one port to another has grown, water ballast tanks have been adopted to compensate for maintaining the trim and stability of the vessel.

The principles behind ballasting are quite simple. Let us assume that the vessel does not have a ballast system on board. In such cases, the following conditions may arise: the propeller may not fully immerse in water, affecting the engine efficiency of the ship; the ship may list or trim as the cargo capacity of the ship is not fully reached; the shear and torsion loads on the vessel may increase the stresses on the ship structure, leading to bending moments and slamming; and or the vessel may face issues of dynamic transversal and longitudinal instability. To compensate for the aforementioned conditions, ballast water is taken on board to ensure a safe operating condition. In other words, ballasting helps reduce stresses on the hull of the vessel. It also provides for transverse stability of the ship. As the propeller is submerged, it aids the propulsion plant in maintaining its efficiency. Ballast helps in immersing the rudder, supporting the manoeuvrability of the vessel, and reducing the exposed hull surface. The ship continually uses fuel and water from its tanks leading to weight loss. The ballast operation helps compensate for this weight loss. As we have already stated, the master and chief officer are primarily responsible for adding or removing ballast water from the ship's ballast tanks, depending on the ship's stability condition. There are three types of ballast conditions: light ballast, heavy ballast, and port ballast. When the ship is heavily loaded, and it does not require any additional ballast, the water ballast tanks are kept empty. This condition is known as *light ballast*. When in a seagoing state, if the ship is not fully loaded, the ship's ballast tanks are filled to their capacity. This condition is known as *heavy ballast*. Many ports around the world have a restriction on the use of ballast water. Dedicated port ballast tanks are provided to correct the trim and list of the ship during loading or discharging operation. When used, this is called *port ballast*. The water ballast tanks are provided at various locations depending on the type of ship. The following are some of the most common locations for ballast tanks on merchant ships (specialist vessels and naval ships may have completely different configurations depending on their design, and function, and roles):

> *Topside tanks.* As the name suggests, these are tanks located in the topside spaces of the ship. The topside tanks are triangular and are fitted with wings on both sides of the cargo holds. They are more common in bulk carrier ships and are constructed using transverse frames arranged in the following ways: (1) deck

transverse, i.e., under the main deck which supports the deck plating; (2) bottom transverse, which forms the part of the supporting frame for the bottom area of the topside tank; (3) side transverse, which forms part of the structure for supporting the side shell plating of the tank. This is kept in line with the side shell frames within the cargo holds (typically in single-skin bulk carriers). These tanks are directly connected to the ship's main ballast pipelines. During the cargo loading and or discharging operation, the volume of ballast water in the topside tank is kept in equilibrium with the cargo weight. The design of the topside tanks helps avoid cargo shift, which is especially useful for ships carrying fluid-type cargoes such as grains and light ores.

Lower hopper tanks. Similar in construction to the topside ballast tanks, these water ballast tanks are located on the bottom wing sides of each cargo hold and are kept in continuation to the double bottom tanks which run through the centre of the vessel. The hopper tanks are designed to function as additional ballast space for the ship. Their design provides a slope in the cargo-hold corners, which eases the collection of cargo in the mid-position of the hold. The adjacent fuel tank plating of the hopper tank forms a slant boundary to carry static and dynamic loads during cargo loading and ballasting.

Double bottom (DB) tanks. The DB of the ship is a safety feature to avoid the ingress of water in the event of grounding or collision. These void spaces are used to store ballast water, which helps stabilise the ship. The DB tanks are located between the forward section (i.e., from the collision bulkhead) to the aft peak bulkhead, dividing the engine room. On some vessels, such as container ships and bulk carriers, the DB space is divided transversely into three sections (instead of two). This is done to provide a cofferdam in the centre known as the duct keel, which is used to carry ballast and bunker tank valves, as well as piping for ships ballast tank and bunkering systems. The construction of the DB tanks is related to the length of the ship. Vessels longer than 120 m (393 ft) will have additional longitudinal framing in comparison to the transverse framing for vessels less than 120 m (393 ft). Unlike upper topside tanks, these water ballast tanks are located adjacent to the fuel oil tanks in the DB. Hence, they are usually not connected to the ballast system to avoid the risk of contamination.

Fore and aft peak ballast tanks. The fore and aft peak ballast tanks are provided to perform precise trimming for the ship. To achieve the required trim these tanks are filled partially to avoid free surface effect from occurring. The construction of the fore and aft peak tanks is different from the ship's other ballast tanks, as their shape is irregular due to their location within the ship's hull, being dependent on the design of the bow and stern. By necessity, the design of these ballast tanks is narrow at the bottom, and as the tank moves upward, the width of the tank increases significantly. The tank breadth corresponds to the breadth of the ships' hulls. The valve used to control the flow of water into the ballast tank can either be a manually controlled butterfly valve or a hydraulically operated remote valve. For the fore and aft peak tanks, only remote-controlled (hydraulic) valves are used due to the sensitivity of their location.

Oil tankers have a separate set of regulations for their ballast tanks. The two main types of ballast tanks found on these types of vessels are the segregated ballast tank and the clean ballast tank. *Segregated ballast tank (SBT)*. As per MARPOL annex 1, regulation 18, every crude oil tanker of 20,000 tonnes deadweight and above, and every product carrier of 30,000 tonnes deadweight and above delivered after 1 June 1982, as defined in regulation 1.28.4, must be provided with segregated ballast tanks. The segregated ballast tanks are dedicated tanks constructed for the sole purpose of carrying ballast water on oil tankers. For protection, they are completely separated from the cargo and fuel tanks. The segregated ballast tanks avoid any risk of mixing oil and water, which usually happens when cargo holds are used to carry ballast water. *Clean ballast tank (CBT)*. Oil tankers often sail without carrying cargo in their holds, which can lead to severe stability issues. This is especially so on heavy seas. Hence, the cargo holds which carried oil in the last voyage must be cleaned and filled with ballast water. During the discharge of ballast water, an oil content monitor control is used. Only effluent which is <15 ppm is permitted to discharge overboard; the remainder is transferred to the ship's slop tanks.

BALLAST TANK MONITORING

There are three types of monitoring that must be conducted on the ballast water tanks. These are level monitoring, atmosphere monitoring, and volume monitoring. *Level Monitoring*. The water ballast tank on the ship is installed with level sensors to control the valves and ballast pumps. This helps ensure safe ballasting and de-ballasting operations. Multiple ballast pumps are provided in the engine room which take suction from the main seawater line (i.e., from the sea chest), and during de-ballasting, they discharge the ballast water overboard. The cargo control room is usually where the ballast tank level monitoring system is located. The pump cutoff is controlled once the water level reaches the sensor level to activate the trip. *Atmosphere monitoring*. On oil tankers, the ballast tank is provided with gas-measuring sensors at various levels. These are typically at the upper and lower levels of the tanks. During a loaded condition of the ship, the ballast tank will be kept empty. In such situations, the three-way valve in the sampling line will be set towards the lower sampling point. When the ship is in a ballast or partial ballast condition, the sampling line is adjusted to activate at the top sampling point. This is done to avoid the egress of water into the analysing unit through the sampling points. *Volume monitoring*. The volume monitoring of the ballast tank is done to achieve the ballast or de-ballast rate of the pumping system. This is done by the load indicator software installed on the ship. The ship's BWM officer manipulates the results displayed by the load indicator to operate the fill or discharge valve of the ballast tank. The change in volume of the ballast tank is used to calculate the ballast pump rate, which in turn, determines the time needed to finish the ballasting or de-ballasting operation. This helps the chief officer to complete the stability operation in time to maintain the ship's estimated time of departure (ETD).

BALLAST TANK PROTECTION

The ballast tank is filled with seawater, which is highly corrosive in nature. When the tank is empty, the damp atmosphere also increases the rate of corrosion on the ballast tank surface. As we well know, corrosion is a major problem on ships and a grave concern for marine engineers. To prevent or, at least, lessen the effect of corrosion, a few techniques may be used, which we will briefly discuss. *Tank coatings*. Coating of the tank surface is the most common protection system used on ships. The advantage of coating is that it protects the entire tank and if the right quality of coating is applied, the ballast tank will resist the effect of seawater and other corrosive elements for a long time. Coating works by providing a protective layer of saltwater-resistant dry film over the steel structure of the ballast tank. The coating dry film can be as thin as 300 μm. The most common type of coating used is heavy-duty, dual-component epoxy. *Anodes.* The use of sacrificial anodes is an immensely popular option for minimising the effect of corrosion. Zinc, aluminium, and their alloys, together with other metals such as tin, are a popular choice of anode. Magnesium anodes are avoided as they tend to generate hydrogen, which can have deleterious effects on some ballast tank coatings. On oil tankers, the use of aluminium anodes is prohibited as they are a spark hazard if dropped from significant heights. *Controlled atmosphere.* If the atmosphere of the ballast tank can be controlled to reduce the oxygen content, the corrosion rate decreases drastically. This system is known as oxygen stripping which is done by introducing inert gas to maintain the oxygen level in the tank at below 4%. This system can be used in tanks coated in a protective layer as well as in tanks fitted with sacrificial anodes (to extend the life of the anodes). When used, the controlled atmosphere technique can reduce corrosion by as much as 84%.

BALLAST TANK INSPECTIONS

The deck and engineering officers must always be aware of the conditions inside the ballast tank. This means regular inspections are required to ensure the tank conditions are kept satisfactory. Primarily, the concern is corrosion. During the inspection, the extent of corrosion of the tank's surface should be recorded, and localised corrosion marked accordingly. If the corrosion of the tank surface exceeds 75% of the allowable margin, immediate[2] repairs such as the renewal of the hull structure must be conducted. Coating plays a significant role in the ship ballast water tank surface protection hence proper inspections of the tank coating need to be performed. Any visible failure of the coating must be recorded together with any indications of rusting of the tank surface. This is especially important along any weld lines and the edges of the tanks. The tank structure must be inspected for signs of cracks or buckling. The strengthening arrangement must be checked for bends or cracks. These should be repaired at the next reasonable opportunity. The accuracy of inspection records is important as the history of the ship (i.e., previous inspection records) plays a critical role in determining the level of response to any faults identified during the tank inspection process. Where a vessel is one amongst two or more sister ships, there is always the possibility that malfunctions or faults in one vessel could affect any of the sister vessels. Timely and accurate record keeping is thus essential to maintaining a safe and seaworthy ship (Figure 18.3).

PERFORMING BALLASTING OPERATIONS

An experienced officer must only ever conduct ballasting and de-ballasting as it is related to the stability factor of the ship. The ballast system often differs from ship to ship but the basics of all ballast systems remain the same; filling, removing, and transferring water from one tank to the other to achieve the required level of stability. Fortunately for us, this is the responsibility of the deck department, so we need not concern ourselves too much with the intricate details of the ballast operation itself; what does concern us, however, is

BALLAST WATER REPORTING FORM
(To be provided to the Port State Authority upon request)

1. SHIP INFORMATION **2. BALLAST WATER**

Ship's Name:	Type:	IMO Number:	Specify Units: M³, MT, LT, ST
Owner:	Gross Tonnage:	Call Sign:	Total Ballast Water on Board:
Flag: Marshall Islands	Arrival Date:	Agent:	Total Ballast Water Capacity:
Last Port and Country:		Arrival Port:	Ballast Water Treatment system
Next Port and Country:			Make/Model:

3. BALLAST WATER TANKS Ballast Water Management Plan on board? YES NO Management Plan Implemented? YES NO

Total number of ballast tanks on board: ______ No. of tanks in ballast: ______ IF NONE IN BALLAST GO TO No. 5.

No. of tanks exchanged: ______ No. of tanks not exchanged: ______

4. BALLAST WATER HISTORY: RECORD ALL TANKS THAT WILL BE DEBALLASTED IN PORT STATE OF ARRIVAL; IF NONE GO TO No. 5.

Tanks/Hold	BALLAST WATER SOURCE					BALLAST WATER EXCHANGE					BALLAST WATER TREATMENT			BALLAST WATER DISCHARGE			
$ (List multiple sources per tank separately)	DATE DDMMYY	Port or Lat/Long	Volume (units)	Temp (units)	Salinity (units)	DATE DDMMYY	Endpoint Lat/Long	Volume (units)	% Exch.	Sea Hgt. (m)	DATE DDMMYY	Volume (units)	Treatment System	DATE DDMMYY	Port or Lat/Long	Volume (units)	Salinity (units)

Ballast Water Tank Codes: Forepeak = FP, Aftpeak = AP; Double Bottom = DB; Wing = WT; Topside = TS; Cargo Hold = CH; Other = O

IF TREATMENT/EXCHANGES WERE NOT CONDUCTED, STATE OTHER CONTROL ACTION(S) TAKEN: ______

IF NONE STATE REASON WHY NOT: ______

5: IMO BALLAST WATER GUIDELINES ON BOARD? YES NO

RESPONSIBLE OFFICER'S NAME AND TITLE (PRINTED) AND SIGNATURE: ______

Figure 18.3 Typical ballast water reporting form.

the machinery and equipment used to perform the ballast operation, as the ship's engineers have responsibility for their upkeep and maintenance. All valves in the ballast system are normally hydraulically operated from the remote operator station in the ship's control centre or in the engine control room in manual mode or in automatic sequence. The ballast pump suction and discharge valves, along with the other valves, have their fail-safe in the OPEN position so that if any valves malfunction or become stuck, they will remain open throughout the ballast operation. The overboard discharge valves are set to their fail-safe position. y position

Ballasting and de-ballasting can be achieved in any one of five ways:

- Transferring water between tanks using gravity.
- Ballasting or de-ballasting tanks from the sea using gravity.
- Ballasting the tanks using the ballast pumps.
- De-ballasting the tanks using the ballast pumps.
- De-ballasting the tanks using the stripping ejectors.

Note: DB tanks should always be filled by gravity.

When ballasting or de-ballasting the vessel, care should be taken to ensure that the tank is not overfilled. Overfilling the tanks will damage the tanks as the pressure vacuum valves have a lower capacity than that of the pump. The filling valves will close automatically when the tanks reach their pre-set point level. Furthermore, care must be taken not to run the pump dry or run the pump with the discharge valves closed. This can be managed through the installation of an automated system, which ensures the pump will not start until all the necessary valves are opened. Valves can be put into auto mode, which ensures that the valve closes automatically once the ballast tank is filled with the required amount of ballast water or once the setpoint is reached. The port and starboard sides are considered two separate systems, each having its own automatic sequence for ballasting and de-ballasting. When filling the ballast tanks with the ballast pumps, it should be ensured the motors are not overloaded. This can be done by checking the current with an ammeter. If this occurs, the number of opened valves to ballast tanks should be immediately reduced (closed) until the current is returns to within the allowable limit. A ballast pump motor overload alarm is given for the safety of the ballast pump. Sometimes during a sea passage, an alarm may trigger indicating the ballast pump suction pressure is high. If this happens, open the suction valve to the sea chest and close it again once the pressure has reduced. The water in the heeling tanks should always be kept at half of their total capacity. If required, the heeling tanks may be used as additional ballast tanks. The ballast pump is used to empty or fill the heeling tank. Also, in some ports, the port authorities may ask for a sample of the ballast that the ship is carrying. In this instance, the sample must be taken from the sounding pipe connection. The locations of all the sounding pipes are provided on the ballast system plan of the ship.

BALLAST WATER TREATMENT

The presence of invasive aquatic species in the ship's ballast water is one of the biggest problems faced by the shipping industry. Posing a great threat to the marine ecosystem, these aquatic species have led to an increase in bio-invasions at an alarming rate. Under the IMO's *International Convention for the Control and Management of Ship's Ballast Water and Sediments*, the implementation of a BWM plan and ballast water treatment system on board ships has become mandatory. To ensure that ships comply with the rules and regulations set by IMO regarding BWM, ship operators have started implementing ballast water treatment systems on their ships. A variety of technologies are available on the market for treating marine ballast water. That said, constraints such as the availability of space, the cost of implementation, and the level of environmental protection afforded by such systems have stymied the wider adoption of ballast water treatment systems. Several factors should be considered when choosing a ballast water treatment system for a ship. These include effectiveness against ballast waterborne organisms, environment friendliness, the safety of the crew, cost-effectiveness, ease of installation and operation, and space availability on board. The main types of ballast water treatment technologies currently available on the market in 2022 are filtration systems (physical), chemical disinfection (i.e., oxidising, and non-oxidising biocides), ultraviolet treatment, deoxygenation treatment, heat (thermal treatment), acoustic (cavitation treatment), electric pulse/pulse plasma systems, and magnetic field treatment. Most typical ballast water treatment systems employ two or more technologies in tandem to ensure the treated ballast water meets IMO convention standards.

Physical separation/filtration system ballast water treatments. Physical separation or filtration systems are used to separate marine organisms and suspended solid materials from ballast water using sedimentation or surface filtration systems. The suspended/filtered solids and waste (backwashing) water from the filtration process is either discharged in the area from where the ballast is taken or further treated on board ships before discharging. The following equipment is used for ballast water filtration: (1) *screens and discs*. Screens (fixed or movable) or discs are used to remove suspended solid particles from the ballast water with automatic backwashing. These are environmentally friendly as they do not require the use of toxic chemicals in the ballast water. Screen filtration is effective for removing suspended solid particles of larger sizes but are not particularly good at removing smaller particles and organisms. Note: it is worth noting that though screens are highly effective in removing most suspended solid particles and organisms from ballast water, they alone are not sufficient to treat ballast water in accordance with IMO standards. Subsequently, a secondary system is required. (2) *Hydrocyclone*. The hydrocyclone is an effective method for separating suspended solids from ballast water. High-velocity centrifugal force is used to rotate the water to separate the solids. As the hydrocyclone does not have any moving parts, it is extremely easy to install, operate, and maintain. Note: it has been found that as the operation of the hydrocyclone heavily depends on the mass and density of the particle, they are not successful in removing smaller organisms from the ballast water; therefore, a secondary system is required. (3) *Coagulation*. As most physical filtration methods are not able to remove smaller solid particles, the method of coagulation is used prior to the filtration process to join smaller particles together to increase their size. As the size of the particles increase, the efficiency of the filtration processes also increase. Such treatments involving the coagulation of smaller particles into small flocs are known as flocculation. The flocs settle more quickly and can be removed easily. Note: some ballast water treatment systems using coagulation and flocculation use ancillary powder (sand, magnetite, etc.) or coarse filters to produce flocs. An additional tank is required for treating ballast water for this process and thus extra space is required onboard ships. *Media Filters*. Physical ballast water treatment systems with media filters can also be used to filter out smaller-sized particles. It has been found that compressible media filters (such as crumb rubber) are more suited for shipboard use because of their compact size and lower density when compared to conventional granular filtration systems.

Magnetic field treatment. The magnetic field treatment method uses coagulation technology. Magnetic powder is mixed with the coagulants and added to the ballast water. This leads to the formation of magnetic flocs, which includes marine organisms. Magnetic discs are then used to separate these magnetic flocks from the water.

Chemical disinfection (oxidising and non-oxidising biocides). Biocides (oxidising and non-oxidising) are disinfectants that remove or inactivate marine organisms in the ballast water. However, it is worth noting that the biocides used for ballast water disinfection purposes must be effective on marine organisms and readily degradable or removable to prevent discharge water from becoming toxic to other marine life. Based on their functions, biocides are divided into two types: oxidising and non-oxidising. (1) Oxidising biocides are general disinfectants such as chlorine, bromine, and iodine, and are used to inactivate organisms in the ballast water. This type of disinfectant works by destroying the organic structures of the microorganisms such as cell membranes and nucleic acids. Some of the processes utilising oxidising biocides on board ships are chlorination (chlorine is diluted in water to kill any microorganisms present in the ballast water), ozonation (ozone gas is bubbled into the ballast water using an ozone generator. The ozone gas decomposes and reacts with other chemicals to kill organisms in the water), other oxidising biocides such as chlorine dioxide, peracetic acid, and hydrogen peroxide may also be used to kill organisms in the ballast water. (2) Non-oxidising biocides are a type of disinfectant which, when used, interferes with the reproductive, neural, or metabolic functions of the organisms. Though there are several non-oxidising biocides available on the market today, only a few such as Menadione or Vitamin K are used in ballast water treatment systems. This is because they tend to produce toxic by-products. Because of this, considerable research is going into developing non-toxic alternatives which are effective against invasive marine species without adversely affecting the wider marine environment.

Ultra-violet treatment method. The ultraviolet ballast water treatment method uses ultra-violet (UV) lamps (Amalgam lamps) that surround a chamber through which the ballast water is allowed to pass. The UV lamps produce ultraviolet rays which act on the DNA of the organisms rendering them harmless and preventing their reproduction. This method has been successfully used globally for water filtration purposes and is effective against a broad range of aquatic organisms. Note: UV systems are the most popular option at present. Efficiency depends on the turbidity of the ballast water as this can limit the transmission of UV radiation. UV systems are suitable for any vessel type but are preferable for those that do not take on much ballast water and have flow rates of up to one thousand cubic metres per hour, such as roll-on, roll-off vessels, container ships, offshore supply vessels, and passenger ferries.

Deoxygenation. As the name suggests, the deoxygenation ballast treatment method involves purging or removing all oxygen from the ballast water tanks to asphyxiate any marine organisms present in the tanks. This is usually done by injecting nitrogen or any other form of inert gas into the space above the water level in the ballast tanks. Note: it takes between two and four days for the inert gas to asphyxiate. Therefore, this method is not usually considered suitable for ships having short transit times. Moreover, such types of systems can only be used on ships with perfectly sealed ballast tanks. If a ship is already installed with an inert gas system, then a deoxygenation system will not require any additional space.

Heat treatment. This treatment involves heating the ballast water to a temperature that will kill the organisms. A separate heating system is used to heat the ballast water in the tanks, or the ballast water can be used to cool the ship's engine, thus disinfecting the water by way of heat from the engine. However, this treatment method usually takes a considerable time to take effect and can increase corrosion in the tanks.

Cavitation or ultrasonic treatment. In this system, ultrasonic energy is used to produce high-energy ultrasound waves to kill the cells of the organisms in the ballast water. Such high-pressure ballast water cavitation techniques are used in combination with other systems.

Electric pulse/plasma treatment. The electric pulse/plasma treatment is still in the development stage. In the pulse electric field system, two metal electrodes are used to produce an energy pulse in the ballast water using very high-power density and pressure. This energy kills the organisms in the water. With the electric plasma system, a high-energy pulse is supplied to a mechanism placed in the ballast water, which generates a plasma arc, thus killing the organisms. Both methods are said to have the same effect.

OZONE GENERATOR FOR BALLAST WATER TREATMENT

After the adoption of the *International Convention for Control and Management of Ship's Ballast Water and Sediment* on 13 February 2004, many different ballasts water treatment methods have been introduced, some of which we have discussed in this chapter. One such system that is worth further discussion is the ozone generator system. Ozone is a colourless oxidising biocide with a pungent smell and is formed naturally in Earth's atmosphere. It has very unstable properties and when injected into water decomposes rapidly. It has been found that ozone is one of the most powerful and quickest-acting oxidisers available and is effective against many types of waterborne bacteria such as moulds, yeasts, organic material, and viruses. The ozone generator is a machine that generates ozone gas by using oxygen from the atmosphere. The principle of the generator is energy from an electric discharge field is used along with ambient air which flows through the ozone cells responsible for converting and increasing the amount of ozone gas. The ozone gas is then injected into the ballast water by an injector unit. The ozone gas dissolves, decomposes, and reacts with other chemicals, killing all organisms present in the water. This makes the ballast safer to discharge overboard. The ozone generator produces ozone gas by taking air or pure oxygen as the feed-gas source; by breaking apart O^2 molecules into single atoms, these then attach to other O_2 atoms forming ozone (O_3). The ozone generator uses two methods to produce ozone gas. The first is the silent corona discharge. These machines use electric discharge to split diatomic oxygen atoms into single atoms. The second method uses ultraviolet radiation. This process is like the process in which the sun's UV rays split oxygen atoms into two. This method is considered less efficient than the silent corona discharge method. Ozone generators are categorised according to their (1) control mechanism (i.e., voltage or frequency unit), (2) cooling mechanism (i.e., water, oil, or combination thereof), and (3) the physical arrangement of the dielectrics (i.e., whether horizontal or vertical). In addition to these three categories, it should be noted that ozone generators may also be categorised in accordance with their unique manufacturer's characteristics.

Because the ozone treatment method is particularly harmful, there are several considerations which need to be factored. As ozone is a strong oxidant and virucide, it provides direct oxidation and destruction of the bacteria cell wall resulting in the leakage of cellular constituents outside the cell wall. These constituents can cause reactions with radical by-products. The use of ozone causes the carbon-nitrogen bond to break leading to depolymerisation. Last of all, ozone causes damage to the constituents of nucleic acids (i.e., purines and pyrimidines). Whilst these are highly effective against waterborne bacteria, they are also extremely hazardous to other aquatic lifeforms. Subsequently, great care must be taken when using ozone treatment systems.

The basic layout of the ballast water system comprises an ozone generator, injector unit, line filter, and neutraliser. A line filter is fitted before the ozone generator and the injector unit, whereas the neutraliser is fitted on the discharge line of the ballast system. The neutraliser comes into use when the water is de-ballasted from the ship. Neutralisation is done as ozone is a toxic gas and is harmful to humans. The gases that are not dissolved thus need to be removed by the neutraliser prior to discharge out to sea. The ozone generator system is more effective for the removal of larger organisms when combined with other treatment methods such as electrolysis or UV methods.

BALLAST WATER SYSTEM MAINTENANCE

Ballasting and de-ballasting operations are frequently performed on board ships. These operations are conducted when the vessel is in port, during ballast exchange, cargo transfer, and when taking on heavy weather ballast. A considerable amount of machinery and effort is needed to keep this essential system in good working order. Therefore, it is always prudent to discuss the ballasting and or de-ballasting plan with the engineering department during the pre-arrival meeting or cargo work briefings. While preparing a ballasting or de-ballasting plan, the critical stages, with respect to ship stability, stresses, and changeover of tanks should be clearly identified and discussed. Some of the main problems that can affect the efficiency of the ballast system are as follows.

1. *Poor familiarisation.* The officers and crew responsible for operating the ballast pumps and valves must be fully conversant with the line-up and meaning of various "valve indications" and symbols provided on the ballast line-up diagrams in the cargo control room and pump room. They should be able to differentiate between the indication of manual valves, hydraulic valves, manual-hydraulic valves, suction and discharge gauges for pumps and the prime mover for ballast pumps, such as steam, electricity, or hydraulic pressure.
2. *Faulty gauges and sounding pipes.* Most of the ballast tank gauges work on a pneumatic pressure difference method, where a measured pressure of air and difference of counter-pressure gives the level of ballast in a tank, and the reading is displayed in the cargo control room through a digital or analogue reading. These gauges should be purged regularly, and the readings compared with manual soundings to eliminate the risk of erratic readings. Sounding pipes are often found choked with rags or sounding rods or tapes. They must be always kept clear to provide accurate manual

soundings to ensure the tank is completely empty or the intermediate readings are correct in the event gauges are faulty. This will prevent the dry running of the ballast pumps. During the topping up of the ballast tanks, if the sighting ports for ballast manholes are kept open, this can help significantly should the tank reach the overflow level in case the cargo control room readings are erroneous.

3. *Failure to check the shore installations.* Before starting a de-ballasting operation, always ensure to check alongside the berth, pier, or jetty for any electrical fittings or electrically operated shore bollards. It is common for vessels to damage these quayside fixtures, resulting in costly repairs.
4. *Pressure surges.* When conducting ballasting or de-ballasting operations, the pumps in use are of a centrifugal type. To start a centrifugal pump, positive suction pressure must be generated. The discharge valve of the pump can be kept up to 30% open to prevent causing damage to the valve body or valve seat ring. The discharge pressures and rpms of the pump need to be increased slowly and gradually to avoid any pressure surges in the lines and load surges in the engine room. Pressure surges are common causes of damaging the lines and valves. They can even cause the load on the boiler or generator to fluctuate abruptly, thereby tripping the plant completely and delaying the operation. During a changeover or brief, idle periods during operation, pumps can be run in a 'sea-to-sea' mode to avoid dry running and overheating the pump casing or causing further damage to the pump seal. Whilst taking ballast by gravity during a loaded passage, the inboard line to the tank should be opened completely, keeping the sea chest valve closed. Only then should the sea chest valve be opened. When the ship is in a loaded condition, due to deeper draught, seawater enters under heavy pressure. If this is not monitored and managed appropriately, the high pressure can damage the line fittings and valves if any valve is closed in between. When de-ballasting by gravity, the line from the ballast sea chest to the tank must be opened completely, keeping the tank valve closed. Once the line is fully open, the tank valve can be opened slowly.
5. *Shallow waters and sediment deposits.* Care should be taken when conducting ballast operations in areas with low under keel clearances (ukc) or with sedimentary bottoms. This is the main reason ballast pump strainers become choked, due to the large sediment deposits that build up inside the tanks after de-ballasting. When these puddles of mud accumulate in the tanks, they can become quite considerable if the tanks are not inspected and cleaned regularly. Open sea ballast water exchange is another effective method for removing sediment from the tanks. Mud and sediments can choke the suction bellmouth inside the tank. Retaining excess ballast on board can be crucial where a vessel is loading to her draught marks.
6. *Discharge pressure in parallel suction lines.* Often during ballasting, when two ballast pumps are running together (if the suction lines are common) one of the pumps can develop a better suction than the other. This is especially the case when one has a direct suction line as opposed to a branched line from the mainline. This should be borne in mind and the discharge pressure of the pumps adjusted accordingly to ensure good suction is provided by both pumps until the tank's water level reduces. As the level falls further, positive use of trim and list can help the pumps to retain suction for a longer period thus reducing de-ballasting time. If, however, one of the pumps loses suction, it can be run in 'sea-to-sea' mode for some time, and as the suction and discharge pressure builds up, the changeover of suction to the tank can be conducted to resume de-ballasting.

MARPOL ANNEX I

The prevention of oil spillage from ships and to keep the sea safe from oil pollution is the responsibility of the deck officers, though like all operations on board today's ships, the engineers have a key role to play as well. Oil pollution from ships can be the result of accidental spills and leaks or through the negligence of the ship's crew. Though less common today, oil pollution may be caused by the intentional discharging of oil-contaminated ballast water overboard. When oil is discharged into open water, it spreads quickly over the surface, although the intensity of the pollution depends on the relative density and composition of the oil. The results can be disastrous. as oil sheen has a significant negative impact on aquatic life as well as on the human environment. Worse still, oil contamination can have long-lasting effects on the coastal environment, especially when oil seeps into the substrate. The immediate effects of oil contamination are toxic smothering leading to the mass mortality of fish and other food species, birds, and so on. This has a knock-on effect for local economies which rely on fishing and tourism for their livelihoods. In part because of the inherent dangers of associated with oil contamination, the occurrence of oil spill incidents at sea has drastically reduced in recent years; however, it cannot be permanently eliminated. The objective of MARPOL annex I, which entered into force on 2 October 1983, is to protect the marine environment through the complete elimination of pollution by oil and other damaging causations, and to reduce the risk of accidental oil discharges into the open sea.

Central to MARPOL annex I is the definition of "oil." As marine engineers, it might seem perverse to define the term "oil," but it is useful to recognise the definition provided by MARPOL. MARPOL defines "oil" as a viscous fluid containing petroleum which may be in the form of crude oil, heavy fuel oil, sludge, oil refuse, and refined products such as marine gas oil. As per annex I, ballast water and tank washing residues originating from ballast tanks and the washing out of cargo tanks on tanker vessels are also included within the definition. In fact, MARPOL annex I covers all fluids which contains oil, and which can be discharged overboard at sea. Even oily water separator which we will cover later in this book) treated discharge water is also included under the annex. Under MARPOL annex I, ships must carry and maintain in a fully functioning condition the following equipment:

All ships:

- Oil filtering equipment
- 15 ppm alarm arrangements
- Standard discharge connection

Tanker specific:

- Oily water interface detector
- Crude oil washing system, if fitted
- Oil discharge monitoring and control system
- Cargo and ballast pumping, piping, and discharge arrangements
- Engine room bilge holding tank to slop tank pumping and piping arrangement

Control of the discharge of oil under MARPOL annex I, regulation 4

Under regulation 4 of MARPOL annex I, any discharge of oil or oily mixture generating from the ship's engine room or cargo spaces on tanker ships is prohibited, except when they fulfil the following criteria[3]:

- The vessel must be on route.
- The vessel is more than 12 NM (13.8 mi, 22 km) from the nearest shoreline.
- The vessel must carry, and use, an approved oil filtering system which complies with the requirements set out in MARPOL annex I, regulation 14, pertaining to the treatment of oily water effluents.
- The oil ppm in the treated effluent without dilution must not exceed 15 ppm.
- The oily mixture being treated by the oil filtering equipment is exclusively drawn from the engine room and not from the cargo holds.
- The oily mixture is not mixed with fuel tank or cargo tank oily residues.

Discharges in special areas

MARPOL annex I prohibits the discharge of oil or oily water mixtures from vessels over 400 gross tonnes deadweight in any mandated special areas. Special areas are marine regions of environmental, scientific, or economic value. The special areas listed under annex I (as of 2022) include the Mediterranean Sea, the Baltic Sea, Black Sea, Red Sea, the Arabian Gulf, the Gulf of Aden, the Antarctic Sea, Northwest European Waters, the Oman area of the Arabian Sea, and the southern waters around South Africa. A vessel may discharge oil or oily water mixtures into any of the previously listed special areas only under very stringent conditions. These are:

- The vessel must be on route.
- The vessel must carry, and use, an approved oil filtering system which complies with the requirements set out in MARPOL annex I, regulation 14, pertaining to the treatment of oily water effluents.
- The oil ppm in the treated effluent without dilution must not exceed 15 ppm.

In the Antarctic Sea special area, any discharge into the sea of oil, oil effluent, or oily mixtures from ships is absolutely prohibited in all circumstances.

Discharges from oil tankers

Discharge from the cargo area of an oil tanker (which includes the cargo tanks, pump rooms, machinery spaces, and bilges mixed with cargo oil residue, etc.) is only permitted under the following circumstances:

- The tanker ship is not inside a special area as defined by MARPOL annex I.
- The tanker ship is more than 50 NM (57 mi, 92.6 km) from the closest shoreline.
- The tanker ship is making progress on route.
- The instantaneous rate of discharge of oil content does not exceed 30 NM/L.
- The total quantity of oil discharged into the sea does not exceed:

- For existing tankers (delivered on or before 31 December 1979): 1/15,000 of the total amount of cargo from which the residue formed a part thereof.
- For new tankers (delivered after 31 December 1979) 1/30,000 of the total amount of cargo from which the residue formed a part thereof.
- The tanker has in operation an oil discharge monitoring and control system and a slop tank arrangement.

Discharges from oil tankers in special areas

Any oily mixture or oil effluent discharge from the cargo area of an oil tanker into the sea which forms part of a special area is prohibited, although the regulation does not apply to the discharge of clean or segregated ballast water. In respect of the special area around Antarctica, any discharge into the sea of oily water or mixed effluent from any ships is prohibited.

COMPLYING WITH MARPOL ANNEX I

Every vessel of 400 gross tonnes deadweight or above, and all tanker ships of 150 gross tonnes deadweight and above, must conduct the surveys to comply with MARPOL annex I: (1) *Initial survey*. This survey is performed before the ship is set into service. In this survey all equipment, machinery, systems, fittings, etc., which are covered under annex I are inspected and tested. (2) *Annual survey*. The annual survey is performed every year within a buffer period of three months prior to or after the anniversary date the IOPP Certificate was issued. (3) *Intermediate survey*. An intermediate survey takes place within a buffer of three months before or after the second-anniversary date or within three months before or after the third anniversary date of the Certificate replacing any one of the annual surveys. (4) *Renewal survey*. The renewal survey is done on or before the five-year period of the certificate expiry date. In this survey, a detailed inspection of all equipment, material, machinery, fittings, etc., covered by annex I is performed. (5) *Additional survey*. If there are significant repairs and renewals conducted on any of the machinery, systems, or fittings, which fall under annex I, an additional survey must be performed which can be general or partial, depending on the circumstances. (6) *Condition assessment scheme survey*. The CAS is done to confirm that the structural strength of single-hull oil tankers is acceptable under the periodical surveys as indicated in the Statement of Compliance (SOC). The first CAS survey is usually done in parallel to the first intermediate or renewal survey after 5 April 2005 or once the vessel completes 15 years of age, whichever occurs later.

Certificates, plans, and records under MARPOL annex I

International Oil Pollution Prevention Certificate (IOPP). The International Oil Pollution Prevention Certificate or IOPP is issued following the initial survey and after each renewal survey. The IOPP Certificate states that the vessel's equipment, systems, fittings, machinery, and so forth (as covered by MARPOL annex I) are compliant with the regulation. The validity of this certificate is never more than five years. Individual administrations may decide to issue the certificate for a period of less than five years, depending on circumstances. If the certificate is on the verge of expiring, and the ship is still out at sea engaged in a voyage, the administration may extend the certificate validity so that ship can complete the voyage and come to a port where the survey can be conducted. In such situations, the maximum extension period permitted is three months from the date of expiration. The validity of the certificate may expire in any of the following conditions: (1) any of the relevant surveys are not conducted under a specified period as stated in the annex; (2) if the endorsement is not completed as per the requirements of the annex; and (3) if there is a change of Flag to another Flag state.

Oil record book. The oil record book is an important legal document under MARPOL annex I. There are two parts to the oil record book depending on the type and deadweight of the vessel. For ships of 400 gross tonnes deadweight or less, only part 1 need be carried. This also applies to oil tankers of 150 gross tonnes deadweight or less. For vessels of 400 gross tonnes deadweight or more, and oil tankers of 150 gross tonnes deadweight or more, part 1 and part 2 needs to be continued board. The oil record book contains the following generic information: the name and IMO number for the ship, the GRT, the owner's details, any official numbers or references, and the period of usage. Part 1 of the oil record book acts as a log of (1) all shipboard operations involving oil, oily mixtures and effluents; (2) dates, geographical positions, quantities, tank identifications, and the durations of operations; (3) ballasting and cleaning of the fuel oil tanks including the discharge of dirty ballast or cleaning water from the oil fuel tanks, disposal of oil residues (sludge), and non-automatic discharges overboard, or disposal otherwise, of bilge water accumulated in the machinery spaces; (4) automatic discharge overboard, or disposal otherwise, of bilge water, collected in the machinery spaces (for example transfers of bilge water to the slop tank); (5) conditions of oil discharge monitor (ODM) and oil discharge control system; (6) any accidental or other exceptional discharges of oil; (7) bunkering of fuel or bulk lube oils; and (8) any additional operational procedures and general remarks. Port State authorities are legally entitled to take copies of any, and all, entries, and if so requested, the master is required to confirm any copies are a true likeness of the original. The oil record book must be retained on board for three years following the last date of entry.

In addition to part 1 of the oil record book, all vessels over 400 gross tonnes deadweight and tankers over 150 gross tonnes deadweight are required to carry part 2, which records all oil-related activities from cargo to ballasting. Part 2 records the following activities: (1) loading and unloading of oil cargo, (2) internal transfer of oil cargo during the passage, (3) cleaning of cargo tanks, (4) crude oil washing COW system (only), (5) ballasting of cargo tanks, (6) ballasting of segregated CBT tankers (only), (7) discharge of dirty ballast water, (8) discharge of clean ballast from the cargo tanks, (9) discharge of ballast from segregated CBT tankers (only), (10) discharges of water from the slop tanks into the sea, (11) condition of the ODM and control system; (12) accidental or other exceptional discharges of oil overboard, (13) additional operational procedures and general remarks, (14) loading of ballast water (tankers engaged in specific trades), (15) location of ballast water within the ship, and (16) ballast water discharged to reception facilities.

Shipboard Oil Pollution Emergency Plan (SOPEP). The SOPEP is a prevention plan which should be available on board all ships of 400 gross tonnes deadweight and above and all oil tankers of 150 gross tonnes and above. In addition to SOPEP, all oil tankers with 5,000 tonnes deadweight or more must have quick access to coast-established computerised damage stability and residual structural strength calculation programmes.

Other essential requirements of MARPOL annex I

In addition to the points already discussed, MARPOL annex I also enforces the following mandatory provisions:

Fuel oil tank protection. All ships which delivered after 1 August 2010, having a fuel oil capacity of 600 m^3 and more must comply with the following regulations: (1) The individual fuel oil tank should not have a capacity of more than 2,500 m^3. (2) For those ships which have a fuel tank capacity more than 600 m^3 but less than 5,000 m^3, the fuel oil tank location should be inboard of the moulded line of the side shell plating, and not less than w, as described by the formula:

$$W = 0.4 + 2.4 \times C/20,000$$

(where the minimum value of $w = 1.0$ metre (3.28 ft))

(3) For ships having total fuel oil tank capacity of 5,000 m^3 and above, the fuel oil tank location should be inboard of the moulded line of side shell plating, and not less than w, as described by the formula:

$$w = 0.5 + C/20,000$$

or

$$w = 2.0 \text{ m} (6.56 \text{ ft}) \text{ whichever is less}$$

(where the minimum value of $w = 1.0$ m (3.28 ft))

In these calculations, C is total fuel oil volume.

Pump room protection. The pump room, which is an essential part of an oil tanker, is located at the bottom part of the vessel. It is provided with a DB and should be designed such that the distance between the bottom of the pump room and the ship's baseline (indicated as L), when measured perpendicular, should not be less than:

$$L = B/15 \text{ or } 2.0 \text{ m } (6.56 \text{ ft}) \text{ whichever is less}$$

Here, B is the breadth of the ship.

Oily water separator (OWS). This is equipment specific to the marine industry which is used to separate oil from water. It only allows effluent of less than 15 ppm to be discharged out to sea. We will cover the OWS in more detail later in this book.

ODM and control system. This system is used on tanker ships to discharge effluent generated from the cargo and ballast tanks. It is not a filtration unit and only monitors the content, allowing discharges to flow overboard only when the ppm content is satisfactory.

Other pollution prevention equipment and tools. The SOPEP locker contains various pollution prevention tools, chemicals, and equipment, which can be used for managing on board oil pollution, and to restrict the oil from going out to sea, as well as for responding to oil spill pollution.

Roles and responsibilities under MARPOL annex I

Ship's staff. Under MARPOL annex I, there are various duties and responsibilities for the ship's staff and Port State Control. The ship's staff plays the most vital role in implementing annex 1 on ships. The main duties that annex I places on the crew are (1) maintaining the OWS, ODM, and control system, and ensuring all other MARPOL annex 1 equipment is kept in good working condition. For crews working oil tankers, additional responsibilities apply, which are managing all oil transfers (internal and external) carefully to prevent any oil spills, and ensuring all bunkering operations are conducted carefully. All records pertaining to the transfer of oil, sludge, bilge, tank washing, etc. must be maintained in the appropriate logbooks and oil record books; the master is required to carry out regular pollution prevention training and drills on board; every member of the ship's crew must be fully familiarised with the SOPEP locker and other pollution prevention equipment available on board; the ship's officers must know how to complete the appropriate oil record books; the master and chief engineer must check the entries for accuracy and countersign the oil record book entries; the crew must, by all reasonable means possible, restrict and prevent any oil from going overboard. If the ship participates in an accident, the ship's crew must try to minimise the outflow of oil into the marine environment; and any accidental oil spills must be immediately reported to the nearest shore administration.

Port State Control (PSC). The main aim of any PSC inspection is to find deficiencies in the vessels they visit so that those defects can be rectified before the ship goes out to sea. Under MARPOL annex I, the following equipment, systems, and logs are usually checked by the PSC inspectors: the oil record book (which is checked for different entries) and the sounding logbook (which is reviewed). The PSC inspector may ask to take actual soundings for cross-checking; the engine room bilges for oil content; evidence of leaks in machinery which may contribute to oily bilges; the discharge pipe of the OWS (before the overboard valve) can be opened and checked for oily layers, which indicates the OWS is not performing as required; the operational log of the oil content metre; the validity of the IOPP Certificate; and the seals on the OWS discharge valve and ODMCS discharge valve. The PSC inspector may randomly pick any member of the ship's crew and ask questions regarding oil spill drills or SOPEP locker location to check the practical knowledge and familiarisation of the crew regarding their ship.

MARPOL annex 1 is one of the first regulations to come into force with the aim of reducing and preventing marine pollution. It also includes provisions for port reception facilities, and their requirements, so that ships can dispose of sludge and bilge waste which cannot be treated and discharged overboard in a safe and environmentally friendly manner.

COMPLYING WITH THE BALLAST WATER CONVENTION

In 1997, the IMO adopted guidelines for the control and management of ship's ballast water to minimise the transfer of harmful aquatic organisms and pathogens from one sea area to another. Later, in February 2004, the Marine Environment Protection Committee (MEPC) adopted the *International Convention for the Control and Management of Ship's Ballast Water and Sediments*. In accordance with this convention, all ships must carry and implement ballast water and sediment management protocols contained with a ballast management plan, maintain a ballast water record book, and conduct ballasting and de-ballasting operations in accordance with the provisions of the convention. There are ten key points which must be followed to ensure compliance with the convention. These are (1) *general methods for managing ballast water whose source is anything other than the deep ocean* and are as follows: (a) proceed to a Port State–approved offshore location to carry out ballast water exchange, (b) sealing of tanks against ballast water discharges when in port waters, (c) pump water to shore reception facilities, (d) prove by laboratory analysis that the ballast water is safe and acceptable, and (e) treat the ballast water in-situ using an approved treatment method. (2) *Important points to consider for taking ballast in port*. To the greatest extent possible, the ship should be slightly ballasted in port in preparation for departure using ballast tanks to allow for safe navigation. The master should ensure that the ship takes on ballast water that is as clean as possible and care should be taken to minimise any sediment uptake. Taking on ballast under the following circumstances should be avoided wherever possible: (a) in shallow waters, (b) in the vicinity of sewage out-falls or dredging operations, (c) within proximity of any known outbreak of diseases communicable through ballast water, (d) in areas with toxic phytoplankton blooms (for example, harmful algae blooms such as red tides), (e) at night when bottom-dwelling organisms may rise up in the water column, (f) where the incoming or outgoing tide is known to be turbid, and (g) where tidal flushing is considered poor. (3) *Important methods of ballast water exchange*. Ballast water exchange operations should be conducted in deep water, preferably in open ocean, and as far as possible from shore. The exchange should be conducted when the vessel is sailing in clean ocean water, which can be found more than 200 NM (230 miles, 370 km) from shore and in water where the depth is at least 200 m (656 ft). There are currently three methods for conducting ballast water exchange. These are the sequential method, the flow-through method, and the dilution method, of which the sequential method is the preferred option.

> *Sequential method.* The ballast tanks with water to be exchanged are emptied and refilled with clean ocean water to achieve at least 95% volumetric exchange. In this method, the ballast tanks are emptied until the ballast pumps lose suction. The tanks are further stripped by eductor systems.
> *Flow-through method.* This is a process by which replacement ballast water is pumped into a ballast tank, allowing the incoming water to overflow through the air vent or dedicated overflow vents. In practice, this means that at least three tank volumes are to be pumped through each ballast tank to achieve 95% efficiency of exchange.
> *Dilution method.* A process in which the replacement ballast water is filled through the top of the ballast tank with simultaneous discharge from the bottom of the tank at the same rate, while maintaining a constant level in the tank during the operation.

(4) *Ballast water sampling point*. The sampling of ballast water may be required by local authorities such as the port Quarantine Officer. Normal ballast water sampling may be done through the sounding pipe. Location of the sounding pipes' heads should be marked and clearly visible on the vessel plan. (5) *Sediment management*. Routine cleaning to remove sediments from tanks which are used to carry ballast should be regularly conducted. This involves (a) flushing the tanks with small quantities of clean water in an empty tank before de-ballasting, (b) manual cleaning by making entry into the ballast tank or tanks. The tank entry is a hazardous operation and safety procedures should be followed before entry. Always inform the port authorities and follow the enclosed space protocols. A permit to work will also be required. (6) *Chain lockers and sea chests*. Subject to practical accessibility, all sources of sediment retention such as anchors, cables, chain lockers, and suction wells should be cleaned out regularly as an additional precaution to further reduce the possibility of spreading contamination. (7) *Disposal of sediments*. During cleaning and de-silting operations, the tank sediments should be safely disposed of and should never be discharged in estuarine or coastal waters. Safe disposal infers the transfer of contaminated water to shore-based facilities or designated landfill sites. It may be possible to discharge ballast into the deep ocean, so long as it is done in accordance with the provisions of the convention. (8) *Record keeping*. All data and records pertaining to ballasting and de-ballasting operations, including ballast water exchanges, must be recorded with the time and position of the vessel in the ballast water record book. When sediment is cleaned from the ballast tanks or chain lockers, this data should also be logged. Accordingly, (9) *designated BWM officer*. Usually the chief officer, the BWM officer is the designated person on board with responsibility for BWM. They must ensure that the procedures laid out in the BWM plan are implemented properly. It is important to plan and conduct ballasting and de-ballasting operations taking into account the stability and trim factors of the vessel and to advise the other officers on conducting the operation safely. The *BWM officer* must also prepare the ballast water declaration form prior to arriving in port; assist PSC or quarantine officers with any sampling that may need to be taken and maintain the ballast water record book. (10) *Crew training and familiarisation*. All officers and ratings should be trained and familiarised with the ship's pumping plan, positions of air and sounding pipes, location of tanks and manholes, compartment and tank suctions, pipelines as well as remote pump operations and sounding equipment. Moreover, all crew should be made aware of the safety precautions and hazards involved in ballast water operations. In addition, officers should be familiarised with recordkeeping. The approved stability booklet, and ballast water-log should be cross-checked every day.

Research is ongoing into refined and enhanced BWM systems, with the aim being to eradicate the transfer of harmful aquatic organisms and pathogens, thus preventing disastrous changes in the marine environment because of the global shipping industry. Plans for making ships without ballast tanks are also underway, which it is hoped will further help in putting an end to marine pollution caused by ballast water. In this sizeable chapter, we have covered a lot of ground regarding the function of ballast water, the regulations governing the use and discharging of ballast water at sea and in special areas, and the procedures for safely performing ballasting and de-ballasting operations. In the next chapter we examine the role and function of the OWS, a key piece of equipment which we have already touched on at points within this chapter.

NOTES

1. Examples of such organisms are Golden Mussels, Zebra Mussels, North American Comb Jellyfish, the Cladoceran Water Flea, and the North Pacific Seastar.
2. In this instance, immediate refers to the first reasonable opportunity where the ship can be dry docked and repairs carried out safely without hindering the safety of the ship and its crew.
3. This provision applies to all ships of 400 gross tonnes deadweight (other than oil tankers) and above sailing outside special areas.

Chapter 19

Oily water separator

Ships produce vast quantities of oil and water mixtures daily. This mixture needs to be separated from each other before it can be legally and safely discharged. If the oily water mixture is not separated prior to discharge overboard, the penalties can be extremely severe. Under annex I of the MARPOL regulations, ships must carry a piece of equipment called the oil discharge monitoring and control system together with oil filtering equipment. Collectively, this equipment is called the *oily water separator* (OWS) (see Figure 19.1). As the name suggests, this equipment is designed to satisfy the requirements of annex I, which limits the oil content in bilge water that vessels can legitimately discharge out to sea. It achieves this by separating the maximum amount of oil particles from the water to be discharged overboard from the engine room or cargo hold bilges, oil tanks, and

Figure 19.1 Typical example of an OWS.

oil-contaminated spaces. As per the MARPOL regulations, the oil content in water processed by the OWS must be less than 15 parts per million of oil. If we recall from the previous chapter, MARPOL annex I, regulation 4, paragraphs 2, 3, and 6 states, '[A]ny direct discharge of oil or oily water mixture into the sea shall be prohibited'. The regulation further explains how an oily water mixture should be treated on board before discharge out at sea: for ships of 400 gross tonnes deadweight and above, discharge of oil mixture can be done under the following conditions: (1) the ship is on route; (2) the oily mixture is processed through an oil-water separator filter meeting the requirements of annex I, regulation 14; (3) after passing through the oil-water separator system, the oil content of the effluent without dilution must not exceed 15 parts per million; (4) the oily mixture must not originate from cargo pump-room bilges on oil tankers; and (5) on oil tankers, the oil-water mixture is must not be mixed with oil cargo residues. If the ship is anywhere within the Antarctic Sea region, any discharge into the sea of oil or oily mixtures from any vessel is prohibited. Furthermore, in accordance with the MEPC 107(49) regulations, the following provisions also apply to vessels over 400 gross tonnes deadweight: (1) bilge alarm or an oil content monitor, which provides for the internal recording of alarm conditions, must be certified by an authorised organisation; (2) the oil content alarm provided with the OWS must be tamper-proof; (3) the oil content alarm must activate and sound an alarm whenever freshwater is used for cleaning or zeroing purposes; (4) the OWS must be capable of achieving 15 ppm on type C emulsions. Provided the vessel meets these requirements, the discharge of oily water emulsion and effluent can be legally and safely discharged at sea (outside any special areas).

DESIGN, CONSTRUCTION, AND WORKING PRINCIPLES OF THE OWS

The OWS consists of three main segments: the separator unit, the filter unit, and the oil content monitor and control unit. *The separator unit.* This unit consists of catch plates which are located inside a coarse separating compartment and an oil collecting chamber. As the oil has a density lower than that of the water, the oil rises into the oil collecting compartment. The remainder of the non-flowing oil mixture settles down into a fine settling compartment, having passed between the catch plates. After a brief period, more oil will separate and collect in the oil collecting chamber. The oil content of the water which passes through this unit is around 100 ppm of oil. A control valve (either pneumatic or electronic) releases the separated oil into the designated OWS sludge tank. A heater may be incorporated into the unit to smoothen out the flow and enhance the separation of the oil from the water. The heater may be incorporated into the unit either at the middle or the bottom of the unit depending on the area of operation and capacity of the separator equipment. The first stage in the oil separation process helps remove physical impurities from the mixture, which then aids fine filtration in the next stage. *The filter unit.* This is a separate unit whose input comes from the discharge of the first unit. The filter unit consists of three stages: (1) the filter stage, (2) the coalescer stage, and (3) the collecting chamber. In the first stage, any impurities and particles are separated by the filter and settled at the bottom of the separation chamber for removal. In the second stage, the coalescer induces a coalescence process through which oil droplets are joined together to increase their size. The coalescer achieves this by breaking down the surface tension between the individual oil droplets in the mixture. These large oil droplets rise above the mixture in the collecting chamber to be later removed when required. The output from this unit should be less than 15 ppm to fulfil the legal discharge criteria discussed earlier. If the oil content in the water is more than 15 ppm, then remedial work, such as filter cleaning or filter renewal, must be conducted. A freshwater inlet connection is also provided to the filter unit to clean and flush the filter. This is usually done before and after the operation of the OWS unit. *Oil content monitor and control unit.* This unit functions together in two parts: (1) monitoring and (2) controlling. The oil content monitor continuously monitors the ppm of oil. If the ppm is high, the monitor will trigger an alarm and feed data to the control unit. The control unit continuously monitors the output signal of the oil content monitor. If an alarm is triggered, the control unit shuts the three-way solenoid discharge valve, which deposits the separated emulsion overboard. There are normally three solenoid valves commanded by the control unit. These are in the first unit oil collecting chamber, the second unit oil collecting chamber, and on the discharge side of the OWS. The three-way valve inlet is from the OWS discharge, where one outlet is to overboard and the second outlet is to the OWS sludge tank. When the oil content manager triggers the alarm, the three-way valve discharges the oily mixture directly into the sludge tank. A small pipe connection for freshwater can be provided to the oil content monitor unit for flushing. Whenever this line is in use, an alarm is sounded and recorded in the oil content monitor log. This ensures an accurate record is kept of the operation of the discharge valve. For most shipping companies, the OWS is only ever operated by the chief engineer. This means, unfortunately, training standards and levels on OWS systems for other engineering department crew members are usually minimal if any is provided at all.

OPERATING THE OWS

The OWS should only be operated when the ship is underway. To meet the requirements of MARPOL, annex I, the OWS must be capable of filtering the oil content of bilge water to 15 ppm of oil. Should the OWS be defective, and oil discharges found to exceed the permitted allowance, the chief engineer can be arrested, fined, and, in particularly egregious circumstances, even imprisoned. Because of such high risks, operating the OWS should be done with absolute care to minimise the risks of marine pollution. Although it is common to find a *How to Operate* guide on or near the OWS, the standard process for operating the OWS is discussed here (see Figure 19.2 for a typical OWS schematic diagram). (1) The OWS overboard manual discharge valve should be kept locked and held by the chief engineer. Before operating the OWS, open the lock and overboard valve. Then open all the other valves of the system. (2) Open the desired bilge tank valve from which the oily water mixture is to be discharged from the OWS. (3) Open the air valve if the control valves are air operated. (4) Switch on the power supply of the control panel and oil control unit. (5) Fill the separator and filter unit with fresh or seawater to clean and prime the system until the water comes out from the vent in the second stage. (6) Start the OWS supply pump, which is a laminar flow pump, and the one that will supply the oily water mixture to OWS. (7) Observe the oil content monitor for ppm value and keep checking the soundings of the bilge tank from where the OWS is taking suction and of the OWS sludge tank. (8) A skin valve or sample valve is usually provided just before the overboard valve (which is a three-way valve). Regularly take samples of the effluent and monitor its clarity. (9) Keep a watch of the ship's side at the overboard discharge valve. (10) Once the operation is complete, switch off the power and shut and lock the overboard valve. Return

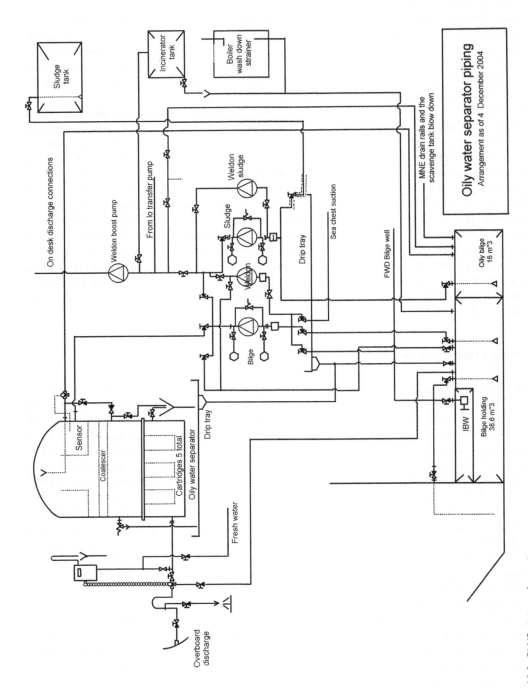

Figure 19.2 OWS piping schematic.

the keys to the chief engineer. (11) An entry must be made by the chief engineer in the oil record book summarising the operation. This record must be signed by the operating officer (if not the chief engineer), the chief engineer (if not already signed) and the master.

The OWS is designed to work under ideal conditions; however, the ship is never an ideal condition, and there are many potential sources of contamination to the bilge water. This bilge water is a mixture of various grades of oil mixed with water, together with suspended solids such as rust, chemicals, detergents, soot, paint chips, cargo dust, etc. All these particles can affect the efficiency and working of the OWS if proper maintenance and operation are neglected.

OWS MAINTENANCE

Every marine engineer should be acutely aware of the importance of the OWS, and the need to keep it in good working order. As with all things on board a ship, this requires regular maintenance, and the OWS is no different. Unfortunately, the OWS has acquired a reputation for being quite temperamental and liable to malfunction. Other than the filter becoming clogged due to continuous use, there are many other reasons for the OWS to suffer inferior performance. In this section, we will briefly cover some of the ways to ensure the OWS is kept functional, and more importantly, operating in accordance with the MARPOL annex I regulations. (1) *Oil in the bilge.* The OWS is suitable for separating small quantities of oil in the bilge, and not the other way round. If there is a mixture consisting of a small quantity of water in oil, it is best not to pass it through the OWS and instead directly transfer it to the waste oil tank for sludge disposal onshore or for shipboard incineration. (2) *Keep the viscosity within acceptable limits.* Highly viscous oil, when supplied for oily water separation, will clog the filter. Therefore, it is important to keep the viscosity of the oil within 1,000 mm^2/s. (3) *Discharge outside the separator.* Never discharge or drain oily water mixtures from the separator outside the OWS, as this will lead to the separated oils adhering to the coalescers, rendering it inoperative. Always ensure to discharge the collected separated oil outside the separator and clean the interior of the OWS by supplying fresh water. (4) *Install a dust filter.* When using the OWS for discharging bilge mixtures containing dust and sand, it will be difficult for the coalescer filter to pass dust and sand due to the size of their particles. This will reduce the operating hours of the filter and, in turn, the efficiency of the OWS. The ideal service life of the filter normally ranges from 12 to 24 months depending on the manufacturing process and accounting for a daily operation of one hour. Therefore, it is strongly recommended to install a dust filter in the inlet line of the OWS. (5) *Use an exchanging probe for fault finding.* In most 5ppm OWS models, the oil level detecting probe and transmitting converter in the first- and second-stage chambers are identical. In the event of operating issues, try exchanging these probes with each other, between the chambers. This will help establish whether the fault lies with the probe or with a different part of the OWS. (6) *Keep a check on the heating device.* If a heating device is provided, ensure it is ON whenever the OWS is in operation and switched OFF before the OWS is stopped. Where the OWS is run for a long time, monitor the heater for potential overheating of coils. If the OWS is overheated, this may lead to the production of combustible gases within the chambers. (7) *Protect the internal coating.* The OWS interior is applied with a tar epoxy coating, which is inflammable. Avoid introducing naked flames or performing hot work, such as welding, over the surface or body of the OWS as the heat generated will damage the coating, making the

OWS susceptible to corrosion. (8) *Monitor the water level.* Always ensure the OWS is filled with seawater before any bilge mixture is supplied. This helps increase the working life of the filters and maintains the operational efficiency of the OWS. (9) *Prevent leaks.* Always avoid any leaks or water flow out of the OWS by siphon effect. Failure to do so will allow the oil to flow unimpeded into the secondary chamber. Doing so will affect the treatment capability of the OWS and lead to the clogging of the second-stage filters. (10) *Monitor the effluent.* Frequent checks on the effluent should be done to assess the performance of the OWS. If the effluent is found to be contaminated, immediately stop the separator, and take corrective actions.

To ensure the OWS works properly, and to keep oily water discharges below the mandated ppm, it is useful to keep in mind the following good practices.

- *Avoid emulsions.* Emulsions are formed when the interfacial tension between two liquids is reduced sufficiently to allow the droplets of one liquid to disperse into another. Mechanical agitation, shearing forces, solvents, chemicals, surfactants, and the presence of particulate matter can all reduce interfacial tension and result in the formation of emulsion.
- *Avoid chemical emulsions.* Chemical emulsion is formed by the addition of chemicals into the water. These chemicals function as surfactants and hold the oil drops together in an emulsified state. These surfactants may be detergents used for cleaning, alkaline chemicals used for boiler cleaning and conditioning, and so forth.
- *Avoid secondary dispersion.* Mechanical emulsions are either primary or secondary types. Primary emulsions are larger drops of oil dispersed in water and are separated within 24 hours. Secondary emulsions are fine droplets of oil that are thermodynamically stable and do not separate. Secondary dispersion is caused by turbulent conditions.
- *Avoid suspended solids.* Suspended solids lead to the stabilisation of emulsion, causing problems in the separation of the oil from the bilge water. Suspended solids may be mud, boiler soot, or cargo residues sucked from the blowers. These suspended solids get coated with oil and stabilise the emulsion. Neutrally buoyant solids that neither rise nor fall are most problematic as it is difficult to remove them. They will also trigger the high ppm alarm.
- *Avoid turbulence.* The OWS requires a stable laminar flow to operate at its most optimum. Avoid using the OWS during heavy rolling and keep all line valves fully open to avoid generating turbulence in the pipes. Rolling motions, retrofitting on old pipelines, and inadequate opening of the suction line valves, can all lead to the generation of turbulent flow within the OWS, resulting in a fall in OWS capacity due to the formation of emulsion. Sometimes, turbulence can cause oil droplets to form which are smaller than eight microns and are affected by the random motion of water particles. This random motion is called as *Brownian motion*,[1] which nullifies the forces of buoyancy, causing the oil droplets to fall.
- *Avoid particulate matter.* Fine particulate matter, such as soot, rust, microbial contamination of bilge water, etc. can also function as an emulsifying agent. Although most soot resulting from boiler washing settles down in the bilge holding tank, fine soot particles (i.e., those of 1μm or less) will give the bilge water a blackish appearance. These particulate matters will not only trigger the ppm alarm but will also function as an emulsifying agent.
- *Optimal use of chemicals.* Sometimes it is necessary to use special chemicals called emulsion breakers to separate the oil from the water. This releases the free oil.

If an emulsion breaker chemical is used, care should be taken to ensure it is used in accordance with the manufacturer's instructions. Using too much emulsion breaker can lead to an increase in emulsion production.

Restrict the drainage of chemicals. Many different chemicals are used throughout the engine room, such as for water conditioning, corrosion inhibition, rust removal, cleaning, degreasing, etc. Care should be taken to collect these chemicals and dispose of them properly. Allowing chemicals to drain free into the bilges is not good housekeeping. Moreover, should the pH of the water rise above 10, or below 4, chemical emulsification can occur.

Detergent disposal. Detergents are often used for mopping and soap washing the bulkheads, deck plates, and other such areas. In most cases, these detergents are the same as those used at homes for domestic cleaning. These detergents are designed to function as surfactants and cause emulsion of oil in water. To avoid this, always dispose of dirty mop water separately.

Avoid prolonged storage. Normally, oily water mixtures, when allowed to stand for some time (say 24 hours), separate into a layer of oil on top of the water. This layer of oil is called *free oil*. This free oil is easy enough to remove, but long retention of contaminated bilge water can cause modification to the properties of free oil caused by oxidation and microbial action. If this modification occurs, it becomes much more difficult to remove the oil.

Conduct proper filtration. If there are copious amounts of solid particles, floating media, jute, etc., present in the bilge water, these should be carefully removed using strainers to avoid fouling the filter media.

Prevent and collect leakages. Always ensure that the absolute minimum of oil reaches the bilge wells. If the oil quantity is more than water in the mixture, pump it into a separate oil holding tank. Always remember the OWS is a separator, not a purifier.

Heat the effluent. Heating the effluent reduces the viscosity of the continuous media. This helps the separation process leading to more efficient OWS operation.

Segregation of wastes. Never mix sludge and bilge. Even a small amount of sludge can contaminate a large volume of bilge water. In some ports even the discharge of treated sewage is not permitted in accordance with local laws therefore in the absence of a dedicated sewage holding tank, treated sewage must be stored in a bilge holding tank. This should be avoided wherever possible as the OWS cannot be operated without first cleaning and sterilising the tank.

Fill up the OWS prior to use. Prior to operating the OWS and allowing the bilge water to enter, always ensure that it is filled up with clean water and all air pockets have been removed. This is important as air pockets can confuse the capacitive sensors leading to the OWS malfunctioning.

Conduct back flushing. Back flushing the OWS should be done in accordance with the recommended frequency provided by the manufacturer. Doing so increases the lifespan of the filter media.

Clean Sensors. Frequent cleaning of the electronic interface sensors helps to ensure the correct operation of the OWS.

Remove accumulated oil residues. Other than the automated oil removal, any other accumulated oil residues should be removed from the OWS chambers regularly to ensure the OWS functions properly.

Follow the proper operating procedures. Always ensure that the correct operating procedures for the OWS are followed. Failing to do so can lead to the OWS malfunctioning, reduced oil filtration, erroneous discharges overboard, and inadvertent triggering of alarms. All these issues are recorded by the oil content monitor, which may be inspected at any time by Port State Control officials.

Conduct proper OWS maintenance. Like the previous point, proper maintenance of the OWS is necessary to keep it in good working condition.

We have now covered some of the main principles of good OWS management, operation, and maintenance. It is always worth keeping in mind that the performance of the OWS is always recorded by the oil content monitor, which may be inspected at any time by Port State Control. Any deficiencies, faults, or abnormalities in the OWS operation are likely to attract the attention of Port State Control. Should the OWS be found to be defective in any way, the consequences can be severe. In the next section of this chapter, we will examine the oil discharge monitor and control system in more detail.

OIL DISCHARGE AND MONITORING AND CONTROL SYSTEM ON OIL TANKERS

Most dry cargo ships do not have oil discharge monitoring and control systems as part of their OWS installation. Oil tankers, however, by the very nature of their cargo, do carry these types of systems. As tankers often discharge their cargo then sail unladen to their next port of call, they require additional ballast to be brought on board to stabilise the vessel during normal sea conditions, and especially so in heavy seas. Where this additional ballast is brought on board, it is usually pumped into the empty cargo holds. Before new cargo can be brought on board, the ballast must be cleaned then discharged out to sea. To enable this, an oil discharge monitoring and control system (ODMCS) is used to prevent marine pollution by oil. In accordance with MARPOL annex I, all oil tankers of 150 gross tonnes deadweight and above must carry an approved ODMCS. The key performance factor with the ODMCS is that they must be able to work manually in the event of a system failure. The generic ODMCS consists of four systems:

1. *Oil content metre.* The oil content metre is used to analyse the content of oil in the water that is to be discharged overboard. This oil is expressed in parts per million (ppm).
2. *Flow metre.* The flow rate of the oily water to be discharged is measured at the discharge pipe.
3. *Computing unit.* A computing unit calculates the oil discharge in litres/nautical miles and the total quantity, together with a date and time identification.
4. *Overboard valve control system.* The auto-control valve is installed at the overboard so that it closes and stops the discharge once the permissible limit has been reached.

The oily mixture is pumped out to sea via the ODMCS by a pump. A sampler probe and a flow metre sensor are connected at the discharge pipe, before the overboard valve, to monitor the oil content and the flow of the mixture. The data provided by the two sensors are fed to a control unit where it is analysed. The same control unit controls the discharge valve. If the control unit detects a rise in the ppm and flow, when compared to the permissible value, it will shut the overboard valve and open the recirculation valve which is connected to ship's slop tank. For the ODMCS to be operated legitimately, there are several regulatory requirements which must be met before the oily mixture can be discharged overboard. These are (1) the vessel must be on route, (2) the vessel must not be in a special area, (3) the vessel must be at least 50 NM (57.5 mi, 80.4 km) from the nearest shoreline, (4) the instantaneous rate of discharge of oil content must not exceed 30 L/NM (6 gal per 34.58 mi, 55.5km), (5) the total quantity of discharge must not exceed 1/30,000 of the total quantity of the residue formed cargo, and (6) the tanker must have an operational and approved ODMCS.

Furthermore, as per MARPOL, annex I, the following inputs must be recorded by the ODMCS:

- The discharge rate of the pump which is discharging the oily water mixture overboard
- The location of the ship in terms of latitude and longitude
- The date and time of the discharge operation
- The total quantity that has been discharge overboard
- The oil content of the discharged mixture in ppm

All records pertaining to the operation of the ODMCS must be stored on board for a minimum of three years.

SLUDGE PRODUCTION AND MANAGEMENT

The operation of the main engine, several types of auxiliary machinery, and the handling of fuel oil results in the production of sludge on board ships. This sludge is stored in various engine room tanks and is either discharged to onshore facilities or incinerated on board. The sludge may also contain effluent from various leaks involving the seawater and freshwater pumps, coolers, and bilges. In this section, we will discuss where and how this sludge is produced, how it is stored in the engine room tanks, what record-keeping requirements are needed, and how the sludge and bilges are incinerated, evaporated, or discharged. To start with, there are several core components which make up the sludge management system. These are the

- Fuel oil purifiers,
- Lube oil purifiers,
- Main engines scavenge drains,
- Main engine stuffing box, and
- Leak trays.

Fuel oil purifiers. The fuel oil purifiers have a designated discharge interval depending on the quality of fuel oil being used. After every set interval, the bowl of the purifier discharges the sludge that has accumulated, into the sludge tank or designated fuel oil purifier sludge tank. This sludge contains oily water and impurities which have been separated from the fuel oil by the purifiers. *Lube oil purifiers.* The lube oil purifiers have a designated discharge interval depending on the quality of lube oil used and the running hours of the main engine and auxiliary generators. After every set interval, the bowl of the purifier discharges the sludge that has accumulated, into the sludge tank or designated lube oil purifier sludge tank. This sludge contains oily water and impurities which have been separated from the lube oil by the purifiers. *Main engines scavenge drains.* When the main engine is running, oil residue in the scavenge spaces is collected from the cylinder lubrication after being scraped down from the liners. This oil is drained through the scavenge drains of each unit of the main engine and is collected in the sludge tank or designated scavenge drain tank. *Main engine stuffing box.* When the main engine is running, oil residue is collected from the stuffing box. This is scraped off the piston rod. The oil comes from the stuffing box drains of each unit of the main engine and is collected in the sludge tank or designated stuffing box drain tank. *Leak trays.* All fuel oil machinery, i.e., the pumps, filters, purifiers, etc., have a tray underneath them which collects any leaks. The drain of the tray goes directly into sludge tanks. *Miscellaneous.* There are various other drains that feed into the sludge tank, for example, the air bottle drains, fuel oil settling tank drains and the service tank drains. All these drains contain oily water which is then collected in the sludge tanks.

Sludge tanks

The number of sludge tanks varies from ship to ship, from which shipyard the ship originates, and the machinery installed in the engine room. Some ships have one common sludge tank, whereas other ships may have individual sludge tanks. In either case, a sludge pump is used to make internal sludge transfers and to transfer sludge to onshore reception facilities. All sludge tanks must comply with Flag state and class regulations. This includes the records which must be kept of every transfer of sludge in the oil record book. All designated sludge tanks and bilge tanks must have an IOPP Certificate. Any transfer from or to the IOPP tanks must be recorded in the ship's oil record book by the chief engineer.

Oily water evaporation and sludge incineration

The sludge generated by ships typically originates in the heavy fuel oil and lube oil purifiers, the heavy fuel oil settling and service tank drains, and the air bottle drains. When this sludge is slated for incineration, the sludge is transferred from the various sludge tanks into a waste oil holding tank pending incineration. Before incineration, the water content in the sludge must be evaporated to enable a clean and efficient burn. When this sludge is transferred from the heavy fuel oil purifier sludge tank, lube oil purifier sludge tank, and the oily bilge sludge tank into the waste oil tank, the steam valves (inlet and return) must be kept open to allow any water to evaporate. The tank temperature is slowly increased to 100°C (212°F). Once the tank temperature goes above 100°C (212°F), this indicates that the water has fully evaporated, and the oil has started to heat up. At this point, the sludge is ready for incineration. The quantity of water evaporated must be recorded in the oil record book. If there is a common sludge tank, then water is allowed to settle for a few days in the bottom of the common sludge tank. Once the water has settled at the bottom, suction from the bottom is taken and transferred to the waste oil tank for water evaporation. Before transferring any sludge into the waste oil tank, the temperature of the waste oil tank must be less than 90°C (194°F) to prevent boil-off in the tank. Boil-off will result in an instantaneous and tremendous rise in pressure within the tank. This must be avoided at all costs. After the water has fully evaporated and the sludge is heated up, it is ready for incineration.

Once the sludge is ready for incineration, the subsequent steps should be followed: (1) Drain and check if any water from the waste oil (sludge) is present in the tank before burning. (2) Agitate the sludge in the waste oil tank (if an agitator is present). This will help in emulsifying the oil into an even mixture for fine atomisation. (3) Warm up the incinerator with diesel oil. The incinerator should be operated by a qualified officer. (4) After warming up the incinerator, open the feed valve for the waste oil from the waste oil tank. (5) Ensure steam tracing is proper for the waste oil line and the strainers are not choked. (6) Adjust the damper and temperature in accordance with the operator's manual. (7) As the waste oil pump will take suction from the waste oil tank, continue burning the waste oil and maintain the correct incinerator parameters. (8) Depending on the capacity of the waste oil pump, compare and check how much waste oil is burning in the incinerator. (9) Record the final volume of sludge to be incinerated in the oil record book. The volume of sludge generated onboard is in proportion to the ship's fuel consumption. In general, average sludge production is considered around 1.5% of total fuel consumption. If sludge generation is more than 1.5%, the sludge production for the ship is excessively high.

ENGINE ROOM BILGE WATER GENERATION

Leaks from the freshwater and seawater pumps and coolers are collected in the engine room bilge wells. The bilge wells are located forward of the bottom platform at the tank top in port and starboard positions. Additional bilge wells include aft of the engine room in the recess under the flywheel and in the shaft tunnel if there is a separate space for the shaft tunnel. All leaks in the engine room bottom platform are collected in these bilge wells and can be transferred to the bilge holding tank via the oily bilge pump. The oily bilge pump may also pump these spaces directly to the sludge tank (via the sludge pump bypass line) and the deck connections for discharge to onshore or barge. The oily bilge pump transfers bilges to the bilge holding tank via the bilge primary tank. The bilge primary tank is usually of smaller capacity and is used to separate oil from the bilges. The bilge primary tank is overflowed to the bilge holding tank. Any oil layers formed on top of the bilge primary tank can be removed.

Bilge holding tank. Bilge from the bilge well is transferred here and stored to be discharged overboard via the OWS and or ODMCS (for oil tankers) or to be discharged ashore. *Bilge primary tank.* Bilge is transferred here to separate oil by gravity. Any oil layer formed on the top can be removed. *Bilge evaporation tank.* This bilge tank is present on some ships in which bilge can be transferred and evaporated by heating. *Air coolers drain tank.* All the moisture from the main engine scavenges air coolers and the generator scavenge air cooler is drained in this tank. There may be some oil, as the engine room air may contain oil vapours. This is then discharged overboard via the OWS or ODMCS. Given that the atmospheric air in the engine room contains moisture, and this air is compressed in the turbocharger and then cooled in the air cooler, the moisture condenses to form water droplets. If these water droplets enter the cylinders with the scavenge air, they can remove the oil film from the liner, resulting in excessive cylinder liner and piston ring wear. Additionally, the removal of water droplets from the air minimises the risk of sulphuric acid formation in the cylinders and uptakes due to the dissolving of acid products of combustion in the water droplets. To prevent these issues from developing, water is removed from the combustion air by water separators fitted after the scavenge air coolers. The water droplets are directed from the air coolers, via drain traps, to the air cooler drain tank. The discharge from this tank is pumped overboard by the air cooler drain discharge pump or bilge pump. The water flowing to the overboard discharge line passes through an oil detector, which monitors the oil content of the water being discharged overboard. It is also possible to pump the contents of the air cooler drain tank to the bilge holding tank using the oily bilge pump.

Whichever method is used, all bilge transfers, bilge discharges overboard or to shore, and bilge evaporation operations must be recorded in the oil record book. Whenever the OWS is operated, the position of the vessel at the date and time of starting and stopping must also be recorded along with duration and volume of bilge discharge. The ppm monitor will not allow any discharge of bilge consisting of more than 15ppm of oil content.

CARGO HOLD BILGE WATER GENERATION

Cargo holds, i.e., those found on container vessels, have bilge wells located at the bottom of each side – port and starboard. The hold bilges are normally pumped overboard through a bilge eductor from the fire and general services pump as they contain only water. However, before pumping out the hold bilge wells, a visual inspection must be conducted of the bilge wells. If any traces of oil are found, then they must be pumped to the hold bilge collecting tank or another designated engine room tank from where the bilge is processed by the OWS. Before any bilges are pumped directly overboard, it must be ensured that no local or international anti-pollution regulations will be contravened. Moreover, the eductor should only be used when the vessel is at sea. The hold bilge line additionally takes suction from the bow thruster room bilge wells, pipe duct bilge wells, chain locker bilge well, and forepeak void space. All the bilge wells valves can be operated remotely from the ship's bridge or the engine control room.

Sludge and bilge management are especially important responsibilities for both the engineering and deck officers. MARPOL rules are stringent and must be followed properly to prevent any pollution at sea from ships. Violations of MARPOL are likely to result in severe punishments, including large fines and even imprisonment.

GOOD BILGE MANAGEMENT PRACTICES HELP IMPROVE OWS PERFORMANCE

As we have covered earlier, the performance of the OWS depends on many factors including the OWS design, operational factors, bilge management, and so forth. In this section, we will discuss some good bilge management practices that can help optimise the performance of the OWS for the benefit of the ship, its crew, and the marine environment. No matter what equipment is installed on board the ship, if the bilge management is not properly maintained and used, OWS is bound to malfunction. (1) First amongst our good practices is effective bilge management. In the engine room, the bilges and bilge wells are located at the very bottom of the engine room. These collect the oil and water from leaks, condensates, and waste so that they can be pumped to the bilge holding tank. Clean bilges are the first line of defence against marine pollution. All seasoned marine engineers know that if the bilges are clean and dry, there is usually little to be concerned about when Port State Control officers come knocking. This is because dirty engine room bilges are one of the main causes of detainable deficiencies during Port State inspections. (2) *Use the bilge well properly*. The following are some of the good practices and tips for ensuring an efficient bilge well. Used or waste oil should never be intentionally drained into the bilges or bilge tank. All oil should be collected and put in the separate oil tank or dirty oil tank. Thereafter, it can either be incinerated or landed ashore. Discarded chemicals should never be disposed of in the bilge tanks, as the pH of this water is often above 10 and below four. This can cause chemical emulsification of the bilge water and difficulties in oil and water separation. (3) *Place drip trays where there are leaks and ensure these are dealt with as soon as reasonably possible.* (4) *Where possible, install an oil-water vertical separator for bilge water.* Always use the clean drain tank effectively. In tropical climates, condensation of more than 1–2 m^3 per day is prevalent. If this water is allowed to go to the bilge tank, this will increase the load of the OWS unnecessarily. As this is mostly clean water, it should not be allowed to drain to the bilge tank; instead, it should be put into the clean drain tank and thereafter disposed of properly. Any leaks from the freshwater and seawater pumps should also be put in the clean drain tank. (5) *Use mechanical seals where possible.* Mechanical seals, though expensive, lead to cleaner engine rooms, as there is minimal (at worst) or zero leakage (at best) from the main engine glands. With conventional gland-type pumps, although the dripping water may appear insignificant, small but persistent leaks can lead to a large build-up of water. (6) *Ensure the pipework is appropriate.* During new building, repair, and retrofitting, it must be remembered that inlet piping should be as smooth as possible and designed with few bends, as these can lead to turbulence. Inlet piping should have the least number of valves, bends, and other fittings. Where possible straight-line valves, such as gate valves, are preferred over angle valves and globe valves to avoid turbulence. The inlet piping just before the entry to the OWS should be straight for a length equal to ten times the diameter of the piping and should be sufficiently sized to avoid pressure drop. Vertical pipelines causing the shearing of upcoming water should be avoided as much as possible. Small diameter inlet pipelines cause shearing of water and force the oil droplets to become smaller. These droplets are difficult to remove later, therefore, the inlet pipeline should be of a proper diameter. (7) *Limit the ingress of air.* Sometimes the ingress of air cannot be avoided, as the positive displacement pumps handle a small volume of air. Any fall in vacuum should be investigated, as these air pockets can make the capacitance oil probes feedback erroneous data leading to the activation of the oil release valves. (8) *Limit the use of bilge cleaning chemicals.* Bilge cleaning chemicals must be OWS compatible. The use of the wrong type of chemicals will make the oil soluble in water, preventing its separation. (9) *Remove dust and cargo residues.* Dust and cargo residues should be manually swept and collected using a brush and pan and not blown by air into the bilges. These particulates can lead to the stabilisation of emulsions. (10) *Separate soot residues.* Soot from the boilers and economisers should be disposed of in a separate tank. They should never be drained to the bilges. (11) *Conduct the correct boiler blowdown procedures.* Whenever a boiler blowdown is conducted, it should be done overboard and not into the bilges. This will prevent the cleaning chemicals from producing chemical emulsions. (12) *Separate condensates from the air conditioning systems.* Condensate from the accommodation and engine room air conditioning systems should not be put in the bilges. Instead, they should be drained into separate tanks or pumped directly overboard. (13) *Do not drain deck water into the bilges.* Mopping water containing detergents as well as hand wash water should not be put into the bilge tanks, as the chemicals can lead to emulsification.

If care is taken in controlling the ingress of water and waste into the bilges, there should not be any problems with the OWS, as it will be operating within the parameters of its design.

CONSTRUCTION AND WORKING PRINCIPLES OF THE WASTE INCINERATOR

We have now touched on the waste incinerator a few times in this chapter. It should be obvious what the waste incinerator is and what it does, but it is worth spending a few moments discussing the construction and working principles of the waste incinerator for the benefit of completeness. In accordance with MARPOL 1973/78, annex I, the regulations for the handling and disposal of maritime waste at sea must be strictly followed. Until about 20 years ago, it was common for vessels to dispose of their waste by chucking it overboard. This of course is a cheap way of disposing of waste (for the ship) but the consequences for the marine environment are horrendous and long-lasting. In response to this, the IMO via the MARPOL regulations has determined that various waste materials such as galley waste, food scraps, accommodation waste, linen, cardboard, oil sludge from lubricating oil, fuel oil, bilge and purifier wastes, and sewage sludge should be disposed of through onboard incineration. This not only helps the environment but also reduces the amount of space needed to hold physical and fluid wastes. As the residue left over from the incineration is ash, this can be disposed of cheaply. For all foreign-going vessels, an incinerator installed onboard ship on or after 1 January 2000 must comply with the requirements of the standard specifications for shipboard incinerators developed under resolutions MEPC.76(40) and MEPC.93(45) (Figure 19.3).

Figure 19.3 Typical marine waste incinerator.

Accordingly, the following materials are not permitted, under MARPOL annex I, II, and III, to be incinerated and must instead be disposed of onshore using an approved waste disposal contractor:

- Polychlorinated biphenyls (PCBs).
- Garbage or waste, as defined in annex V of MARPOL, containing more than traces of heavy metals and refined petroleum products containing halogen compounds.

The incineration of sewage sludge and sludge oil generated during the normal operation of a ship may also take place in the main or auxiliary power plant or boilers, but in those cases, it must not take place inside ports, harbours, and estuaries. Furthermore, the temperature of the flue gases must be monitored and should not be less than 850°C (1,562°F) for continuous feed and must reach a minimum temperature of 600°C (1,112°F) within five minutes (the time may vary depending upon the capacity of the incinerator) for a batch feed.

There are several types of marine incinerators. The horizontal burner type and the vertical cyclone type are the most common. *Horizontal burner type.* The set-up is like a horizontal fired boiler with a burner arrangement horizontal to the incinerator combustion chamber axis. The ash and non-combustible material remaining at the end of the operation must be cleared out manually. *Vertical cyclone type.* With this type, the burner is mounted on the top and the waste to be incinerated is introduced into the combustion chamber at the same location. A rotating arm device is provided to improve combustion and remove ash and non-combustibles from the burner surface. The important parts of the incinerator are the combustion chamber with diesel oil burner, sludge burner, pilot fuel heater, and the electric control panel. A flue gas fan may be fitted with a flue gas damper or frequency inverter. In addition, a sludge service tank with a circulating pump and heater, sludge settling tank with filling pump and heater (optional), a water injector (optional), and a rotating arm for removing ash and non-combustibles (for vertical cyclone type incinerators) may also be fitted. When the incinerator is operated, the sludge burner is placed in the incinerator to burn and dispose of sewage, sludge, and waste oil. An auxiliary oil burner is also fitted to ignite any garbage. Automatic controls are provided for the system to secure the igniter when the refuse starts burning without the need for the igniter. Combustion air is supplied with the help of a forced draught fan. A loading door, typically pneumatically operated, is provided to load the garbage. An interlock is also provided with the burner and forced draught fan, which trips when the loading door is put into the open condition as part of the incinerator's safety features. Solid waste is fed from the loading door, and the incineration process starts after the door is closed. Any liquid waste is fed into the system when the refractory of the incinerator becomes hot.

On completion of the incineration process, the incinerator must be allowed to cool down, and any residues such as ash and non-combustibles are removed by pulling the ash slide door. The rotating arm in the vertical cyclone type scrapes off the entire solid residue in the ash box, which can then be disposed of easily. During incineration, it is important to control the exhaust temperature, which should not be extremely high or too low. Elevated temperatures can lead to melting of metal and can cause damage to the machinery, whereas temperatures that are too low will not be able to burn the residue, sterilise the ash, and remove any odours. This temperature control can be achieved by introducing cold-diluted air into the exhaust stream at the point which is close to the incinerator discharge (Figure 19.4).

Figure 19.4 Typical marine waste incinerator.

Good practices for the incinerator

To keep the incinerator working efficiently, the following good practices should be adhered to: (1) Always keep the incinerator chamber inlet, outlet, and burner parts clean. A daily inspection should be conducted as part of the morning rounds. (2) Do not throttle the air/steam needle valve more than three-fourths of a turn closed. If the pressure increases above the defined limit, clean the sludge burner nozzle. (3) Do not turn off the main power before the chamber temperature is down below 170°C (338°F). Keep the fan running to cool down the chamber. (4) If, when operating the incinerator, any problems are experienced regarding high temperatures in the combustion chamber, flue gas, or control of sludge dosing, replace the dosing pump stator. (5) Do not transfer sludge to the service tank during sludge burning in a single tank system, as this can damage the refractory. (6) It is always recommended to heat the sludge overnight, without starting the circulating pump. Drain off the free water and start the sludge programme before performing the incinerator operation. (7) Never load glass, lithium batteries, or copious quantities of spray cans into the incinerator. Avoid loading substantial amounts of oily rags or filter cartridges, as these may damage the flue gas fan. (8) Inspect the cooling jacket at least every six months (open the cover plates) and clean as required with steam or hot water. (9) Always read the instruction manual, and never change any settings unless instructed by the manufacturer. Further to the points listed above, the following should also be noted: never incinerate metals such as soft drink and food cans, flatware, cutlery such as serving spoons or metal trays, hardware such as nuts and bolts, structural pieces, wire rope, chains, or glass such as bottles, jars, and drinking glasses, as loading glass into the incinerator will result in a rock-hard slag, which is hard to remove from the refractory lining. Moreover, flammable materials such as bottles or cans containing flammable liquids or gases, including aerosol cans, must not be incinerated. If a blackout is experienced during the operation of the incinerator, and the incinerator temperature is above 220°C (428°F) start the flue gas fan as soon as possible to prevent damage to the incinerator through accumulated heat in the refractory lining.

SHIP OIL POLLUTION EMERGENCY PLAN (SOPEP)

If you are a maritime professional working on board a ship, the SOPEP locker (room) is one of the most important locations you will be required to familiarise yourself with, together with the SOPEP Plan, within the first few days of joining the vessel. When an oil spill occurs at sea, it tends to spread over the surface of the water, leaving a thick sheen which is fatal to marine life and birds and is deleterious to the shoreline. The cost of cleaning up oil spills depends on the quantity and quality of the oil discharged and is calculated based on factors such as legal claims, money paid as penalties, the loss of the oil itself, any repairs and clean-ups, and, most importantly, the loss of marine life and the effects on human health. As prevention is better than cure, an oil spill prevention plan must be carried on board all ships. This plan is known as SOPEP or the ship oil pollution emergency plan. SOPEP is a requirement under MARPOL 1973/78, annex I, and applies to all vessels over 400 gross tonnes deadweight. For oil tankers, the deadweight tonnage is reduced to 150 gross tonnes. The ship's master is the ultimate authority for the SOPEP and carries responsibility for its implementation, along with the chief officer and chief engineer. The SOPEP sets out what actions the ship and its crew will take in the event of an inadvertent discharge of oil (whether accidental or as the result of maloperation, malfunction, or poor maintenance).

The essential SOPEP requirements for the ship are (1) the SOPEP must be written in accordance with the provisions of MARPOL 1973/78, annex I, regulation 37; (2) the approved plan provides a guide for the master and senior officers on board the vessel in the event an oil pollution incident occurs or the ship is at risk of such an incident; (3) it is a requirement under MEPC circular no. 256 that the SOPEP contains all the information and operational instructions related to the emergency procedures and SOPEP equipment provided in the SOPEP locker; (4) the plan must contain important telephone numbers, telex numbers, and the names of responsible contacts in the event of an oil pollution incident (or risk of one); (5) a recognised authority must approve the SOPEP, and there must be no changes or revisions made to the SOPEP without the prior approval of the vessel's Flag state administration; (6) if, however, any changes are made to the plan which are non-mandatory, these do not necessarily need to be approved by the Flag state administration; and (7) the shipowner and ship manager must update the appendices each time a non-mandatory change is made to the plan where Flag state administration approval is not sought.

The SOPEP contains (1) guidance, instructions, and contact details for personnel who need to be informed in the event of an actual oil pollution incident, or in the event of a near-miss oil pollution incident. In addition, the SOPEP contains (2) the duties of each crew member at the time of the spill, including the emergency muster and individual actions; (3) the SOPEP must contain general information about the ship, as well as her owners and operator(s); (4) steps and procedures for containing the discharge of oil into the sea using the SOPEP equipment on board; (5) an inventory of the SOPEP material provided for pollution prevention such as oil absorbent pads, sawdust bags, and booms; (6) onboard reporting procedures and requirements in the event of an oil pollution incident; (7) details of the competent authorities to be contacted and the reporting requirements in the event of an oil pollution incident; this includes authorities such as the nearest Port State Control, oil clean-up contractors, and salvors; (8) details and drawings of all fuel lines, oil lines, the positioning of vents, save trays, etc.; (9) the general arrangement of the ship including the location of all oil tanks together with their individual capacities, contents, etc.; (10) the location of the SOPEP locker and its contents; (11) guidance on the records to be made and statutory retention periods (for liability, compensation, and insurance purposes); (12) material for reference from essential organisations such as guidelines issued by the ICS, OCIMF, SIGTTO, INTERTANKO; (13) procedures for testing the various plans described in the SOPEP; (14) procedures for maintaining records as required by the competent authorities; and (15) details of when and how to review the SOPEP.

The general duties of the ship's crew under SOPEP are (1) The master. The master is in overall charge of any incidents related to oil spills and is required to inform the competent authorities as soon as the incident becomes apparent. The master is also required to ensure all crew members comply with the plan and that the necessary records are maintained without exception. (2) The chief engineer. The chief engineer is responsible for bunkering operations and is required to instruct their subordinates to prepare the SOPEP kit in advance of any oil-related operations, including sludge transfers, lube oil bunkering, and fuel oil bunkering. The chief officer should also ensure the master is informed and updated on the bunkering progress, as well as any incidents arising from the bunkering operation. (3) The chief officer. The chief officer always oversees the complete deck operations. This includes preventing any oil spills, or in the event of an oil spill, informing the master of the incident and taking all appropriate and reasonable steps to manage and mitigate the incident. (4) The officer of the watch. The duties of the officer of the watch are to assist the chief officer in the deck watch and to alert the chief officer and chief engineer of any potential oil spill situation. (5) The engineer of the watch (EOW). The EOW has responsibility for assisting the chief engineer during any oil transfer operations, which includes the preparation of SOPEP material and the readiness of firefighting equipment. (6) The duty rating(s). The responsibilities of the duty rating(s) are to assist and alert the duty officer and engineer of potential oil leaks and to immediately assist, by all means, the mitigation of oil spill incidents. They should also bring the SOPEP materials to the location of the incident to prevent oil from reaching the ship's railing.

Importantly, SOPEP not only provides details for preventing and fighting during an oil spill but also acts similarly to any other regulations covered by SOLAS to the extent that it details how to save the ship and crew in the event of any incident involving oil, such as fires, collisions, or listings.

This completes the chapter on oily water management. In the next chapter, we will look at and discuss wastewater management on ships.

NOTES

1. Brownian motion, or pedesis, is the random motion of particles suspended in a medium. This pattern of motion typically consists of random fluctuations in a particle's position inside a fluid sub-domain, followed by a relocation to another sub-domain.

Chapter 20

Wastewater management

Discarding the sewage produced on board a ship is one of many tasks that requires the absolute care and attention of the ship's engineers. As the sewage cannot be stored on the ship for exceptionally long, it must be discharged out to sea. Although sewage can be legally discharged into the sea, ships are not permitted to do so directly, as there are regulations regarding the discharge of sewage that must be followed. Ship sewage is that waste which is produced from toilets, urinals, and WC scuppers. The regulations which apply to the management and discharge of shipboard sewage can be found in annex VI of the MARPOL 1973/78 regulations. The regulations state that shipboard sewage may be discharged into seawater only after it has been treated and the distance of the ship is at least 4 NM (4.6 mi, 7.4 km) from the nearest shoreline. Where the sewage is untreated, the vessel must be a minimum of 12 NM (13.8 mi, 22.2 km) from the nearest shoreline. Moreover, the regulations state the sewage must not contain any visible floating solids nor should it cause any discolouration of the surrounding water body. Ships prefer to treat sewage before discharging overboard to save themselves from any type of financial penalty and media embarrassment. There are various methods available for treating sewage, though the most common method is the use of bioagents or aerobic bacteria. This system consists of holding tanks where the sewage is stored. Aerobic bacteria are pumped into the tank which then feeds on the waste. As the aerobic bacteria neutralises the harmful bacteria in the waste, the waste decomposes into a watery sludge. Although this method is considered an effective and environmentally friendly approach to handling shipboard waste, it is important to note that the bacteria produce harmful gases such as hydrogen sulphide (H_2S) and methane (CH_4), both of which are harmful to aquatic life. Whichever system or method is employed, the ship's Class must certify it as performing to the required standards and regulations.

DESIGN AND CONSTRUCTION OF THE SEWAGE TREATMENT PLANT

Sewage treatment plants on ships are standard from one vessel to another, in the sense that they consist of the following primary components: (1) *The screen filter*. The screen filter is a mesh fitted on the first tank near the inflow sewage valve. It helps remove non-sewage components such as toilet paper, plastic paper, and other solids which cannot be broken by the aerobic bacteria and would otherwise clog the system if allowed to pass through the filter. (2) *The biofilter*. The biofilter forms part of the aeration chamber which treats the sewage as it passes from the screen filter. The biofilter reactor, with the help of fine air bubbles supplied from the blower, disperses the sewage into a finer mist. This helps aerate the sewage with oxygen which in turn promotes aerobic bacteria growth. (3) *The settling or sedimentation chamber*. Once the sewage has passed through the biofilter reactor, it is fed into a separate chamber for settling. The mixture is then separated into two further tanks: one for high-grade water and the other for sediment. Clarification is achieved in the clarification hopper, which has sloping sides to prevent sticking and the accumulation of sludge. Any untreated sludge is settled at the bottom of the sedimentation tank, whereafter it is returned to the biofilter. The process then begins all over again. (4) *Activated carbon*. Activated carbon is fitted post the settling chamber to remove chemical oxygen demand (COD) through a filtration and absorption process. The carbon also helps to treat the biological oxygen demand[1] (BOD) and suspended solids. (5) *Chlorinator*. The chlorinator is fitted within the last chamber to treat the final stage water before discharging overboard. The chlorinator can be of a tablet dosing type or a chemical injection type. Inside the tablet-based chlorinator, clean water comes directly into contact with the chlorine tablets, forming a chlorine solution. The chlorinator comprises cylinders for filling the chlorinator with tablets. With the chemical pump type, a measured set quantity of sodium hypochlorite (NaOCl) is injected into the sterilisation/chlorination tank using a diaphragm-type reciprocating pump. (6) *The air blower*. There are usually two air blowers installed in the sewage treatment plant: one which supplies air (via air bubbles) to help aerate the sewage, and one which is kept on standby. The air blowers also help transfer the sludge from the sedimentation tank, supply air to the activated carbon tank, and back flush the sludge. (7) *Discharge pump*. The discharge pump is provided in a duplex format which are mounted on the last compartment of the sewage treatment plant. The discharge pumps are centrifugal and of a non-clog type which is coupled to their respective motors. The pump is run on auto mode, which is controlled by level switches installed in the sterilisation tank. The pump is usually operated on manual mode when taking out sludge from the compartments after cleaning the tank interiors. (8) *Piping*. The inlet pipe which carries the sewage to the plant is installed with a proper slope to prevent clotting and condensation. The sewage pipe is arranged in such a way that the inside holes are accessible for cleaning during maintenance. The overboard discharging outlet should be placed at least 200~300 mm lower than the lowest water level within the discharge tank. To prevent back flushing, the discharge pipe is provided with a non-return valve. (9) *Floats and level switches*. On most sewage treatment plants, there are three float switches – namely, the elevated level, low level, and high alarm level. These are fitted to the chlorination and sterilisation chamber. This chamber is also fitted with level switches to control the start-stop operation of the discharge pump.

> *Operation of the biological sewage treatment plant*. The basic working principle of the biological sewage treatment plant is bacterial decomposition of raw sewage. This process is achieved by aerating the sewage chamber with fresh air. The aerobic bacteria survive on this fresh air and feed on the decomposing raw sewage. Once the sewage has been fully treated, it is safe to discharge overboard. Oxygen is particularly important to the functioning of the biological sewage treatment plant because, if oxygen is not present, it will lead to the growth of anaerobic bacteria, which produce toxic gases that are hazardous to health. Furthermore, sewage that has been treated with anaerobic bacteria will produce a dark black liquid which causes the discolouration of water. As we mentioned earlier, this is not permitted under annex IV of the MARPOL regulations. Subsequently, with the biological sewage treatment plant, the

main provision is the introduction and maintenance of fresh air flows. The biological sewage treatment plant is divided into three chambers. The chamber is the aeration chamber. This chamber is fed with raw sewage which has been ground into small particles. The advantage of breaking the sewage into small particles is that it increases the respective area of each particle. This allows an increased volume of bacteria to attack the sewage particles simultaneously, which aids the decomposition process. The sewage is decomposed into carbon dioxide, water, and inorganic sewage. The air is forced through the diffuser into the air chamber. The pressure of air flow plays a significant role in the decomposition of the sewage. If the pressure is high, then the mixture of air and sewage will not take place properly and it will pass through without having any positive effect. To avoid this, the pressure must be controlled to around 0.3–0.4 bar. Once the liquid and sludge are mixed, it is passed to the settling tank from the aeration chamber. In the settling tank, the sludge settles at the bottom with a clear liquid forming towards the top of the chamber. The sludge, present at the bottom of the chamber, is not permitted to remain inside the settling tank, as this will lead to the growth of anaerobic bacteria. The sludge is then recycled with incoming sludge where it is mixed before passing once more through the aeration process. In the chlorination and collection chamber, the clear liquid that is produced in the settling tank is overflown and treated with chlorine disinfectant. This is done to combat the presence of e-Coli bacterium present in the liquid. To reduce bacteria to acceptable levels, chlorination is performed. Moreover, to reduce the levels of e-Coli further, the treated liquid is kept in the chlorine suspension for a period of at least 60 minutes. With some plants, disinfection may also be achieved with ultraviolet radiation. Once the liquid has been treated, it is discharged overboard or pumped into the settling tank depending on the ship's location. If the ship is in restricted waters or in proximity to a shoreline, the sewage must be discharged into the holding tank; otherwise, the sewage is discharged directly out to sea. A low-level switch will activate when the level inside the discharge tank reaches the low-level parameter.

Precautions for the efficient operation of the sewage treatment plant. The aeration blower is designed to run continuously as it helps the aerobic bacteria to grow and expand in number. For this reason, it is important to never switch off the blower, as this will cause the death of the aerobic bacteria. This will impact the efficiency of the sewage treatment plant; it can take as much as several weeks for the aerobic bacteria to reproduce to an efficient number. Never dispose of foreign substances such as cigarette butts, paper, rags, or female personal hygiene products into the lavatory, as these can block the pipeline or hamper the filtering operation of the treatment plant. Toilet tissue used on board ships should be free of vinyl, as this affects the growth of aerobic bacteria. To avoid killing the aerobic bacteria, only use approved chemicals and detergents when cleaning the lavatories, sinks, and showers. At the point where the sewage is pumped overboard, the pH level must be within a range of 6–8.5 and the nitrite content must not exceed 10 mg per litre of nitrogen dioxide (NO_2).

SPECIAL AREA REGULATIONS

We have already touched on the special area regulations stipulated under annex IV of the MARPOL regulations, but it is worth spending a few moments expanding on this very important aspect of wastewater management. As of 2022, only the Baltic Sea is recognised as a special area under annex IV. The discharge of sewage from passenger ships within special areas is prohibited, except where the ship has an approved sewage treatment plant installed which meets the nitrogen (N) and phosphorus (P) removal standards. In accordance with resolution MEPC.275(69), the discharge requirements for special areas under regulation 11.3 of the MARPOL annex IV for the Baltic Sea Special Area apply to (1) all new passenger ships from 1 June 2019; (2) all existing passenger ships, except for those specified in (3) from 1 June 2021; and (4) all existing passenger ships on route directly to or from a port located outside the special area and any existing passenger ship on route directly to or from a port located east of longitude 28°10′E within the special area, and where such vessels do not make any port calls within the special area.

MAINTENANCE AND CHECKS

An efficiently operating sewage treatment plant requires periodic maintenance and daily checks of the system. Failure to do so can lead to products that cannot be legally discharged out to sea, blockage of pipelines, and the malfunction of critical parts. There are several factors that aid in the smooth operation of the sewage treatment plant, which this section will now briefly discuss.

> *Routine checks.* During daily rounds, the pressure of the system should be checked it is within acceptable limits as set by the manufacturer. The airlift return should be checked to ensure the system is working properly. This is usually done by confirming the flow through the transparent plastic pipe which is present in the installation. If a clear sludge can be seen flowing through the tubes to the aeration chamber, this means the plant is working as expected. Over extended periods, the sludge content in the aeration tank will increase due to the recycling of the sludge from the settling tank and the introduction of fresh sewage. This sludge content or suspended solids are measured in mg per litres. The method for checking this content is to take samples using a conical flask provided by the manufacturer and filling it up to the 1,000 ml mark. The sample is then allowed to settle before a reading of the sludge content is recorded. The sludge content should not be above the 200 ml mark. If it is above the 200 ml mark, the tank should be emptied as soon as practicable. On some ships, this may be checked by filtering the sample through a pre-weighted pad which is dried and then re-weighed. Whichever method is used, the check must be conducted weekly. A new bio-pac should be added to the plant every week. The bio-pac contains the aerobic bacteria which is activated when mixed with warm water. The level of BOD is also checked to ensure it is no higher than 50 mg per litre. The sample is checked by incubating the sample at 20°C (68°F) and then oxygenating the sample. The amount of oxygen absorbed over a period of five days is measured. This is done to check the volume of oxygen required to achieve a full breakdown of sewage products once it has been treated with aerobic bacteria. The internal coating of the sewage treatment plant should be checked for signs of cracking and blistering. If any form of damage requires the affected tank to be emptied before any necessary repairs are conducted, special precautions should be taken before entering the tank, as it will contain toxic gases that cause suffocation. The presence of toxic gases should be checked using a Dragor tube.[2] When the toxic gas level inside the tank has been reduced to acceptable levels (usually through manual ventilation) entry may be made donning full personal

protective equipment. After completion of the repair work, the area around the tank and the repair team must be thoroughly disinfected. If the sewage treatment plant is fitted with ultraviolet disinfectant systems, as opposed to the standard chlorination system, the ultraviolet lamp should be evaluated and, where necessary, changed as recommended by the manufacturer. The high- and low-level limit switches should be checked for auto cut-in and cut-out of discharge to the overboard pump. Last of all, always ensure the standby sewage discharge pump is switched onto auto operation whenever the sewage treatment plant is operated.

In the event of a blockage in the sewage line, there is a connection for back flushing, which uses seawater. This is used to unclog the sewage pipelines; however, it should be noted that only the necessary valves are left open; otherwise, sewage may back flush into the lavatories and showers. As a general note, stewards are usually instructed to use approved chemicals provided by authorised suppliers such as Drew Marine and Unitor. Even so, it is important not to overuse even approved chemicals, as this can lead to the destruction of aerobic bacteria. Only ever use approved chemicals, and in the amounts specified by the manufacturer.

REDUCING MARINE POLLUTION FROM SHIPS

The marine industry contributes many thousands of tonnes of garbage and waste products every year from day-to-day operations. Sadly, much of this ends up in the world's seas and oceans causing horrendous marine pollution. The most potent indication of this problem is the colossal Pacific Garbage Patch, which is visible even from space. Nobody knows how much detritus and refuge has accumulated in this area, but one established fact is that as much as 20% of all marine pollution comes from ships. This is clearly an unsustainable problem that requires radical and innovative solutions. Despite that, the answer also lies in changing seafarer attitudes towards disposing of shipboard waste overboard. MARPOL has done more to combat marine pollution by forcing the marine industry to adapt and change the way the industry manages waste on ships. Unfortunately, however, we as a planet are still far off from achieving the ambition of a pollution-free maritime industry. To that end, even small adjustments to individual attitudes can make an enormous difference. Most of the waste and garbage that is generated on board ships consists chiefly of inorganic plastics, dunnage and packing materials, cleaning materials and rags, paper products, food wastes, the remains of paints, solvents, and chemicals. Proper handling of these waste products is critical to reduce, and prevent, further marine pollution. A top-down approach is often considered the most effective, where senior officers are seen to lead by example. Every effort should be made to reduce waste and to encourage recycling wherever possible. Companies should implement garbage reduction strategies, which are then proactively enforced by the senior officers. Any waste, such as plastics, metals, glass, batteries, medical wastes, oily rags, sludge, and waste oils which cannot be disposed of safely at sea should be incinerated or stored on board for later disposal onshore. Compactors should be used to reduce the volumes of plastics and other waste materials which can be compressed. Modern techniques which separate glass from mercury and metal should be followed on board. Finally, as far as is possible, every effort to reduce the production of oily waste and sludge should be made. We all share planet Earth and as such, we each hold a stake in its future. Few industries are as polluting as the marine industry, but this does not need to be so. Bold action, investment in technology, and acceptance of collective and personal responsibility are all key to managing the very fragile resource that is the world's seas and oceans. As professional seafarers, we owe it to ourselves, our colleagues, our families, and future generations to safeguard our world.

We have now completed part 4 of this book. In the next part, we will look at the main fire and emergency response systems in the ship's engine room.

NOTES

1. Biochemical oxygen demand (BOD) is a test to identify biological decomposable substances and to test the strength of the sewage. BOD depends on the activity of bacteria in the sewage. These bacteria feed on and consume organic matter in the presence of oxygen. BOD can also be defined as the amount of oxygen required by the microorganisms in the stabilisation of organic matter. The results are generally expressed as the amount of oxygen taken by a 1 litre sample (diluted with aerated water) when incubated at 20° for five days.

    BOD of raw sewage is 300–600 mg/l. IMO recommends BOD of less than 50 mg/litre 25 Q_i/Q_e mg/l (updated by MEPC159), and the chemical oxygen demand (COD) does not exceed 125 Q_i/Q_e mg/l. The test method standard should be ISO 5815 1:2003 for BOD5 without nitrification and ISO 15705:2002 for COD, or other internationally accepted equivalent test standards after treatment through the sewage treatment plant.
2. The Dragor tube is a special tube which analyses for various types of gases.

Chapter 21

Freshwater generation

The freshwater generator is one of the critical machineries in the ship's engine room. In fact, without the freshwater generator, it would be almost impossible for the ship to function at all. Freshwater produced by the freshwater generator is used for myriad reasons, including drinking, cooking, washing, and operating other machineries and systems which use freshwater as a cooling medium. Freshwater is produced on board using the evaporation method. The evaporation method uses two items which are readily available on every modern ship: seawater and heat. The freshwater generator works by converting seawater into freshwater by evaporating the seawater using any available heat source. The evaporated seawater is then cooled by fresh seawater, and the cycle repeats itself. In most cases, the heat source is taken from the main engine jacket water, which is used for cooling the main engine components such as the cylinder head or liner. The normal operating temperature of the jacket water is approximately 70°C (178°F). But, as we know, water will only evaporate at 100°C (212°F); therefore, the water must be pressurised to achieve the required temperature. To produce freshwater at 70°C (178°F), we need to reduce the atmospheric pressure, which is done by creating a vacuum inside the evaporation chamber. As a result of the vacuum, the evaporated water can also be cooled at a lower temperature, which makes the process much more efficient. Once the freshwater is cooled, it is collected and transferred to the freshwater tank. An alternative method, which is primarily employed on passenger vessels which have substantial freshwater demands, is reverse osmosis. We will cover reverse osmosis later in this chapter, but first, we will examine the arrangement and operation of the freshwater generator.

The main component of the freshwater generator is a large cylindrical body which consists of two sections or compartments. The first compartment is the condenser, and the second is the evaporator. The freshwater generator also has an educator which helps generate the required vacuum. The freshwater pump and ejector pump transfer water to and from the freshwater generator. Before starting the freshwater generator, we must check that the ship is not sailing in congested waters, in a canal or restricted waterway, and is a minimum of twenty nautical miles (23.01 mi, or 37.04 km) from the nearest shoreline. This is necessary as inshore waters tend to be contaminated with effluents from factories and raw or processed sewage which is discharged straight out to sea. We also need to confirm the main engine is running above 50 rpm. The simple reason for this is that at low rpm, the temperature of the jacket water will only be around 60°C (140°F), which is insufficient for the water to evaporate, even with the help of the condenser. Next, confirm the drain valve at the bottom of the generator is in the closed position. Open the suction and discharge valves of the seawater pump. This will provide the seawater for evaporation and cooling and to the eductor for the creation of the vacuum.

OPERATING THE FRESHWATER GENERATOR

To start the freshwater generator, begin by opening the seawater discharge valve; this allows any excess seawater to return to the sea after circulating inside the freshwater generator. Close the vacuum valve situated at the top of the generator. Now, start the seawater pump and check the pressure of the pump. The pressure should be ideally 3–4 bar. Wait for the vacuum to build up. The vacuum should be at least 90%, which can be seen on the generator gauge. On most models, the time taken to create a satisfactory vacuum is about 10 minutes. Once a satisfactory vacuum is achieved, open the valve for the feedwater treatment. This prevents the formation of limescale on the interior plates. Open the hot water (jacket water) outlet and inlet valves slowly to about halfway. Always open the outlet valve first, followed by the inlet valve. Slowly increase the opening of the valves till they are fully open. By now we should be able to see the temperature increasing as the vacuum begins to drop. Once the vacuum has fallen to about 85% it should be evident that evaporation has begun. Open the valve from the freshwater pump to the drain. If not already switched on, activate the salinometer. This is usually set to auto-start. Start the freshwater pump; this will begin the flow of freshwater out of the condenser. When the production of freshwater is in full operation, there should be a slight but noticeable drop in temperature as the vacuum begins to return to normal pressure. Sample the water coming out of the salinometer to verify it is not salty; also, check the reading of the salinometer. This is done to confirm the salinometer is working properly to prevent saltwater from contaminating the freshwater. The value of the salinometer should be below 10 ppm. Having checked the salinity of the freshwater, open the valve for the freshwater pump and close the drain valve. To stop the freshwater generator, close the jacket water inlet valves. The inlet valve is always closed first followed by the outlet valve. Close the valve for the feedwater treatment. Stop the freshwater pump. Switch off the salinometer. Stop the seawater pump (also referred to as the ejector pump). Open the vacuum valve. If required, close the seawater suction valve and overboard valve. This is not needed, as both valves are non-return type valves.

REVERSE OSMOSIS

Every ship is fitted with a freshwater production plant which produces freshwater from seawater. As we mentioned at the start of this chapter, there are two methods for producing freshwater: the freshwater generator and reverse osmosis. We have already covered the use and operation of the freshwater generator, so now we can briefly discuss the principles and process of generating freshwater through reverse osmosis. Reverse osmosis is a modern process which, like the freshwater generator, produces fresh potable water from seawater. Unlike the former, reverse osmosis does not use heat to desalinate the seawater. Instead, as the name suggests, it reverses the osmosis process. When a chemical solution is separated from pure water by passing it through a semi-permeable membrane (which allows the flow of water but not the chemical solution) the pure water flows through the membrane leaving the chemical solution behind. Reverse osmosis is the use of this phenomenon but in the reverse direction. This results in water being forced through the membrane from the concentrated solution toward the more diluted solution. This is achieved by applying osmotic pressure to the concentrated solution. The osmotic pressure of seawater is about 28 bar. But to overcome system losses and the fact that the seawater concentration increases as it passes through the length of the membrane, much higher pressures of around 40–70 bar, depending on the plant size, are required. A triplex plunger pump is commonly used to produce high pressure across the membrane. The membrane used has an exceptionally fine barrier of dense holes which only allows water molecules and gas atoms to pass through while preventing the passage of solutes such as salt and other impurities. The freshwater produced after this stage is treated with chemicals and ultraviolet treatment to make it drinkable and safe for other uses.

In this chapter, we have briefly examined the freshwater production system on ships. In the next chapter, we will look at the various types and designs of pipes, tubes, bends, and valves that marine engineers must know.

Chapter 22

Pipes, tubes, bends, and valves

Pipes, tubes, bends, and valves are among the most crucial components in any ship's machinery. Ships carry hundreds, if not thousands of miles worth of pipes along their lengths. In this chapter, we will very briefly examine some of the most common types of pipes found on ships, together with their associated bends and valves. Before we do, though, it is worth discussing the difference between pipes and tubes. From a technical perspective, the issue of pipes and tubing is a matter of nomenclature. On the one hand, a pipe is rigid and resistant to bending, whereas tubes, such as copper tubes and brass tubes, can be flexible. However, in many structural projects, tubes are equally as rigid. Pipes are classified by schedule and nominal diameter. For example, a 250 mm nominal diameter and schedule 80 pipes. Tubes are classified by outside diameter and thickness. For example, 10 mm copper tube 2 mm thickness. With pipes, all the fittings can be matched by nominal size and schedule. This means a schedule 40 one-inch pipe will have fittings specified by the same dimensions. These pipe fittings would not fit a 1in diameter tube. Pipes are always round or cylindrical. Alternatively, tubes may be square, rectangular, and cylindrical. Pipes start from ½ inch to exceptionally large sizes. Tubes, however, are of small diameter only. We may use a 10-inch pipe, but never a 10-inch tube. Tubes are used in applications where the outside diameter must be precise, such as in cooler tubes, heat exchanger tubes, boiler tubes, etc. Pipes are used to carry and contain fluids and have pressure ratings and hence are scheduled. In tubes the thickness increases in standard increments such as 1 mm thick, 2 mm thick. With pipes however the thickness depends on the schedule of the pipe and there is no fixed incremental increase in size. Pipe joining is more time-consuming, as it requires welding, threading, flanges with bolts, and so forth. On the other hand, tube joining is much faster and easier as flaring, brazing, and couplings can be used instead. One decisive point to note is that tube dimensions are actual dimensions. This means that a 1" tube will actually have an overall dimension of 1". Pipe dimensions are only nominal, which means the 1" schedule 40 nominal size pipe has an internal diameter of 1.049", an overall dimension of 1.32", and a wall thickness of 0.133". Tube fittings are compression fittings, which include ferrule and union nuts, flared fittings, biting fittings, or mechanical grip-type fittings. Pipe fittings on the other hand incorporate pipe-to-pipe butt welding, threaded pipe fitting connectors, flange-to-flange bolted fittings, and so forth. In the field of marine engineering, we use schedule 40 pipes for light duties and schedule 80 pipes for heavy duties. There are however many other schedules which have been incorporated in part due to improvements in metallurgy and in part due to requirements in increased pressure demand. The schedule of a pipe refers to its pressure rating. The higher the schedule the higher the pressure it can contain. Schedules are normally 5S, 10S, 10, 20, 30, 40S, 40,

60, 80 100, 120, 140, and 160. As the schedule increases the wall thickness of the pipe increases, which results in the internal diameter of the pipe deceasing.

BENDS AND ELBOWS

As we have discussed the differences between pipes and tubes, and the meaning of nominal diameter and pipe schedules, we can move on to discuss the several types of bends, elbows, and mitre bends. There is often doubt regarding the use of the terms 'bends' and 'elbows' on ships. For our purposes, however, we can refer to the following differences. A bend is a generic term for any offset or change of direction in the piping. It is a vague term that also includes elbows. An elbow is an engineering term, and they are classified as 90° or 45°, and short or long radius. Elbows have industrial standards and have limitations to their size, bend radius, and angle. The angles are usually 45° or 90°. All other offsets are classified as bends. Bends are made or fabricated as per the need of the piping; however, elbows are prefabricated as standard and are available off the shelf. Bends never consist of sharp corners, whereas elbows do. Pipe bending techniques have constraints as to how much material thinning can be allowed to safely contain the pressure of the fluid. As elbows are prefabricated, cast, or butt welded, they can be as sharp as right angles, with some return elbows being as much as 180°. These are called return elbows. Whereas the elbow is a standard fitting, bends are custom fabricated. In bends, as the pipe is bent and there is no welding involved, there is less pipe friction meaning the flow is smoother. In elbows, the welding can create some friction. In summary, the basic difference is the radius of curvature. Elbows have a radius of curvature between one to twice the diameter of the pipe whereas bends have a radius of curvature more than twice the diameter of the pipe.

Short-radius and long-radius elbows

As we know, elbows are classified as either long radius or short radius. The difference between them is the length and curvature as demonstrated in (Figures 22.1 to 22.8). A short-radius elbow will give the pipe a sharper turn compared to a long-radius elbow.

With long-radius elbows, the radius of curvature is 1.5 times the nominal diameter. In a standard elbow, the radius of curvature is 1.0 times the nominal diameter of the pipe. This means the long-radius elbow provides less frictional resistance to the fluid when compared to short elbows. Moreover, long-radius elbows create less pressure drop than short-radius elbows. Because short elbows are compact and can fit into smaller spaces and crevices, they are less expensive than long-radius elbows.

Figure 22.1 45° short-radius elbow.

Pipes, tubes, bends, and valves 287

Figure 22.2 90° short-radius elbow.

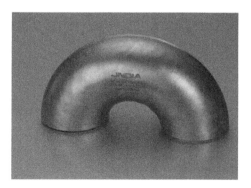

Figure 22.3 180° return elbow.

Figure 22.4 Mitre bend.

Mitre bends

Another type of bend is the mitre bend. A Mitre bend is a bend that is made by cutting the pipe ends at an angle, which are then joined. A true mitre bend is a 90° bend made by cutting two pipes at 45° angles and then joining them by welding. Similarly, three pipes cut at 22.5° will give a 90° mitre bend.

PIPE FITTINGS

Pipe fittings are special parts that are used for joining two pipes together. This may be to join a smaller size pipe to the existing piping or to regulate the flow of a fluid. There are many diverse types of pipe fittings, which are provided in Table 22.1.

Table 22.1 Different formats of pipe fittings

Nipples	The nipple is a pipe with male threads on each side to facilitate the joining of two pipes or fittings with similar female threads. It is made by cutting threads on both sides of a pipe with a die or suitable process. The nipples may be short or long.

Figure 22.5 Nipple.

Long Pipe Nipples	A long pipe nipple is like a standard nipple but is longer in length. It is used when there is more distance between the fittings or when a fitting like a valve is to be inserted at a distance.

Figure 22.6 Long pipe nipple.

Hexagonal Nipples	A hexagonal nipple is a pipe connector with male threads on both sides and a hexagonal nut in between for easy installation and screwing. Both sides are to be screwed into the pipe or fitted with reciprocating female threads.

Figure 22.7 Hexagonal nipple.

(Continued)

Pipes, tubes, bends, and valves   289

Table 22.1 (Continued)

Hexagonal Reducer Nipples	A hexagonal reducer nipple is a hexagonal-shaped nipple with two varied sizes of threads on each side. The purpose of the hexagonal reducer nipple is to connect the pipes of two varied sizes together.	 *Figure 22.8* Hexagonal reducer nipple.
Hexagonal Long Nipples	A hexagonal long nipple is used where the distance between the pipe and the fittings is to be joined. They are otherwise like the hexagonal nipples apart from the length.	 *Figure 22.9* Hexagonal long nipple.
Close Nipples	A close nipple is completely threaded and there is no unthreaded area. This means that there is no hexagonal nut of plain pipe for putting the spanner. These can be damaged by using spanners. Therefore, a special tool called a nipple spanner, which holds the nipple from the inside, is used for fitting.	 *Figure 22.10* Close nipple.
Couplings	A pipe coupling is a connector with female threads on both sides that allows two pipes or fittings with male threads to be joined. Couplings can be a normal pipe coupling or a hexagonal coupling. Pipe couplings are fittings that help to extend or terminate pipes.	 *Figure 22.11* Coupling.
Reducing Couplings	A reducer coupling is a coupling with two varied sizes of threads on each side. They can be a plain reducer coupling or a hexagonal reducer coupling. They are used to connect two varied sizes of pipes or fittings and sometimes also for flow control.	 *Figure 22.12* Reducing Coupling.

(*Continued*)

Table 22.1 (Continued)

Adapters	Pipe adapters are fittings that adapt to changes and are therefore used for joining diverse types of pipes such as a pipe to a hose. Adapters are used to extend or terminate the piping. They are also used to connect dissimilar pipes. There are male-to-female adapters, parallel-to-taper thread adapters, and pipe-to-hose adapters. Pipe adapters have a male or female thread on one side and an opposite-gender thread on the other side.	 Figure 22.13 Adapter.
Reducing Adapters	A reducing adapter is used for joining a pipe to a hose or tube, as well as to help in flow control. In doing so, it controls the pressure acting on the hose at the end. It is used in applications where a copper tube is joined to the main steam pipe for trace heating.	  Figure 22.14 Reducing Adapter.
Gauge Adapters	A gauge adapter is used for fitting pressure gauges and instrumentation fittings. They are also used for fitting pressure-measuring instruments such as pressure gauges, gauge cocks, and shut-off valves. The diverse types of gauge adapters are male to female, female to female, male to male, self-sealing nipples, left-hand–right-hand unions, union nuts with nipples, compression fittings with ferrule, swivel adapters, and so forth.	 Figure 22.15 Gauge adapter.
Hose Adapters	A hose adapter is used where a hose is connected to a pipe. It may be the termination of the piping, such as a garden hose at the end, or a high-rating hydraulic hose used to give flexibility and movability to the system.	 Figure 22.16 Hose adapter.
Reducing Bushings	A reducing bushing is a pipe fitting with both male and female threads that joins two pipes of different diameters.	  Figure 22.17 Reducing bushing.

(Continued)

Table 22.1 (Continued)

Pipe Caps	A pipe cap is a fitting to seal the end of a pipe. It acts like a plug but has female threads and screws on the male threads onto the end of a pipe or adapter.	 Figure 22.18 Pipe cap.
Pipe Plugs	A pipe plug is a fitting to close or seal the end of a pipe. The plug has male threads and screws onto the female threads on a fitting.	 Figure 22.19 Pipe plug.
Female Elbows	A female elbow is a pipe fitting that is used to change the direction of the piping and has female threads on both sides that allows for fixing pipes or fittings with male threads. They come in various angles such as 45° and 90°.	 Figure 22.20 Female elbow.
Male Elbows	A male elbow is a fitting with male threads on both sides and is used for changing the direction of the piping. They fix onto female threads of the fittings on both sides. They come in various angles such as 45° and 90°.	 Figure 22.21 Male elbow.
Street Elbows	A street elbow is different from the male and female elbows in the sense that it has a male thread on one side and a female thread on the other side. The advantage of using the street elbow is that it can directly fix to the pipe without connecting a nipple. They come in various angles such as 45° and 90°.	 Figure 22.22 Street elbow.

(Continued)

Table 22.1 (Continued)

Reducing Street Elbow	A reducing street elbow is a pipe fitting that has a male thread on one side and an unequal female thread on the other side. They are used for altering the direction of flow as well as for joining pipes of two varied sizes. They come in various angles such as 45° and 90°.	 *Figure 22.23* Reducing street elbow.
Female Tee	A female tee is a pipe fitting that joins one pipe to another pipe in a perpendicular direction. The female tee has female threads on all ends, and pipes or fittings with male threads fix into it.	 *Figure 22.24* Female tee.
Male Tee	A male tee is a pipe fitting that joins two perpendicular pipes together to make a "tee." It is like a female tee but has male threads on all ends and joins to fittings with female threads.	 *Figure 22.25* Male tee.
Street Tee	A street tee is a pipe fitting which joins one pipe to another in perpendicular. In addition, it has male threads on one end and female threads on the other two.	 *Figure 22.26* Street tee.

(Continued)

Pipes, tubes, bends, and valves 293

Table 22.1 (Continued)

Reducing Tee	A reducing tee has two openings of the same size and one of a smaller size. These are also used to join two pipes perpendicular to each other as well as to obtain flow control.

Figure 22.27 Reducing tee.

Cross Tee	A cross tee is a pipe fitting that joins four pipes at 90° each. There is either one inlet or three outlets or vice versa.

Figure 22.28 Cross tee.

VALVES

A variety of valves are used in the piping and machinery systems of ships, as per the need and flow pattern of the fluids. All shipboard valves share the same basic functionality of regulating the flow of a fluid through the pipage. Valves are used for every machinery system. Although valves are often referred to as efficiency-decreasing devices, as they reduce the energy in the liquid flow, their use is imperative in applications where flow limitation is required. Therefore, any aspiring or serving marine engineer needs to know the construction and working principles of the several types of valves used on ships.

Gate valve

To begin with, we will start with the most used valve, which is the *gate valve*. The gate valve is one of the simplest valves in terms of design and functionality. As the name suggests, the gate valve consists of a simple mechanism called the 'gate'. This is a valve disc, which performs the function of regulating the flow of fluid through the valve junction. It should be noted that the gate valve only has two settings: full-flow and zero-flow, and subsequently can only be operated in one of two positions. The valve provides a full-bore flow without initiating a change in flow direction. The gate valve is not suitable for operations

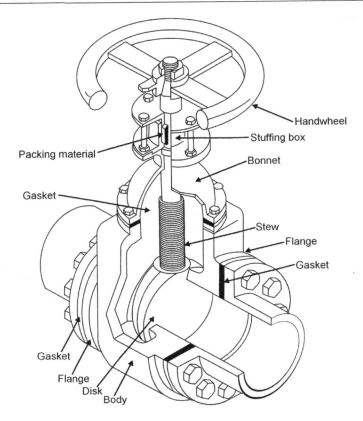

Figure 22.29 Gate valve cross section.

which require a partially open operation. The main parts of the gate valve are indicated in Figure 22.29.

The working principles of the gate valve are quite simple as they do not involve any complex mechanism. The spindle wheel, which is attached to the spindle rod, is rotated to move the 'gate' at right angles to the flow of the fluid. The screwed spindle works a nut which lifts the valve to open or close the 'gate' between the circular openings furnished with seats. The valves and seats may be either tapered or parallel to their facing sides. Gate valves are classified according to their internal operation and the stem type. There are two important types of gate valves with respect to the stem operation: (1) Rising stem type. With the rising stem gate valve, the stem has threads, which are mated with the integral thread of the yoke or within the bonnet. When the valve is operated, the stem rises above the actuator and the valve attached to the stem opens. (2) Non-rising stem type. With this type of gate valve, the gate or the valve disc is itself internally threaded and connected to the stem. As the stem thread mates with the disc, the valve will open or close without raising the stem as is the case with the rising stem type (Figures 22.30 and 22.31).

As the valve disc (gate) directly works against the flow of the fluid, the metal surface undergoes wear and tear. This often results in the valve leaking. Gland packing is used within the gate valve to stop any kind of water leaking through the space around the spindle. Over time, the gland packing may also become damaged. To prevent leaks, the gland packing should be replaced as per the preventative maintenance schedule. The valve seat of

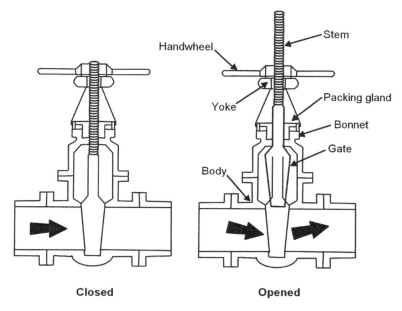

Figure 22.30 Rising stem type.

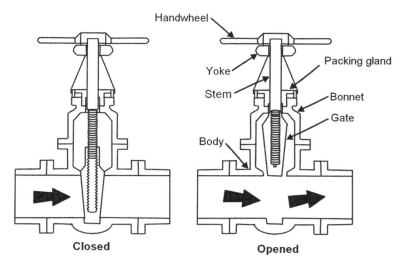

Figure 22.31 Non-rising stem type.

the gate valve can also cause problems and thus should be checked for clearances during the standard maintenance routine.

Globe valve

The glove valve is a type of valve commonly used on board ships in places such as the bilge suction lines. A linear motion valve, which regulates the flow of the fluid, the valve has a globe (bulbous) shaped body that houses the valve seat and the disc. The general

Figure 22.32 Globe valve.

arrangement of the globe valve consists of the valve seat and the disc arranged at right angles to the axis of the valve. The perpendicular movement of the disc, away from the valve seat, causes a space between the disc and the disc ring, which opens the valve. This characteristic provides good throttling ability to the globe valve, which is extremely helpful in regulating the flow of the fluid. The globe valve is used not only in starting and stopping the fluid flow but also for regulating the fluid flow (Figures 22.32 and 22.33).

There are three main designs of globe valves:

1. Z-body type valve
2. Y-body type valve
3. Angle valve

Z-body type valve. This is the simplest and most used design of globe valves and is favourable for applications involving water flow. The 'Z-body' shape is derived from the Z-shaped diaphragm or partition in the bulbous-shaped body. The stem and the disc move at right angles to the valve body, aligning with the horizontally arranged seat. The stem passes through the bonnet attached to a wide opening at the top. *Y-body type valve.* The Y-body type valve design is a solution to the problem of pressure drop which is commonly found in globe valves. With this design, the valve seat and stem are angled at approximately 45° to the valve axis. This results in a much straighter path for the fluid flow, providing improved pressure resistance inside the valve. Y-body valves are suited for high-pressure

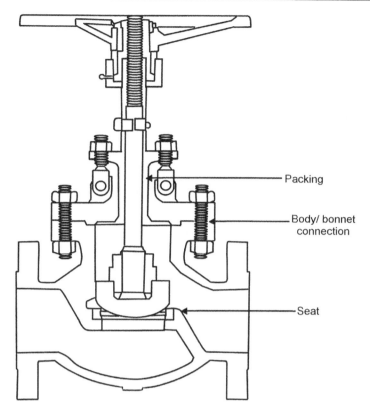

Figure 22.33 Global valve cross section.

applications where problems associated with pressure drop are common. However, in the case of throttling fluids, the pressure resistance might not work effectively, especially if the valve is too small. This is also because, with smaller valves, the flow passage is not as streamlined as those in larger valves. *Angle-type valve.* The angle globe valve is a modification of the basic Z-shaped glove valve. The inlet and outlet ends are placed at right angles and the diaphragm is a simple flat plate. The fluid flows through the valve body in a right-angle flow path and is discharged downwards in a symmetrical manner. These characteristics are extremely important in applications involving high pressure. However, in moderate conditions, the valve functions like a Z-shaped valve.

Globe valves can be used for both low and elevated-temperature applications. In low-temperature applications, the globe valve is installed so that the pressure is under the disc. This helps provide easier operation of the valve and protects the packing and prevents erosion to the seat and disc. For elevated-temperature applications, the valves are installed so that the pressure is above the disc. This prevents the steam below the disc from contracting upon cooling, which tends to lift the disc off the seat.

Both the valve seat and the disc are arranged at right angles to the axis of the valve. The seat can either be screwed into the valve chest or form part of the body. The seating can either be flat or mitred and the stem either threaded above or below the stuffing box. The spindle is held in the valve disc by a nut. Leakage is prevented by the stuffing box, which is packed with packing material and glands.

There are three main types of discs used in globe valves:

1. Ball disc
2. Replaceable composition disc
3. Plug disc

Ball disc. The ball disc is used for applications involving low-pressure and low-temperature systems. The ball disk is so arranged that it fits on a tapered, flat-surfaced seat. Though the ball disc can be used for throttling purposes, the efficiency may not be as strong as required. *Replaceable composition discs* are designed with a hard, non-metallic insert ring on the disc which helps in creating additional tightness. Such rings are used in steam and hot water applications as they help prevent erosion, which can cause damage to the globe valve from solid particles. *Plug discs* are used for applications which require excessive throttling. The disc provides better throttling than the ball and replaceable composite discs because of its long and tapered design.

The stem of the globe valve connects the disc in two ways:

4. T-slot construction
5. Disc nut construction

The only difference between these two construction types is that with the T-slot construction, the disc slips over the stem, whereas with the disc nut construction, the disc is screwed directly to the stem. The valve seat can form either an integral part of the valve body or be separately screwed into the body. Most valves have a backseat arrangement. This is a design function which provides a seal between the stem and the bonnet. The disc sits against the backseat when the valve is fully open. This helps prevent system pressure from building against the valve packing. The gland packing of the valve is subject to wear and tear and thus needs to be changed regularly as the sealing quality will reduce over time. Moreover, continuous opening and closing of the valve leads to damage to the metal of the valve seat. Lapping paste can be applied to the seat to ensure the surface remains smooth and clearances are removed.

In addition to the gate and globe valves discussed earlier, the third most common type of valve used on ships is the pressure relief valve. The relief valve is used in every machinery system which operates with high pressures. These high pressures often tend to exceed the predetermined limits and the overpressure created is the pressure just above the maximum allowable working pressure (MAWP) or designed working pressure. The relief valve operates through the action of a spring which is determined by an operating pressure. The opening pressure of the spring can be managed by an adjusting screw provided on top of the relief valve. The spring acts in the opposite direction of the pressure and thus during normal operation of the machinery, the spring tension will not allow the valve to operate. When the pressure acting on the valve seat increases to above normal and equalises with the force of the spring acting downwards, the relief valve will lift and release the excess pressure until a level of equilibrium is reached. The lifting pressure of the valve can vary from ~8% to ~15% of the working pressure for unfired systems, but this depends on the manufacturer's design. As soon as the pressure of the system returns to normal, the spring, which was set to lift at a particular pressure, will close the valve because of the spring tension, and the machinery or the system will return to its normal operation. Some valves are designed to have a full lift to release the excess pressure when the system is continuously operating in an overpressure condition. Such types of valves are called safety valves.

The relief valve is a critical part of the ship's safety system and is integrated directly into the ship's machinery. The various parts of the relief valve are as follows: (1) *The body*. The body of the relief valve is normally made up of cast steel. It incorporates all the parts such as the valve spindle, valve, spring, and seat. It must be strong enough to withstand the high pressure when the valve opens to release the excessive pressure through the body. (2) *Inlet and outlet connection*. The relief valve inlet is connected to the machinery or system, and its outlet connection can be opened to the atmosphere near the system only in the case of an air system driven through a duct outside the engine room normally for a steam system or connected to the inlet or to some reservoir for a hydraulic system. (3) *Diaphragm*. The diaphragm acts as a seal between the inlet-outlet connection and valve body so that media does not leak through the valve body when the relief valve operates. (4) *Valve seat*. The seat must be soft enough so that it should not damage the valve and yet durable enough for the operating conditions of the valve; otherwise, the media for which the valve is used will leak. The seat is normally made of stainless steel and coated with soft metal to prevent pressure erosion and corrosion. (5) *Valve*. The valve plays a key role in the controlled operation of the relief valve. Any malfunction of the valve will lead to leakage of media from the machinery or system. It is normally made of stainless steel. (6) *Spindle/plunger*. The spindle/plunger, also referred to as the valve stem, has a valve attached at the bottom on which the spring acts on top of it. The force exerted by the spring is transferred to the valve through the spindle. The material used for the spindle is stainless steel; (7) *Spring and adjusting nut*. The helical spring should have proper elastic strength so that the valve seat can open and close at the correct set pressure. Adjusting bolts are located on the top of the body. By rotating the screw, the lifting pressure of the valve can be adjusted. The adjusting screw and the spring are made of steel alloy.

When choosing the type and design of a relief valve, a variety of parameters must be considered, including the following: (1) *The required relief pressures*. The expected relief pressure is a crucial factor, which in turn will depend on the type of system and or machinery the relief valve is to be fitted. (2) *Fluid properties*. The chemical properties of the fluid should be considered for determining the best materials for valve construction. Each fluid will have its own unique characteristics and may react differently with different metals and materials. Care should be taken to select the most appropriate body and seal materials, as these will be in constant contact with the fluid. The parts of the pressure relief valve which will meet the fluid are referred to as "wetted" components. It is important to select the correct valve material and design appropriate piping for flammable or hazardous fluids to prevent inadvertent sparking or heat friction. (3) *Materials*. As discussed in the previous point, the fluid properties and system application will determine the selection of materials used for the construction of the pressure relief valve. Common pressure relief valve component materials include brass, aluminium, and various grades of stainless steel. The springs used inside the relief valve, which act as the driving component of the valve, are typically made of carbon steel or stainless steel. Brass may also be used as it is cheaper. Where there are weight constraints, aluminium is typically used depending on the type of fluid the pressure relief valve will handle. For hazardous and corrosive fluids, stainless steel alloys are a popular choice. They also operate well when the operating temperatures is high. Equally important is the compatibility of the seal material with the fluid and with the operating temperature range. (4) *Flow rate*. Once the relief valve is lifted, how much pressure must be released from the system is determined by the flow requirements. This is known as the flow rate. Piping arrangements, porting configurations, and effective orifices are designed based on the flow rate requirement. (5) *Size and weight*. With many applications, space and weight are limiting factors. For example, even high-pressure machinery like the air compressor will require small-sized pressure relief valves when compared to other machineries which are much larger but have a lower pressure relief requirement. It is also important to carefully consider the port (thread) sizes, adjustment styles, and mounting options, as these will influence size and weight. (6) *Temperature*. The materials selected for the pressure relief valve not only need to be compatible with the fluid but must also be able to function properly at the expected operating temperature. Here, the primary concern is whether the elastomer will function properly throughout the expected temperature range. Additionally, the operating temperature may affect flow capacity and or the spring rate in extreme applications.

In this chapter, we have discussed the four main types of valves found on board ships. In the next chapter, we will discuss the general emergency drills, alarms, and emergency systems for which all members of the ship's crew must be familiar, and for which the marine engineers are responsible for maintaining in good working order should the worst happen.

Part V

Engine room tanks and bunkering operations

Chapter 23

Main fuel, diesel, and lube oil tanks on ships

Modern ships are massive storage floating structures which are primarily used to store and transfer liquid or dry types of cargo from one port to another. A ship requires many tonnes of oil – both fuel and lubricating types, together with water to ensure the propulsion and other auxiliary systems operate efficiently. To store these different types of oils and water, ships are designed and built to incorporate many different types of tanks, some of which we have discussed already. In essence, these tanks help to store fuel oil, lube oil, hydraulic oil, potable water, freshwater, ballast water, and so forth. The size, number, type, and location of ship's tanks depends on many factors but the predominant are: (1) the size and type of ship; (2) size and type of propulsion plant, auxiliary engines, and the design of the ship's other machineries; and (3) where the vessel is expected to operate and the areas of passages (for example, on global service, in coastal waters, on inland waterways). The ship's tanks are spread throughout the ship from the bow to the stern, and from port to starboard. In this chapter, we will very briefly examine the role and function of the main types of tanks found on the majority of ships today, starting with the fuel and diesel oil tanks (Figure 23.1).

Figure 23.1 Typical bunker station.

FUEL AND DIESEL OIL TANKS

Bunker tanks. These are the biggest tanks in terms of capacity on board the ship. They are used to store fuel and diesel oil which is received in bulk during bunkering. The locations of the bunker tanks are normally outside the engine room, and they are generally wing or double-bottom tanks. Low-sulphur oil and marine gas oil is bunkered in separate dedicated bunker tanks to avoid mixing the fuels. This is important if the vessel is expected to enter an emission control area or special area where the use of high-sulphur fuel oils is prohibited. *Settling tanks.* Generally, more than two settling tanks are present and located on a ship as part of the bulkhead of the engine room. Oil from the bunker tank is transferred into the settling tank. The diesel oil settling tank may be located as a double-bottom tank in the engine room. Settling tanks for low-sulphur oil and marine gas oil are kept separate from other fuel oils for the same reasons as the bunker tanks. *Service tanks.* The service tanks on board ships are used to store and supply treated oil to the main engine, the auxiliary engine, and the marine boiler. The number of such tanks can be one or more. Fuel oil and diesel oil service tanks are normally positioned as part of the bulkhead of the ship's engine room. Low-service fuel oil and marine gas oil tanks are dedicated tanks to avoid mixing. *Overflow tank.* The overflow tank is provided for both the fuel and diesel oil systems in the engine room. They provide a receptacle for collecting overflowed oil from the bunker tanks. Return lines and leak-off lines may also connect to the overflow tank. It is a normal practice to have a common overflow tank for high- and low-sulphur fuels. *Emergency generator diesel oil tank.* Fuel for the emergency generator is supplied from a separate diesel oil tank. The capacity of this tank must be in accordance with the SOLAS regulations regarding the size and type of vessel. The location of the tank is in the emergency generator room, which is located outside the engine room.

LUBRICATING OIL TANKS

It is almost impossible to conceive of any machinery operating without the use of lubricating oils. For this reason, various grades of lube oils are stored on board ship. For this reason, the main types of lube oil tanks present on board are the *main engine crankcase (MECC) oil tank*. The MECC oil is stored in one or several tanks, and low-sulphur system oil is kept in separate tanks. There are no other settling or service tanks in the lube oil system as the oil is taken directly from the main tank. *Main engine cylinder oil tank.* The main engine cylinder oil is used inside the combustion chamber between the piston and the liner and is stored in the cylinder oil tank. The bulk oil is bunkered directly into these tanks. Low-sulphur oil is kept separate in different tanks. *Main engine cylinder oil daily tank.* The daily tank is located in the engine room. The oil is transferred from the storage tank to the daily tank. The capacity of the daily tank is as per main engine cylinder oil daily consumption. *Main engine turbocharger oil tank.* If the main engine comprises a turbocharger system with forced lubrication, a turbocharger lube oil storage tank is provided. This ensures sufficient lube oil is available to keep the turbocharger operating safely. *Main engine turbocharger daily lube oil tank.* A daily oil tank is provided in the engine room. Oil, as per the daily consumption of the turbocharging system, is transferred from the storage tank to the daily lube oil tank. *Auxiliary engine lube oil tank.* Auxiliary engines are four-stroke engines, therefore no separate cylinder lube oil is used. Only auxiliary engine main lube oil is bunkered and kept in this storage tank. One or several tanks may be present as per the vessel's requirement.

TANK INSPECTIONS

The tanks on ships, like every other part, must be regularly and appropriately inspected and maintained to ensure the structures are sound and any faults or defects are found before they turn into serious issues for the ship. Many of these tanks are 'enclosed spaces', compartments or holds where access is strictly regulated as the internal environment of the tank(s) is irrespirable. These tanks require special permission to enter, which comes in the form of a Permit to Work (PTW). PTWs are issued on a case-by-case basis, are timed, and specific for one tank, and one tasking only. Failure to comply with the terms of the PTW puts crew members at risk and must be followed without deviation. Because steel deteriorates over time, its condition needs to be regularly checked for signs of corrosion. If left unattended, the corrosion can eat through the body of the tank eventually leading to structural failure. If the tank contains flammable liquids such as lube oils or fuel oils, this presents a major fire and explosion hazard for the ship. Failed ballast and sludge tanks can cause marine pollution, which may result in the prosecution of the shipping company and the ship's engineers. Larger tanks that fail can cause flooding leading to a loss in stability caused by free surface effect. In summary, the importance of detailed and regular tank inspections cannot be overestimated or understated. In this section, we will look at some of the main inspections that should be carried out on the ship's tanks. Before we go into specific tank failures, it is worth summarising the effects of some of the more general defects that affect steel tank structures. These are corrosion, deformation, and fractures. (1) *Corrosion.* Sometimes referred to as 'material wastage', corrosion is the lead cause for structural deformations and fracturing. It is by far the most 'common' of all defects directly related to steel and its components. If left unattended, corrosion is a disaster waiting to happen; either by cargo or fuel oil contamination, structural loss, marine pollution, or, in the worst cases, even the loss of the ship itself. (2) *Deformation* is a sub-component defect which is caused by damage to the tank's steel plating or material failure. It may manifest as a change in shape or as a physical disfigurement of steel that is caused by implosion (due to vacuum build-up in the tank) or explosion, excessive dynamic moments (wave bending/loading), as well as static stress (for example ship hogging and sagging conditions) and strains on the steel structures, etc. It should be noted that deformations observed on the ship's hull are likely to induce structure deformations on the hull interior too. (3) *Fractures.* Fractures are caused by the propagation of cracks through the steel plating, which has been left unattended. Most of it occurs due to excessive stress concentrations on weakened steel plates throughout the tanks' structure. There have been many cases where inspectors have arrested ships, especially bulk carriers, where imminent cracks (mainly due to concentrations of stress) through the cargo areas have been observed. Welding defects are also known as potential causes of fractures.

As we have said earlier, entering a tank or enclosed space encompasses certain mandatory procedures, which should be followed. These are the issuance of a PTW in enclosed spaces, personal protective equipment suitable for the task and the tank, high-beam lighting, oxygen and or gas detection metres, and suitable means of communication. Once these preliminary requirements have been fulfilled, a competent officer should lead the inspection process. The inspection must be pre-planned, organised, and timely. The most effective way for performing tank inspections is to follow a checklist. This provides both a process to be followed and a record of the actions undertaken inside the tank. When carrying out a tank inspection, there are several key points to look for. These are

- Assessing the overall condition of the tank;
- Corrosion levels;
- Condition of 'sacrificial' anodes fitted inside the tank;
- Evidence of damage, cracks, and or deformation;
- Evidence of pitting corrosion and blister formation;
- Condition of the tank gauging system;
- Condition of the safety devices;
- Evidence of mud or sludge accumulation; and
- Condition of cargo equipment (if relevant).

Assessing the overall condition. Immediately upon entering the enclosed space, we can often determine the condition of the tank by considering the state of the accessways and the ladders, paint coatings, and by closely observing areas susceptible to corrosion such as around weld joints. Rungs, stepways, and ladders are often the first to deteriorate through oxidation. The competent officer should examine the material wastage throughout the access ways and related components. An overall study of the paint coating usually allows an experienced engineer to estimate how the tank has reacted to general corrosion. For easy identification of loopholes, the paint applied on the surface is generally light in colour. Thereby, recoated areas can be easily spotted and should be rechecked for signs of coating failure, or for scaling or pitting.

Condition of corrosion levels. General corrosion is the effect of non-protected oxidation that tends to develop homogeneously on the internal surfaces of holds or tanks which are uncoated. The corroded scale frequently breaks off, revealing the bare metal, which is susceptible to corrosive attack. In tanks and holds that have been coated, corrosion starts affecting the moment the coating starts to break down. Determining thickness reduction in the steel plates is difficult unless excessive shrinking has occurred. For example, corrosion on the inner surfaces of the liquid cargo tanks (on crude oil tankers) is mainly due to the mixture of corrosive gases, crude oil acids, and seawater (through crude oil washing). This, along with the fluctuations in temperatures within the tanks and structural flexing can, over time, shrinks the thickness of the steel plating and associated supports, ultimately leading to failure of the steel structure. Careful examination should be carried out in vulnerable areas, such as the vicinity around the sounding pipes and striker plates; openings for air vents and tank gauging; internal piping, including expansion joints, dresser couplings, and related fittings; joints; and clamps near to operational valves within the tanks, bilges, and tank top areas; the underside of hatch coamings and tank openings; bulkheads in general; joints associated with girders; and web frames.

Condition of 'sacrificial' anodes fitted inside the tank. Typically, sacrificial anodes are made from zinc, among other elements, and provide excellent preventative measures to protect against corrosion within the tanks. This is especially the case with ballast tanks. Due to their sacrificial nature, the anodes, over time, waste away. Hence, to maintain their integrity, the anodes should be checked and inspected closely for evidence of excessive degeneration. A record of material wastage should be maintained for future examinations. It is also important to determine whether the anodes are well secured to the brackets provided.

Evidence of damage, cracks, and or deformation. Adequate lighting in the tanks is necessary for the inspection work and for identifying deformations or surface dents. Shadows are one of the best indicators to highlight buckling or cracks within the tanks. However, this may not be the case for darker paint coatings (for example, coal tar epoxy), where the tanks must be lit up to the maximum to locate any defects. Deformations are generally not readily evident when viewed over a large area. To avoid missing potential defects, it is good practice to highlight each area using a high-beam torch by projecting it parallel to the tank's surface. Where it is difficult to identify defects in a straight line by the torch, a length of string or rope may be used for determining any obscured deformations on the surfaces. Buckling is another symptom of large deformations which can be caused due to a diminutive increase in loads. Permanent buckling may arise due to overloading weak structures, through corrosion or contact damage.

Evidence of pitting corrosion and blister formation. Pitting corrosion is often observed in the bottom plating of ballast tanks, especially near the 'bell-mouth', near the 'bell-mouth' in a liquid cargo tank, or next to the suction wells associated with the submerged pumps fitted within the tanks. Pitting corrosion begins with the local breakdown of coatings, which exposes the bare metal, and accentuates oxidation and galvanic reactions in the area. Blister formation is a common sight in areas where the surface preparation is inadequate before the application of paint coats or where the coating has failed to adhere to the surface. The inspecting engineer must be on the lookout for these bumps on the tank surfaces, as they may be symptomatic of greater decay beneath the top surface.

Condition of the tank gauging systems. Gauging systems that include gas measuring gauges, pressure gauges, temperature gauges, remote level sensing metres, sounding pipes, and striker plates should be checked for functional abnormalities. Rusting is often found underneath the tank top near the conduits that encompass the gauges. If possible, it is always good practice to try and clear out any debris (for example, mud and oil deposits) manually from the remote measuring sensors before attempting to operate them. During the inspection process, physically testing the 'remote' gas measuring device may be worthwhile. The gauges fitted inside or outside the tank must be calibrated during significant inspections (i.e., during drydock) or at intervals defined by the manufacturer.

Condition of the safety devices. The safety devices fitted to the tanks are critical for providing the operator with a remote indication of any hazards that may be manifesting in the system. For instance, water ingress into the bilges of the cargo holds may be due to the cargo sweating. The importance of such devices is high and should be regarded as a priority for visible examination. Given the bilge high-level and low-level alarms of critical spaces (such as in the ship's chain locker, dry cargo holds, and void spaces) are infrequently used, these should be manually tested and scrutinised closely for operational deficiencies.

Condition of mud or sludge accumulation. Accumulation of mud and oily sludge in the tanks is highly detrimental in terms of hiding serious defects and also promotes the development of structural deterioration beneath the horizontal and parallel surfaces.

Therefore, it is strongly advised to remove any excess debris prior to any tank inspection; this means washing down the oil tanks enough to be able to visibly locate defects or physically hosing down mud accretion in the ballast tanks. This also aids in identifying any bottom shell pitting corrosion or deformations.

Condition of cargo equipment. Cargo equipment within the tanks includes heating coils, cargo pumps, crude oil washing machines, remote gauging systems, temperature and heat sensors, etc. A leak test using compressed air or steam should be carried out on the heating coils. Moreover, the pipework and steam traps within the tanks must be thoroughly inspected for faults and leaks. The inspecting engineer should also physically ensure the optimal operation of all the cargo equipment fitted internally. This could be done by remotely operating the system from a control room, with feedback confirmed from within the tank. Any irregularity in the equipment's operation must be recorded.

In addition to the aforementioned, several other defects and issues should be looked for during the tank inspection, which includes the following:

- Ballast tanks that are bordering the hot engine room spaces
- Ballast and void tanks neighbouring the heated fuel oil and cargo tanks
- Tanks that are in the vicinity of areas where vibration levels are high
- Side shell spaces between the loaded and light draughts
- Tanks adjacent to external tug contact points
- Spaces in the forward part of the vessel, especially to be considered after heavy weather

Therefore, to detect and identify where a fault has occurred in the enclosed space, most of the aforementioned factors need to be considered. For evidence and record keeping, using an intrinsically safe camera (or any camera with a certified explosion-proof housing) is highly recommended. Once the inspection is complete, all findings and observations must be entered into the ship's official Tank Inspection Record. If deemed necessary, a copy of the record should be sent to the vessel's onshore manager or technical superintendent. It is strongly advised to keep a correspondence file with records of all subsequent modifications or repairs carried out within the tanks.

TANK CLEANING

Ships use heavy fuel oil, which has very high viscosity. When stored in fuel tanks, this oil tends to stick to the inside of the tanks forming layers of semi-solid residues. Moreover, many impurities of the oil settle down and cling to the surface of the tanks. It is therefore imperative that the fuel oil tanks are cleaned on a regular basis. Generally, fuel oil tank cleaning is performed during dry dock or whenever the inspection of the fuel tanks is due. In addition, internal tank cleaning is also carried out for surveyor inspections or whenever there is any work to be done inside the tanks such as fixing cracks in the fuel tank or attending to leaking steam lines. When cleaning the tanks, various safety precautions must be followed due to the presence of volatile and flammable vapours. Though not exhaustive, the steps laid out in the following section illustrate the general preparations to be taken before cleaning the ship's tanks.

Preparations to be done before cleaning the ship's tanks

The following steps are recommended to be followed prior to starting the tank cleaning process. Empty the tank as much as possible; strip the tank by trimming the ship forward or aft depending on the location of the suction valve outlet. When the ship is going in for dry dock, the keel plan should be sent to the shore facility so that they do not put any keel blocks in the way of the plug in the bottom shell plating. The tank must be properly ventilated to remove any flammable vapours. Ensure the steam connections are closed and signs and placards are prominently displayed to prevent the opening of the steam valves during cleaning. Check the tank for the presence of flammable vapours. Check the tank's oxygen content using an oxygen analyser. Drain off any leftover oil by removing the tank plugs; the location of plugs can be found in the shell plating diagrams. In most cases, the plugs are covered with cement and in line with the shell plating. Complete the enclosed space entry checklist and double-check nothing has been missed or left out. Once these steps are complete, the cleaning of the tank can begin.

During the tank cleaning

Entry is only to be made inside the tank if the oxygen level is at a minimum of 21% by volume and all flammable vapours have been vented. One person should always remain on standby outside the tank manhole door, remain in constant communication with the individual inside the tank, and feedback to the duty officer. If any hot work is to be performed inside the tank, the Port State Authority must be informed prior to commencing the hot work. The hot work may only begin once the Port State Authority has given its permission. In the event hot work is to be carried out, a fire line must be brought inside the tank, in addition to a portable fire extinguisher. The tank is then manually cleaned using soft-hair brushes, rags, and non-flammable detergents. The oxygen content must be continuously monitored; in the event the oxygen content alarm indicates a low level, the space must be evacuated immediately. Remember to remove any equipment when exiting the tank.

After the tank cleaning

Once the tank cleaning task is complete, ensure no tools are left inside the tank, as these may get stuck in the valves or damage the transfer pump. The location of any crack repairs should be checked for leaks. If the task was to repair a steam leak, the coils need to be checked for steam leaks. In the case of cracked or plate renewals, the tank must be pressure tested and checked for leaks. If the repair was major, it should be inspected by a class surveyor before being returned to operation. Close the manhole after the inspection, repairs, and cleaning are complete. Remove the signs and placards. Finally, close out the PTW.

TANK MAINTENANCE

As we know, the ship's engine room is provided with various large and small tanks. Of these, the larger tanks are used daily for the storage of oils, whereas many of the smaller tanks, which are of equal importance, are used for storing waste residues from machinery and other systems. Tanks such as fuel oil sludge tanks, lube oil sludge tanks, scavenge drain tanks, fuel oil filter tanks, dirty oil tanks, and lube oil drain tanks are all essential tanks that need to be kept in good condition. This requires constant oversight and maintenance. The first and foremost aspect of the maintenance of these tanks is that they should be sounded daily, especially so during heavy seas. The ingress of the under-piston scavenge drain tank should be checked daily to assess whether more content is coming into the tank when the vessel is underway or in port. Regular checks should also be done on the other tanks, such as the sludge tank in the ship's engine room. One way to do this is to check if the purifier malfunctions as more sludge is pumped in. Water can enter the under-piston space if the liner rubber rings are leaking or, in rare cases, when there is a hole or crack in the water-cooled turbocharger casing. When opening the under-piston spaces for cleaning, it is very important to check the spaces before cleaning so that their condition can be assessed indicating any blow past, cylinder oil excess consumption, stuffing box leakage into the space, partially burnt fuel leaks, water ingress, blocked drains, etc. If any of these problems are identified, immediate action needs to be taken to rectify the source of the malfunction. It is also important to physically trace out what pipelines in the ship engine room are feeding these small tanks (which are listed in the International Oil Pollution Prevention (IOPP) certificate), which pipes lead out from the tank, which pipes go to which valve and pump. It is also good practice to lay catchment trays below the fuel oil and lube oil transfer systems and to keep them clean and painted, and their drains clear up to the tank where they lead to. It is important to check that the sludge pump does not run dry to prevent pump damage. The scavenge drain tanks should be checked at regular intervals to prevent the contents from spilling out into the engine room bilges or through the air pipes on deck. On some ships, the sludge tank in the ship engine room is located directly below the purifiers, and thus there is no blockage in the discharge of sludge content. However, on ships where the pipe from the sludge tank leads down with a few bends, it becomes necessary to clear the pipe, especially when operating in cold areas when there is no steam tracing line attached to the pipe. Similar difficulty may be experienced when transferring sludge when the suction and discharge pipes are partially clogged due to thickened sludge. External heating by steam and taping of the pipes may be required to loosen up the sludge content.

The ship's tanks are a vital part of the ship's infrastructure, without which the ship cannot function. It is the responsibility of the ship's engineers, in conjunction with the deck department, to oversee and manage the safe operation and maintenance of these tanks. Whilst this chapter has not been exhaustive, it should hopefully provide sufficient information for the reader to recognise just how integral, though often ignored, the tanks are. In the next chapter of this book, we will look at the bunkering process, or the procedure for loading fuel oil and lube oil on board.

Chapter 24

Bunkering operations

Fuel oil bunkering is a critical operation on board ships which involves pumping oil into the ship's tanks without causing any overflow. Bunkerage may be fuel oil, sludge, diesel oil, cargo, and so forth. Due to the nature of the product, bunkering procedures are exceedingly dangerous and pose extreme hazards and risks to the ship and its crew. The word 'bunker' originates from military parlance and is used to define an area to store and safeguard personnel and supplies (such as fuel, ammunition, food). It was derived from the Scottish word 'bunk', which means a reserved seat or bench. In the shipping industry, the word 'bunker' is used for fuel and lube oils, which are stored on a ship and used for machinery operation only. If a vessel is carrying marine fuel or lube oil for discharge in another port, it is not called 'bunker', being 'cargo' instead. If the vessel is carrying the oil to transfer it to another ship for use in its machinery, it is referred to as 'bunker', and the operation performed to transport the oil is known as 'bunkering'. Hence, bunker fuel or bunker oil on a vessel is marine fuel or lube oils which are carried in separate storage tanks and are popularly known as bunker tanks for sole consumption by the ship's machinery. The different types of bunkers which are supplied to a commercial or passenger vessel may include any of the following:

- Heavy fuel oil bunker
- Diesel oil bunker
- Marine gas oil bunker
- Lube oil bunker
- LNG fuel bunker

The bunkerage can be supplied to the ship in different ways. The mode or method may vary depending on the grade or type of fuel being delivered to the vessel, and there may be different types of bunkering facilities which supply the required marine fuel or lube oil to the ship. A small barge or ship carrying bunker fuel can be used to transfer marine fuel oil (such as heavy fuel oil) to the vessel. If the quantity of oil is small (e.g., lube oil, liquefied natural gas, or marine gas oil), it can be supplied directly to the vessel using quayside lorries. The bunkering procedure can be divided into three important stages:

1. *Preparation.* This entails preparing the vessel for the bunkering operation and involves the readiness of bunkering equipment, storage tanks, and bunkering safety.
2. *Performance.* Performing the bunkering operation in real time as per the predetermined procedure and receiving the marine fuel in accordance with the bunker plan.

Figure 24.1 Ship bunkering.

3. *Closeout.* Closing out the bunkering operation with absolute safety and ensuring the correct amount and quality of bunker fuel has been received on board from the bunkering facilities (either a bunker ship or quayside lorry).

The quantity of fuel oil to be received is determined in consultation with the master and the navigation officer. In accordance with the planned passage, adequate reserves of oil are brought on board, usually to last between three and five days. In the event of a long passage (for example, across the Atlantic or Pacific Oceans, where it is not feasible to refuel on route, additional bunkerage must be accounted for). The tanks, into which the oil is to be bunkered, are discussed with the chief officer to ensure proper draught and trim of the ship is maintained, and also to avoid mixing oils as much as practicable (Figure 24.1).

OIL BUNKERAGE PROCEDURES

In preparation of bunkerage, and before the bunkering operation commences, the chief engineer must calculate and check which bunker oil tanks are to be filled once they have received confirmation from the shore office about the amount of fuel to be accepted. It may be necessary to empty some tanks and transfer any leftover oil from one tank to another. This is required to prevent the mixing of two oils and prevents incompatibility issues between the existing oil and the new oil. The sounding of other fuel storage tanks (i.e., those not being used in the bunkering operation) should also be taken to maintain a record of the quantity of fuel already present on board. This helps the ship's officers in the event any valves are found leaking. Finally, a meeting should be arranged and held between the ship's staff to be involved in the bunkering process. This meeting should include clarification of the following:

- Which tanks are to be filled
- Sequence order of tanks to be filled
- How much bunker oil is to be taken
- Bunkering safety procedures
- Emergency procedure in the event an oil spill occurs
- Responsibilities and duties of each officer explained

Once the responsibilities and duties have been allocated, an accurate sounding of the tanks must be taken and recorded. All deck scuppers and save trays are plugged to prevent oil from leaking overboard. An overflow tank must be provided in the engine room. This is connected to the bunker tank and bunker line. It is important to ensure the overflow tank is kept empty for the transfer of excess fuel from the bunker tanks. Adequate lighting at the bunker station and sounding positions must be provided, together with clearly visible *No Smoking* notices at and around the bunkering station. Onboard communications, signs, and signals to stop the operation between the people involved in bunkering should be confirmed. A red flag and or light is present on the masthead. The opposite side bunker manifold valves must be closed and blanked. The vessel's draught and trim must be recorded before bunkering commences. All equipment in the SOPEP locker should be checked and kept near the bunkering station. When the bunker ship or barge is secured to the ship's side, the person in charge on the barge must be advised of the bunker plan. The bunker supplier's paperwork is then checked to confirm the oil's grade and density as per the specification provided by the chief engineer. The pumping rate of the bunker fuel is agreed with the bunker barge or bunker lorry. The hose is then connected to the bunker manifold. The condition of the hose must be checked properly by the ship's staff. If it is found to be defective or otherwise unsatisfactory, the chief engineer must be notified immediately and the bunkerage operation halted. Most bunker suppliers use their own crew to connect the bunker oil pipeline coming from the bunker barge or lorry to the receiving vessel. The ship's staff are strongly advised to recheck the flange connection to eliminate any risk of leakage. Once the connection is made, the chief engineer will ensure all the line valves which will lead the bunker fuel to the selected bunker tanks are open, whilst keeping the main manifold valve shut. For safety, it is critical that proper communications between the bunker barge and the receiving vessel are established and followed scrupulously. In the event of any leaks, all radio communications must cease immediately with manual communications (flags and hand signals) used instead. Most bunkering facilities (ships, barge, terminals, lorries, etc.) provide an emergency stop switch which controls the bunkering supply pump. Ensure this is checked before commencing the bunkerage operation. Once all these procedures are complete and the master and chief officer are satisfied, it is safe to proceed; the manifold valve is opened for bunkering.

At the start of the bunkerage, the pumping rate is kept low; this is done to check that the oil is coming to the tank to which the valve is opened. The ship's staff must track the sounding of the selected bunkering tank and the other tanks not involved in the operation, to ensure the incoming oil is being fed exclusively to the selected tank. After confirming the oil is feeding the correct tank, the pumping rate is increased accordingly. Generally, only one tank is filled at a time, as gauging more than one tank increases the chances of overflow. The maximum allowable limit to which the receiving tank can be filled is 90%. When the tank level reaches the maximum level (or close to it), the bunkerage barge should be instructed to change the flow rate to a low pumping rate to top up the tank. At this point, the valve of any other receiving tank may be opened. During bunkering, soundings are taken regularly, and the frequency of sounding is increased when the tank is nearly full. Many vessels have tank gauges which show the tank level in the engine control room, but this should only be relied on if the system is working correctly. The temperature of the bunker oil must also be checked; generally, the barge or bunkerage supplier will provide the bunker temperature. Temperature is a critical parameter, especially for bunker fuel such as heavy fuel oil, as any deviation in the temperature value may lead to a shortfall in bunker supply. A continuous sample is taken during bunkering by way of the sampling cock located at the manifold.

Once the bunkering operation is complete, it is a general practice to air blow the bunkering supply line to discharge any oil trapped in the pipelines. At this stage, ensure the sounding pipe caps are closed and maintain a watch on those storage tank vents which are at their maximum limit. Avoid opening the bunkering supply line connecting the bunker barge and the receiving manifold. In the event of any discrepancies, the supplier may agree to compensate the shortfall and resume the bunkering operation. Again, the ship's draught and trim are checked. Soundings of the tanks bunkered are carried out. The volume bunkered should be corrected for trim, heel, and temperature correction. In general, for each degree of increase in temperature, the density should be reduced by 0.64 kg/m^3. During bunkering, no less than four samples should be taken. One is kept on board, one is provided to the bunker barge, one is submitted for shore laboratory analysis, and one for Port State Control. Once the master and chief engineer are content the bunkerage operation is complete, the chief engineer will sign off on the bunker delivery note (BDN), which confirms the quantity of bunker received. If there is any shortfall of bunker received the chief engineer can issue a note of protest against the barge/supplier (in the event the deficit is disputed by the bunker supplier). Assuming the volume of bunkerage is correct, the hose connection is disengaged from the manifold and removed. The chief engineer will record the bunkerage operation in the oil record book together with the BDN. The new bunker should not be used until the report from the onshore laboratory is received, confirming the new bunker is satisfactory for use.

LNG BUNKERING PROCEDURES

With the stringent sulphur regulations having come into force in 2020, ships operating with liquified natural gas (LNG) fuels have become increasingly popular. Many different types of ships, including cruise and container ships, are fitted with engines which can burn LNG fuel as these reduce harmful emissions from the ship's exhaust. LNG fuel has very different properties when compared to conventional heavy fuel oil or low-sulphur fuel oil. Due to its cryogenic temperatures, which enables LNG to be stored and transferred from one location to another, the procedure for transferring LNG fuel into the ships requires a different approach compared to other fuel oils. For LNG fuel bunkering, it is important to follow the regulations laid down by local port authorities. This often means securing special permission in advance. The International Code of the Construction and Equipment of Ships Carrying Liquefied Gases in Bulk (IGC Code)[1] must be followed by the bunker ship carrying the fuel. For the ship which will receive the LNG bunker fuel, it too needs to follow the *International Code of Safety for Ship Using Gases or Other Low-flashpoint Fuels* (IGF Code). The master is required to check the weather and tide forecast for the area where the LNG bunkering operation is to be carried out. As far as possible, the operation should be conducted in daylight. When it is a ship-to-ship operation, the bunkering station and nearby areas of the ships must be treated as an explosive classified (EX) area during the bunkering period for both ships. The LNG bunkering operation comes with its own set of issues and hazards such as the following: (1) *Leaks and accidental spills.* Like any other fuel used on board ship, the chance of leaks and spills of LNG is omnipresent. (2) *Cryogenic hazards.* When LNG is carried as a fuel, it is stored at very low temperatures (approximately −162°C or −259.6°F). In the event of direct human contact, it can cause cryogenic burns. (3) *LNG fire and explosion hazards.* LNG is transported in a liquid state so that it can be carried in the maximum quantity possible. This also means that it can only be ignited when in a gaseous state, and the concentration of gas vapours is within its flammable range. Subsequently, there are five explosive hazards which can manifest during the carriage and transfer of LNG fuel: (a) *Flashfire.* This occurs when a cloud of gas burns in an open area without any increase in pressure. (b) *Jet fire.* This occurs when the LNG fuel is released at high pressure from a pressurised vessel. As the velocity of the fuel is extremely high, substantial structure damage can be caused to the pressure vessel and surrounding infrastructure. (c) *Pool fire.* As the name suggests, pool fires occur when the spilt LNG evaporates into a vapour, which then ignites either on land or on water. (d) *BLEVE.* Boiling liquid expanding vapour explosions are a dangerous situation which occur when LNG in closed containment is heated up. Any rupture of the containment system will lead to an explosion. (e) *Rapid phase transition (RPT).* When the LNG contacts another source, leading to heat transfer, this can cause a phase transition from liquid to vapour at an accelerated rate. The heat generated by the transition can ignite the vapour, leading to an explosion. (3) *Bunker line contamination.* The bunker lines are open to the atmosphere. Should they become crusted with salt or contaminated with water or CO_2, the contaminates can freeze, causing ice to block the lines. (4) *Trapped LNG.* Once the bunkering operation is complete, it is possible that some LNG fuel may be left in the pipe, which will convert and expand into a vapour. This can cause high pressure to develop inside the pipe leading to rupture. (5) *Rollover.* Rollover occurs when there is a rapid release of LNG vapours from the storage tank. It is caused by the stratification of vapours when two separate layers of different-density LNG develop inside the tank. The main cause for this stratification is temperature differences in the tank surface at the tank top and tank bottom. (6) *Oxygen deficiency.* If LNG is spilt in a confined space, it can lead to oxygen depletion as the liquid vapourises, displacing the air present in the chamber. This can lead to suffocation and death.

Hence, the crew handling the LNG bunkering operation must be acutely aware of the different hazards related to bunkering LNG fuel. They must be trained and competent and follow detailed instructions. These instructions must be kept or displayed prominently so that they are always readily available. The crew should undergo regular bunker training to maintain competence and currency. Moreover, the crew should not be reticent in using the emergency shutdown device or switch which must be present on both vessels (i.e., the bunker barge and the receiving ship). The emergency shutdown device or switch should be activated whenever there is a loss of either of the ship's power, in the event of a loss of communication between the two ships or their parties, the mooring lines between the two ships become loose or fail, weather conditions make the bunkerage operation too risky or hazardous or where there is a wave condition. LNG bunkering between ship to ship or shore to ship is achieved by using a QCDC coupling. QCDC is the abbreviation for 'quick connect/disconnect'. The coupling used in the LNG bunker hose is usually a dry-break coupling which comprises a female (hose end) fitting and a male (tanker fitting) which are connected disconnected to form a liquid and pressure-tight coupling. Dry-break couplings have a valve arrangement that shuts down the liquid flow prior to the disconnection of the male and female couplings. Hence, the coupling cannot be disconnected while in a live flow condition. Another advantage of the QCDC is that it has a fast connection and disconnection time which allows a quicker bunkerage operation. Nominal size couplings of up to six works for flows up to 650 m^3/h, with a maximum flow rate of 10 m/s. LNG bunker hoses must be clearly colour-marked to prevent the accidental coupling of different-sized hoses.

Prior to bunkering LNG fuel, the engine room and deck crew must prepare the vessel for the LNG bunkering operation. When the LNG bunker barge is in position, it is moored alongside. The pre-bunkering checklists of both ships should be signed off by both parties prior to the start of the operation. The master of the receiving ship and a representative of the LNG bunker barge must agree, in writing, to the transfer procedures, i.e., the maximum loading or unloading rates, modes of communication, etc. Once the mode of operation has been agreed upon, the bunker pipe is transferred and secured. If the vapour return line is provided by the bunker barge, this must be connected between the suppling and receiving vessels. The bunker pipe flanges are then checked for compatibility with the receiving manifold. Communication between both ships is then established; this includes manual emergency communications such as flags and hand signals. Warning and instruction signs must be posted around the operations area. The emergency shutdown must be made available on both vessels and checked for operation. The SOPEP locker should be checked and its contents confirmed with the spill containment equipment kept near the bunker station. Firefighting equipment on both ships must be checked and readied for use. Appropriate personal protective equipment must be worn such as full wrap-around glasses, cryogenic gloves, and safety boots. The hoses with couplings should not touch any unearthed parts prior to connection to avoid possible electrical arcing. The temperature and pressure of both the supplying and receiving tanks must be checked and confirmed within safe limits. Any variation in temperature may lead to vapourisation during the start of the bunker procedure, leading to pressure increases in the receiving tank. It is not compulsory for the bunkering vessel to provide a vapour return line, as the requirement for a vapour return line will depend on the receiving vessel. It may not be needed if

- The ship has a Boil-Off Gas Management System (BOGMS) installed on board,
- The vessel has the ability to cope with vapour pressure created by boil-off hydrogen (BOH), or
- The LNG is supplied in such a way that the boil-off gases are cooled down without generating vapours.

Once the LNG bunkering operation starts, it is important to keep checking the bunker station and filling tanks for the flow rate or pumping rate. Any abnormalities should be reported to the supplier, who will adjust the pump flow rates as necessary. The pressure rate will also depend on the type of filling mechanism, i.e., top spray or bottom flow.

The transfer rate should be within the determined limit throughout the entire operation. To prevent any accidents or leaks from occurring, maintain a vigilant watch over the filling tank condition, temperature, and pressure. Only authorised personnel should be allowed to enter the safety zone during the bunkering operation. All other crew members not directly involved in the bunkerage should keep a minimum safe distance from the bunker station. All naked flames must be extinguished. Continuous level monitoring of the ship's tanks should be carried out and filling should not exceed the maximum allowable limit as determined by the master and the supplier. In the event the high-level alarm is triggered, the emergency shutdown must be activated immediately and all bunkerage operations stopped. It is important to keep checking the mooring lines and tighten as necessary. Ensure the security and safety zones are continuously monitored and should the watchkeeping officer or crew detect any issues, the bunkering operation must be stopped immediately either through communication or by activating the emergency stop. Once the supplier has confirmed the quantity of LNG fuel has been transferred, the pumping will stop completely. If the tank is being topped up, the pumping rate will be reduced and then stopped. The LNG lines should be allowed to vapourise, displacing the remaining liquid back to the suppliers tanks. The manifold valve for the bunkering and vapour line on the receiving ship is shut and the pipes and hoses used for bunkering inerted. The bunker barge will confirm when the bunker hoses, manifold, and piping are sufficiently drained and free of residual LNG. At this point, the bunker pipe and vapour line hose are disconnected and returned to the bunker barge. The bunker hose manifold is blanked and all SOPEP locker equipment is returned. The quantity of bunker fuel exchanged is checked as per the tank levels and the BDN. Provided everything is present and correct, the chief engineer will sign the BDN signalling the cessation of the bunkering operation. The mooring lines are then opened and the bunker barge cast away. The operation must be recorded in the ship's logbook and the Port State Control and port authorities informed that the bunkerage operation is complete.

BUNKERAGE DISPUTES

Bunkerage pilfering and misrepresentation are ongoing major concerns for the global shipping industry. Thousands of vessels worldwide bunker every day. Even relatively minor incidents can quickly add up to tens of thousands of dollars lost to theft.

Bunkering disputes are very common between the receiving ship and the bunker supplier mainly because of differences between the quantity or quality recorded on the BDN, and the actual quantity or quality received. Whenever a dispute arises, a note of protest is issued by the receiving vessel to the bunkerage supplier. The note of protest sets out the details of the dispute, any evidence supporting the claim, and what actions are expected to satisfy the dispute.

Coriolis flow meter

One way of combatting this situation is the adoption of the mass flow metering system, commonly referred to as the *Coriolis flow meter*. This system helps improve existing measurement technologies and prevents fuel pilferage. The system is based on the Coriolis Principle, named after the French mathematician, Gaspard-Gustave de Coriolis (1792–1843). To understand how the Coriolis flow meter works, we first need to understand the

Coriolis Principle itself. The Coriolis flow meter works on the basic principle of *Coriolis Effect* or 'Force'. This is essentially a veering or deflection of a moving object when viewed from a rotating reference frame. In the Northern Hemisphere, a moving object will appear to deflect to the right (facing the direction of motion) and vice versa in the Southern Hemisphere. It should be noted that Coriolis Force is not actually a force per se, but as the objects tend to veer to the left or right of its path, it is assumed that a force has acted upon it to cause the deflection, because of Earth's movement beneath it. A good example of this would be to consider an aeroplane flying from Madrid to New York. Both places share a similar latitude so it would make sense for the pilot to set a course for due west (2700). However, because of the Coriolis Effect, if constant course corrections are not applied the plane will probably end up somewhere in Canada (i.e., right to the intended path in the Northern Hemisphere). Hence, mass flow meter are based on this principle. To simplify how the mass flow meter works, imagine a jet of water, for example from a pressure hose, is pointed straight ahead. The jet of water will move in a straight line but when viewed from a rotating reference, the jet will appear to be veering to the right or left depending on the hemisphere. If the same jet of water were now enclosed in a measuring tube rotating around a fixed point with its axis perpendicular to the direction of the flow of water, the measuring tube would appear to twist/deform due to the change in angular velocity caused by the Coriolis Effect. However, as it is not practical to make the tube rotate, instead, the tube is oscillated electromagnetically causing it to vibrate. This achieves the same effect as if the tube were rotating. This twisting/deformation of the tube results in a phase difference (time lag) which is registered by use of special sensors. This then forms the basis of the mass flow measuring system. The mass flow meter come in different varieties such as a straight tube type, twin tube type, bent tube type, each having its own advantages and disadvantages (though the discussion on these types is beyond the scope of this book) (Figure 24.2).

In summary, the introduction of Coriolis meter will most certainly reduce the amount of bunker quantity disputes, as these meters are less prone to tampering and can be used

Figure 24.2 Coriolis flow meter twin tube.

as an anti-pilferage tool by increasing transparency during a stem operation. The present method of manually gauging the tanks and calculations is not only significantly prone to error but is also time-consuming, especially in an event of a dispute. As technology advances and mass flow meter are perfected, vessel operators and bunker suppliers will see tangible benefits in the form of increased transparency, improved efficiency, and faster turnaround times.

FUEL OIL BUNKERING MALPRACTICES

Unfortunately, several dubious practices are often employed by bunker fuel suppliers during typical bunker stem operations. These malpractices are more prevalent in Asian ports than in North America or Europe, where standards are more stringent. Having said that no matter which part of the world the vessel is in, the importance of accurately measuring the barge fuel tanks before and after delivery is a crucial phase in any bunker stem operation. It is therefore very important that the vessel's bunker operation team methodically take the barge tank measurements, applying the correct trim and list before and after bunkering, recording the actual temperature of the bunker fuel before and after delivery, etc. Proper temperature measurement alone can save thousands of dollars in pilferage. Disputes can arise either by innocent mistake or through deliberate short supply by the barge. One common technique is to introduce air to froth up the fuel (called the *cappuccino effect*) or by submitting incorrect temperatures, and so forth. Also, when bunker fuel is being transferred from a refinery to a storage tank, then to the barge, which is then delivered to the vessel, there is substantial scope for errors and deliberate manipulations that can result in differences (sometimes quite significant) between the quantity claimed to be supplied and the quantity received by the ship. If this is due to an innocent mistake, then it should be relatively easy to locate the 'missing' bunker, provided the supplier cooperates with the receiving shipping company or their agents. However, often this is not the case, and experience tells us that when disputes do arise over the quantity transferred, any 'post-delivery' investigation on quantity shortages is often inconclusive, especially if the shipboard personnel involved in bunkering operation have neglected the basic principles of safeguarding the owners or charterers' rights by way of collecting and preserving evidence. Protests, legal fees, etc., all add to the costs. It is not unusual that, despite an investigation, neither party actually concludes with any certainty what transpired on board. A successful bunker dispute claim will largely depend on the detailed and contemporaneous written evidence collected by the shipboard personnel at the time the supply was made. Indeed, considering the present cost of bunker fuel, 'bunker stem surveys' should be considered a necessary expense to ensure that the quantities stated on the BDN are true and accurate. However, there are many ship operators who leave the bunkering procedure to the chief engineer to save on survey costs. As the chief engineer is most often concerned with the safe operation of the bunkering procedure, rather than the accuracy of the bunkering procedure, it remains quite easy for a vessel to end up paying for an incorrect supply of bunker. It is important to note that even when a surveyor is appointed by the vessel owner or charterer to oversee the stemming operation, the master and or chief engineer is still in charge of ensuring the proper steps have been taken to prevent malpractice and that the surveyor assists the chief engineer and not the other way around. As a final thought, the ship owner and the charterer both share responsibility for the provision of bunker – in a time charter, the charterers will provide bunker, whereas in a voyage charter, the owners will normally supply bunker. Therefore, it is important for both the owners and the charterers to be aware of the tricks of the trade during bunker stem operations.

Tricks of the trade

In this section, we will briefly examine some of the main 'tricks of the trade' related to the bunkerage of marine fuel, starting with the fuel density and weight relationship.

Understanding the fuel density and weight relationship. Marine fuel is always sold by weight (mass) and delivered by volume. For this reason, bunker receipts must always be signed 'For Volume Only' and add the words 'weight to be determined after testing of the representative sample'. Never sign for weight if uncertain about the density. What many bunker surveyors do not realise is that the density given in the supplier's BDN may not be true, and thus the weight determined by calculation should be considered as the 'preliminary' weight of the fuel transferred to the vessel. The actual weight is only determined after the density is verified by an independent fuel testing authority and then factored into the final recalculation of the actual weight of the fuel delivered on board. That is why the importance of accurately obtaining bunker samples both on board the vessel and the bunker barge cannot be stressed enough. After the bunker samples have been analysed by the vessel's chosen independent fuel testing laboratory, a revised bunker survey report is issued. This can then be used to calculate the actual volume of bunker supplied. To put this into context, we can refer to Table 24.1, which sets out a typical scenario of how density can affect the weight of fuel transferred on board. In this example, the ship owner or charterer has a fleet of 20 vessels bunkering an average of 1,000 metric tonnes (MT) each month:

Now imagine the ship owner or charterer has a fleet of 50, 70, or even 100 vessels. The commercial loss would be equal to millions of dollars every year! The key point to note here is if the density of the fuel cannot be verified on board, or independently verified at the time of bunkering, the BDN should be signed only for 'volume' and not for weight. Remember, whenever in doubt, issue a letter of protest.

Understanding the fuel temperature and volume relationship. Petroleum products have a high rate of thermal expansion which must be accounted for when several thousand MT of fuel is transferred or purchased. The bunker barge will often try to under-declare the temperature during the opening gauge and over-declare during close out. This malpractice is quite common in day-to-day bunkering. To avoid falling prey to this, the ship's officers responsible for the bunkering operation must be extra vigilant and check the temperatures

Table 24.1 Typical scenario of how density can affect the weight of fuel transferred onboard

Fuel coast $ USD/MT	650.00	$ USD
Bunker stemmed per month × 20 vessels	20,000.00	MT
Density of fuel @ 15°C (BDN value)	0.9889	
Density of fuel @ 15°C (tested value)	0.9865	
Density differential	0.0024	
Short delivery per vessel per month (approx.)	−2.50	MT
Commercial loss per vessel per month	−1,625.00	$ USD
Fleet commercial loss per month	−32,500.00	$ USD
Fleet commercial loss per year	**−390,000.00**	$ USD

of all bunker tanks during the opening gauge and thereafter periodically check and record the temperature of the fuel as it is pumped on board. The temperatures should be checked both on the bunker barge and at the ship's manifold. If temperature gauges are provided, it is prudent to take photographs, where permissible. If it is suspected that the bunker barge is trying to under-declare the temperature during the opening gauge and over-declare the temperature at closing, always verify the temperatures of all bunker tanks during the opening gauge and thereafter periodically check and record the temperature of the fuel as it is pumped on board. The temperatures should be checked both on the bunker barge and at the ship's manifold with an average of all readings taken during the final calculations. Where temperature gauges are provided, it is prudent to take photographs where permissible. Also, note that the existing flow measurement systems will have separate temperature and pressure gauges. These are easily tampered with. Furthermore, non-aqueous liquid-filled gauges may be purposefully rendered inaccurate with glycerine and silicone oils, or broken sight glasses. The purpose of a liquid-filled gauge is for the liquid to absorb vibrations, thus providing a dampening effect to enable accurate readings and also to reduce wear and tear by lubricating the moving parts. In other words, this affects the integrity and reliability of the gauge readings over time. There have been cases where the glass in the mercury cup case thermometer has been gently heated to create a bubble effect to prevent the correct registering of the temperature of the fuel oil. This malpractice is illustrated by the following example (Table 24.2).

As with the previous example, with a large fleet, losses could run into millions of dollars a year! To avoid these attempts to under-deliver and overcharge, always check and record the temperatures of the fuel tanks before and after bunkerage and periodically during the bunkering operation. It is strongly recommended to carry an infrared laser temperature gun as part of the ship's bunkerage equipment. Remember, whenever in doubt, issue a letter of protest.

Cappuccino bunkers (also sometimes known as the Coca-Cola effect). The cappuccino or Coca-Cola effect may be described as the frothing or bubbling of the bunker by blowing compressed air through the delivery hose. The aerated bunkers when sounded will give the impression that the fuel is delivered as ordered, whereas in fact after the trapped air in suspension settles out of the fuel oil, the oil level will drop and a shortfall will be discovered. With large bunker deliveries, this could be considerable, with significant financial implications. Some of the precautions that the receiving vessel can take against cappuccino bunkers include at the time of opening the gauge, the fuel oil should be observed from the ullage hatches for any foam on the surface of the bunker. Foam may

Table 24.2 Example loss calculation

At opening gauge (under-declared temperature)					
Actual temperature	**53.0°C**				
Declared temperature	**40.0°C**				
GOV m^3	Density @ 15°C (g/ml)	Temp °C	VCFT (548)	GSV m^3 @ 15°C	Weight (MT in air)
1,000	0.9889	53.0	0.9738	973.8	961.92
1,000	0.9889	40.0	0.9828	982.8	970.81
Loss or gain					−8.89
Approx. commercial loss					**−$5,778.50**

also be detected on the ullage tape. If there is no foam, then the oil level on the tape should appear distinct with no bubbles. If by observation it is suspected that air is trapped in the bunker fuel, obtain a sample of the fuel by lowering a weighted bottle into the tank. Pour the sample into a clean glass jar and observe carefully for any signs of foam or bubbles. If these observations show air in the fuel, the chief engineer should not allow the bunkering to start. They must also notify the owners and or charterers immediately. The bunker barge master should be issued with a letter of protest and a copy sent to the ship's agent. If the bunker barge master decides to disconnect from the ship and go to another location, then the agent should immediately inform the Port Authority[2] and try to establish where the bunker barge has gone. All relevant times and facts should be recorded in the ship's logbook. If the chief engineer has not observed any trapped air during the initial barge survey, it is still possible that air can be introduced to the barge tanks or the delivery line during the pumping period, for example, by introducing air into the system by cracking open the suction valve of an empty bunker tank while pumping from the other tanks. Subsequently, it is important for the chief engineer to continue gauging the ship's receiving tanks while the bunkering is in progress, as air bubbles will be readily seen on the sounding tape. Stripping the bunker barge tanks can also introduce air. Therefore, stripping should only be performed at the end of the delivery and for a short period of time only. The bunker barge master must agree to inform the chief engineer when they intend to start stripping and when the stripping is complete. The receiving vessel's crew and surveyor need to be alert during the bunkering procedure and check for any of the following indications:

- Bunker hose jerking or whipping round
- Gurgling sounds when standing in the vicinity of the bunker manifold
- Fluctuations of pressure indication on the manifold pressure gauge
- Unusual noises from the bunker barge

Even after the fuel transfer is complete, it is still possible to introduce air into the delivery line during blowing through at high pressure. Therefore, it is imperative that the bunker barge informs the ship before and after blowing through is completed so that the ship crew can be extra vigilant during this operation. In addition, the ship's bunker manifold valve should be checked and shut before gauging the vessel's tanks.

Identifying cappuccino bunkers is relatively easy if looking for the right signs. These include any or all of the following:

- Signs of froth and foam on the surface of the fuel in the bunker barge tanks during the opening gauge
- Excessive bubbles on the sounding tape prior to, during, and after bunkering
- Bunker hose jerking or whipping around
- Slower delivery rates than what has been agreed
- Gurgling sounds in the vicinity of bunker manifold
- Fluctuations of pressure on the manifold pressure gauge
- Unusual noises from the bunker barge

Important: It is worth noting that hose jerking or evidence of sporadic bubbles that are superficial in nature following line blowing or stripping of tanks is fairly common and should not be construed as evidence of malpractice.

Fuel delivered with high water content. In most cases, traces of water in bunker fuel is normally very low at about 0.1%–0.2% by volume. ISO 8217:2010 Fuel Standards for 'Marine Residual Fuels' provides a maximum allowable water content of 0.5% v/v. Water can originate from any number of sources, such as heating coil damage causing leaks and tank condensation; however, a deliberate injection cannot be ruled out. In the event a large quantity of water is found, a letter of protest should be issued immediately. It should be noted that the exact quantity of water can only be determined after the settlement phase when the water has settled to the bottom of the bunker tank. High water content causes other issues such as removal costs onshore if the oily water separator is unable to filter the water out. This will also adversely affect the vessel's specific fuel efficiency rating. Fuel samples provided by the bunker barge may not have any traces of water, as the samples may have been taken prior to the addition of water to the bunker. Always ensure that the fuel samples are collected during bunkering and not before or after. For these reasons, never sign labels in advance or sign for samples of unknown origin. Samples should only be signed when the sample collection was witnessed. Remember, whenever in doubt, issue a letter of protest.

Inter-tank transfers (gravitating of fuel). During the opening gauge, the fuel could be transferred from a high level to a low level (or empty/slack tank) by gravity. For instance, the bunker barge may have four tanks 1P/1S, 2P/2S, 3P/3S, and 4P/4S. The opening gauge starts from, say, aft tanks 4P/4S. While the gauging is underway, the tank level of 4P/4S could be easily dropped under gravity to a slack or empty tank forward, say, 1P/1S. Thus, essentially the same fuel quantity is measured twice. This method is still in use, and if not detected, the bunker barge can claim that the full quantity was delivered to the vessel. The receiving vessel, however, will suffer a substantial shortfall. Once the bunkering has commenced it is too late to rectify, and it will be virtually impossible to trace the 'missing' fuel. A thorough investigation will be needed to determine the exact stock control quantity and full disclosure from the supplier, which can take many months or even years of legal action. Even then the true state of affairs may not be resolved. It is imperative that the attending surveyor or vessel's representative regauges the tanks in the following sequence: (1) if the initial gauging was forward to aft, then after gauging the last aft tank; the surveyor or vessel's representative should regauge all tanks from aft to forward; the readings should be the same; (2) as an additional precaution, at the commencement of the bunker transfer, the surveyor or vessel's representative should regauge the first tank(s) used to transfer the oil to the vessel. The reading should match that taken during the initial gauging. In essence, the only effective way of dealing with this dubious practice is resounding the tanks as aforementioned before bunkering commences. Remember, whenever in doubt, issue a letter of protest.

Flow meter/pipe work tampering. Bunker barges fitted with a flow meter should be checked for proper functioning by sighting a valid calibration certificate and ensuring the seal is intact. There may also be unauthorised piping (bypass lines) fitted to the flow meter running into the pump suction side. This will register the throughput of fuel twice through the flow meter. To counteract this illegal practice, verify the flow meter seal is intact, verify the validity of the calibration certificate and ensure it is for the same type of flow meter, look out for any suspicious bypass lines running from the flow meter, consult the bunker barge piping diagram if in doubt. If in doubt, issue a letter of protest.

Quantity measurements by flow meter only. The bunker barge may claim that the soundings and ullage ports have been sealed by customs or seized, or indeed because of any other imaginative reason forcing the receiving vessel to go by the volumetric flow meter only. Remember that this may be just the first indication of an unscrupulous bunker barge master and as such be wary of other malpractices. Never ever agree to go by the flow meter only for fuel delivery. If in doubt, issue a letter of protest.

Pumping and or mixing slops into bunkers. Though seldom practiced today as a result of more stringent sampling procedures, introducing slops and thus contaminants into the fuel delivery will reduce the actual fuel amount and can also cause engine problems later. Unfortunately, this scam cannot be detected until the representative fuel samples have been tested by an independent fuel testing facility. A typical scenario where this malpractice would be carried out is after an argument over short supply; the bunker barge would pump in sludge/water to make up for the short supply. As the sample collection would have been completed, it is imperative that, if allowed, a second pumping resample is performed both on the barge and the receiving vessel. Always witness and collect samples through the continuous drip method, i.e., allow the sample to be drawn continuously throughout the bunkering delivery period. It should be good practice on board to isolate the fuel delivered to separate tanks and not to consume the bunker until such time as the fuel testing report gives a clean bill of health. In the event second pumping re-sampling is carried out on both the vessel and the bunker barge, ensure no contaminants such as sludge and water are delivered to the vessel. Fuel contamination, amongst other things, can create problems with the fuel injection system and exhaust valves resulting in costly repairs. Remember, whenever in doubt, issue a letter of protest.

Questionable tank calibration tables. Verify that the sounding/ullage tables are approved by Class (i.e., Class Certified – with endorsement). Carrying more than one set of sounding books is not uncommon and having the tables modified to the supplier's advantage is always a possibility. Inserted pages, corrections, different print or paper types are all potential indications of tampering. Sometimes the bunker barge may have a new calibration table (with the old one being obsolete). This could be following the modification of the tank's internal structure during a dry dock repair or simply because the original calibration is incorrect. Always establish the reason for the new calibration table and ensure it is Class Certified. The same should be said for the list and trim correction tables, which can be easily modified again to the supplier's advantage. If in doubt, issue a letter of protest.

Tampering with gauging equipment. Always verify the condition of the sounding tape. Sounding tapes could be tampered with in many ways, such as the following:
- Tampering with the gauging element
- Deliberately altering the sounding tapes and or using the wrong-sized bobs
- Switched the sounding bobs from different tapes
- Cutting and re-joining the sounding tape resulting in a non-linear tape

To avoid falling foul of these tricks, check the calibration certificate for the gauging equipment. Use a ruler to ascertain the precise sounding/ullage when below the 20-centimetre (7.87 in) mark. Use the ship's own sounding/ullage tapes to confirm the measurements taken by the bunker barge. Pay particular attention to 'millimetre' soundings, especially when the tanks are full and taking ullages, as even small errors will have a significant impact on the total bunker quantity. If in doubt, issue a letter of protest.

Empty tanks – unpumpable fuel (zero-dip volume application). In the event of a short delivery, be wary that the empty tanks may not be empty, even with a zero dip, and that substantial pumpable volumes may exist. Verify the tanks that are claimed to be empty are indeed empty – never take the supplier's word for it. The bunker surveyor or the vessel representative should notify the barge representative that the zero-dip volume of the tank(s) shall be included in the bunker tanker calculations. The condition is considered applicable when the closing gauge indicates no oil cut, whereas the visual inspection of the bunker tanker cargo tank indicates free-flowing oil at the aft of the tank. To avoid zero-dip volume application, sufficient bunkerage should be retained in the cargo tanks such that it touches all four sides of the tank. To apply zero-dip correction, it should be assumed that the tank is rectangular and the sounding is not constrained by a sounding pipe. The sounding should be taken in an 'open sounding' position (from the hatch) where the sounding tape bob is free to travel with the trim of the barge and is not restricted by the sounding pipe. If, however, the tape is inside a sounding pipe, then this correction will be invalid. Liquid cargo should only be trimmed and or list corrected if the liquid is in contact with all bulkheads. When the liquid is not in contact with all bulkheads, a wedge correction should be applied. Never assume any tanks are empty even when reaching the stripping level. Always check the tank calibration tables to verify if they are truly unpumpable. Always apply the correct list/trim corrections during calculations. If in doubt, issue a letter of protest.

Inflated or deflated tank volumes. The level of oil on the tape/bob should be clearly identifiable (i.e., the same colour and viscosity as the rest of the oil in the tank). Soundings can be inflated during opening gauging by pouring diesel oil into the sounding pipe just before gauging. Another method of inflating the sounding is high-pressure compressed air being injected directly into the sounding pipe, pressurising the pipe, and thus causing the level of oil to rise, giving a higher reading without even frothing or creating bubbles. This would be done on route to the vessel and just before delivery. The reverse can also be done – that is the soundings can be deflated during closing gauging by pouring copious amounts of paint thinner into the sounding pipe just before gauging. The thinner washes off the oil level marking on the sounding tape to indicate there is less oil. Always check the level on the sounding tape and if in doubt regauge the tank. As always, if in doubt, issue a letter of protest.

Under-declaring the actual ROB and deliberate short supplying of fuel. The malpractices we have discussed thus far during bunkering operations are quite prevalent amongst unscrupulous bunker suppliers, but on some occasions, it is the receiving vessel that is engaged in illegal or questionable practices. One such example is under-declaring the fuel quantity bunkered on board which is then either sold back to the barge supplier or simply kept hidden on the vessel until an opportunity to profit from the ill-gotten supply arises. For instance, a vessel places an order for 1,000 MT of fuel oil at the next bunkering port. The vessel knows it has ordered an excess of 50 MT (which is undeclared). When the bunker barge comes alongside (through prior negotiations) the vessel is deliberately short-received (or the bunker barge will deliberately short supply) the extra 50 MT of fuel. In other words, the actual quantity supplied is 950 MT, yet the BDN will report 1,000 MT, for which the ship owner or charterer will be invoiced. The short-received (or short-delivered bunker) profit is then shared between the bunker supplier and the receiving vessel. With this loss, the operator suffers a loss twice by (a) paying for the initial 50 MT and (b) by having to backfill the missing 50 MT at the next bunkerage. This is just one example of the many types of scams that are easily perpetrated when there is lax supervision and too much reliance placed on the ship's staff. The problem is further exacerbated by poor bunker stem auditing (which admittedly involves elaborate investigative work carried out by an independent third-party surveyor) and ignoring non-nominating (i.e., non-receiving) tanks to be included in the overall tank measurements during stem operations. Most shipping companies will engage the services of an independent surveyor to protect their interest whenever there is a large discrepancy in the final figures between the bunker barge and the vessel; however, how many companies provide clear instructions to the attending surveyor to measure all non-nominated tanks (non-receiving tanks)? Or how many surveying firms carry out the measurements diligently? Failing to do so leaves the operator vulnerable to deceitful practices as discussed earlier. This problem is further illustrated in Table 24.3.

The excess 53 MT of fuel oil will be in favour of the owners with a loss to the charterers (Table 24.4).

To avoid falling foul to under-declared or over-declared fuel, regularly carry out 'bunker stem audits'. In a large fleet this is an indispensable loss control tool. Measure all

Table 24.3 Scenario 1 – under-declaring to the shipowners advantage

Bunker stemmed by the vessel operator	1,500.00	MT
ROB as per logbook (arrival bunkering port)	350.00	MT
Undeclared fuel onboard	**53.00**	MT
Actual bunker stemmed	1,500.00	MT
Quantity declared on BDN	1,500.00	MT
Final ROB declared in logbook after bunkering	1,850.00	MT
However: actual ROB would be:	1,903.00	MT
Fuel cost $USD / MT	650.00	$ USD
Losses for the ship operator	−34,450.00	$ USD

Table 24.4 Scenario 2 – under-declaring with the aim of profiting for personal gain

Bunker stemmed by the vessel operator	1,500.00	MT
ROB as per logbook (arrival bunkering port)	350.00	MT
Undeclared fuel onboard	**53.00**	MT
Actual bunker stemmed (deliberate short supply)	1,447.00	MT
Quantity declared on BDN	1,500.00	MT
Final ROB declared in logbook after bunkering	1,850.00	MT
However: actual ROB would be:	1,850.00	MT
So where did 53 MT disappear to?	**You guessed it!**	
Fuel cost $ USD / MT	650.00	$ USD
Losses for the ship operator	−34,450.00	$ USD
• For under-declared fuel	−34,450.00	$ USD
• For the short supply fuel	−68,900.00	$ USD

non-nominated tanks prior to stemming operations and again after bunkering is complete. And lastly, always engage the services of a reputable bunker stem surveying firm during stem operations.

FUEL OIL STORAGE ON SHIPS

Marine fuel oil is one of the most important factors that influence the overall efficiency of ships and the shipping industry. The high cost of marine fuel oil requires maritime professionals to bunker, store, and use the heavy-sulphur fuel oil, diesel oil, and marine gas oil very carefully. It is critical to exercise due diligence to keep track of consumption. On many older vessels, it is not uncommon for the heavy-sulphur fuel oil flow meter to not function correctly as it is regularly serviced or calibrated. That said, maintenance of the heavy-sulphur fuel oil flow meter should form a part of the preventative maintenance schedule. Understanding the correct procedure for bunkering is extremely important for the safety of the vessel and for preventing oil spills. Companies and port authorities must also provide necessary training and guidance to ensure safe bunkering procedures. For instance, the Maritime and Port Authority of Singapore has issued instructions and guidance notes for bunkering at Singapore, and these should be studied and followed meticulously. Although fuel quality has rarely been an issue involving Singapore bunkers, there have been occasions where disputes have arisen in response to allegations of cappuccino bunkering. Sounding pipes must be used properly to prevent errors with the tank readings. In many cases we are not able to properly put the tape into the sounding pipes, resulting in erroneous readings. If the sounding pipe is straight, try using a rod to take ullage, especially in cold climates. Also, in some cases, the calibration booklets are not correct on certain trim conditions and thus the quantity of fuel cannot be gauged accurately. Always aim to avoid such problems. One tip is to put a small quantity of diesel oil (no more than 1–2 L) in the sounding pipe of the heavy-sulphur fuel oil tanks one day prior to the scheduled bunkering. This is especially useful in colder regions. From the ship's staff perspective, it is always advisable to learn the tank characteristics from previous crew experiences. These can sometimes help explain any abnormalities or anomalies, before pointing the finger of suspicion at the bunker barge. In addition, there are global attempts to modernise the maritime bunkering process by introducing new regulations coupled with technology. For example, the Maritime and Port Authority of Singapore has made it mandatory for bunker barges to install approved flow meters which utilise modern technology to ensure bunkering is accurate and fair. The International Bunker Industry Association is also taking steps to develop bunkering facilities at various ports around the world, both for heavy- and low-sulphur fuel oils, low-sulphur gas oil, and LNG bunkering. Singapore, Rotterdam (Holland), Fujairah (UAE), and Houston (US) are all major bunkering ports for various qualities of fuel oils, whilst Rotterdam, Seinehaven (both in Holland), Port Fourchon in Louisiana, USA, and Shwinaouski in Poland are being developed specifically for LNG bunkering.

Always ensure correct sampling and expeditious dispatch of fuel samples for laboratory testing. Recently there have been disputes regarding the quality of fuel bunkered from Ukrainian ports (water and total sediment potential) and from Houston in the US. On bulk carriers, which carry dry dust cargoes, we must carefully examine the air pipes and sounding pipes which pass through the hatches (hidden behind the structural protection) to ensure there is no structural damage which can cause dust and other solid contaminants to foul the fuel tanks. A somewhat disturbing trend has emerged over the past decade or so where tank cleaning has been neglected. In the worst cases, it has been found that tanks were cleaned only once every five years (which is in line with the typical dry dock schedule). Moreover, it was also found that the tank or tanks not being used on some ships were still carrying remnants of between 15 and 18 MT of fuel. In many cases the heating coils were found to be defective or leaking, in one instance, sludge had even been pumped into one of these tanks. Even the service and settling tanks are not being cleaned and internally examined at least once every five years. This is a serious issue, and proper steps must be taken to ensure tank cleaning and inspections are performed at regular intervals.

Due in part to improvements in technology, the quantity of diesel oil carried on board ships is significantly reduced by modern standards. Whilst this reduces bunkerage costs and increases vessel efficiency, it does mean that where the vessel is operating in cold areas for prolonged periods, condensation can begin to develop in the tanks. To prevent this, it is necessary to check the diesel oil tank drain and clean the diesel oil line filter regularly to check for the presence of water. The indiscriminate mixing of fuels of various origins (even if the grade, say, 380 CST, is the same) may also cause problems. Adequate purification and filtration help ensure a good quality burn, resulting in reduced fuel consumption. The purifiers and filters should be cleaned regularly. In a similar vein, maintaining the correct temperature and viscosity for the injector (13–15 CST with a temperature around 135°C or 275°F for 380 CST oil) will ensure a good burn. The air pipes and sounding pipes of the fuel oil tanks on deck should be checked to ensure they are structurally sound and that any wire mesh on the air pipes is not damaged or blocked. When the exhaust gas economiser is cleaned, the boiler pressure increases fast as the ship sets sail. In these situations, if the dump steam condenser is not kept clean, much steam can enter the bunker tanks, causing the fuel oil tank high-temperature alarm to activate. This can damage some of the cargo carried on the other side of the bulkhead. It is important to remember the dump steam valve is typically kept on auto; as the low-temperature cooling water temperature rises, the dump condenser is low-temperature water cooled. Many times, we needlessly delay ships because we are trying to sort out the issue of quantity received. This also happens when the vessel is not able to determine the correct trim and when she goes down by the head upon completion of bunkering. A balanced approach at such times helps to settle this issue quickly. Fortunately, these issues will diminish with time as suppliers install technologically sound flow meters on their bunker barges.

FUEL OIL CONSUMPTION CALCULATIONS

Fuel oil consumption calculations and record keeping on board ships are some of the most critical tasks that the chief engineer oversees. Fuel oil is provided by the charterers of the vessel, and the chief engineer has to report to them every day with fuel oil consumption reports, fuel remaining on board, and requirements for the next passage. The applied method for measuring fuel oil consumption is briefly described in this section. The description explains the procedure for measuring data and calculating annual values, the measurement equipment involved, and so forth.

Measuring and reporting fuel oil consumption

Where a flow meter is fitted on a pipeline supplying fuel to an emission source (i.e., the main engine, diesel generator, auxiliary boiler, etc.), flow meter readings are the principal means of determining fuel consumption. Flowmeter readings and fuel temperatures must be recorded daily at 1200 hours ship's meantime, as well as at the time of arrival (as noted in the arrival report) and on departure (as noted in the departure report). Where the figures are calculated using an Excel spreadsheet, the following formula.

$$(\text{Corrected density} = \text{density at } 150°C \times [1 - \{(\text{fuel temperature}(0°C) - 150°C) \times 0.00065\}])$$

should be used to obtain the corrected density at the recorded fuel temperature. In addition to reporting fuel consumption every noon, on arrival, and on departure, it is also necessary to record flow meter readings at the following events:

- At the end of each sea passage
- At the start of each sea passage
- Whenever a fuel change operation is performed

Fuel transferred from the fuel oil drain tank or fuel oil overflow tank back to the fuel/settling tank should be noted in the position, arrival, and departure reports. This amount is automatically subtracted from the voyage fuel consumption. For emission sources that are not fitted with flow meters or when the flow meters are not operational, manual bunker fuel tank monitoring must be carried out instead. In this method, tank readings of all the fuel tanks relevant to the emission source, using tank soundings/ullages or level gauge readings, are noted in the engine room sounding log. Consumptions are to be recorded in the Excel spreadsheet. In addition, fuel quantities in all fuel tanks on board the ship are determined periodically, and at least as per the following schedule (quantities may be determined by use of the fixed gauging system where available or by manual soundings) (Figure 24.3):

- At every arrival of the vessel in berth and at every departure from berth (this may vary according to the company's policy)
- Prior to undertaking bunkering or de-bunkering operations
- After bunkering or de-bunkering operations
- At least once every seven days

Figure 24.3 Bunker vessel.

Position, arrival, and departure reports

The position, arrival, and departure reports in the company's reporting infrastructure for ships are the primary means of reporting information, including fuel consumption, transport work, and other voyage-related data. A position report must be submitted each day at 1200 hours ship's mean time, irrespective of whether the ship is at sea or in port. There must never be a gap of more than 24 hours (in the ship's mean time) between

- Two position reports, or
- Between a position report and an arrival report, or
- Between a departure report and the next position report, or
- Between a departure and arrival report.

Generally, if the gap is more than 24 hours the user will not be able to submit the latest report and will instead need to submit the missing report (with a gap of less than or equal to 24 hours) first. An arrival report must be submitted for the first arrival in port. 'First arrival in port' means the first time (for a specific port or location) that the vessel is

- All fast to a wharf or buoy mooring or single buoy mooring (if berthing directly, without anchoring), or
- Anchored (i.e., 'brought up to anchor') within port limits, or
- Anchored (i.e., 'brought up to anchor') outside port limits, or
- Anchored at a lighterage area, or
- All fast to a lighter vessel[3] (if berthing alongside the lighter vessel directly, without anchoring), or
- Arrival at a lighterage area (if drifting, without anchoring, while awaiting the lighter vessel).

A departure report must be submitted for the final departure from port. 'Final departure from port' means a departure from the last

- Wharf or buoy mooring or single buoy mooring (all lines cast off), or
- Anchorage within port limits (anchor aweigh), or
- Anchorage outside port limits (anchor aweigh at an offshore location), or
- Lighterage location (all lines cast off from lighter vessel and anchor aweigh).

An arrival report for a specific port or offshore location must be followed by a departure report from the same port or offshore location. It will not be possible to submit the departure report if the name of the port or offshore location is different from that in the arrival report. In addition to the position, arrival, and departure reports, other relevant periodic reports, including the noon reports, monthly reports, and quarterly reports, must be filed as per the company's reporting infrastructure.

Determining the fuel bunkered and fuel in tanks

The quantity of fuel bunkered, as stated in the BDN must be checked by gauging all fuel tanks on board, prior to and after the completion of bunkering. This is achieved by applying the appropriate correction factor to density for temperature and obtaining quantities in MT before and after bunkering. The ship's figure of fuel bunkered is the difference between the fuel quantity before and after bunkering. The ship's figure is regarded as the authoritative quantity of fuel bunkered and is the quantity entered by the ship's staff in the departure report. Written records showing the soundings, before and after, of all fuel tanks and details of the calculations showing the ship's figure in MT of quantity bunkered are to be retained on board. The temperature of the fuel in the tanks is to be obtained from tank temperature gauges if provided or by using portable temperature gauging devices. If no gauges are provided, the temperature of the fuel in tanks may be determined by measuring the temperature of the tank sides using an infrared thermometer or else estimated by taking the weighted average of the best estimate of the temperature of the fuel in tanks before bunkering and of the fuel bunkered in each tank. The density of fuel bunkered is to be obtained from the BDN. The density of comingled fuel in tanks is obtained by calculating the weighted average of the density of the fuel remaining in the tanks before bunkering and of the fuel bunkered in each tank. The density of fuel should be corrected using an appropriate temperature correction factor obtained from the ASTM Petroleum Table 54B or equivalent, or from computer software incorporating these tables, or by applying the formula:

$$\text{Corrected density} = \text{density (in the air) at } 15.0°C \times [1 - \{(T0C - 15.0°C) \times 0.00065\}]$$

- Where T0C is the temperature of the fuel in degrees Celsius.

In the case of bunkering from a bunker barge, all tanks on the barge are to be sounded before and after bunkering by a responsible officer. The barge tanks are also to be checked for the presence of free water. A written record is to be made of the results of these soundings and free water checks. The chief engineer is responsible for checking the fuel quantity bunkered. Fuel quantity (in tonnes) in all bunkered tanks is to be re-checked 24 hours after completion of bunkering or else just prior to commencing use of the newly bunkered fuel (if within 24 hours of bunkering) to account for possible settling of fuel. Prior to entering an emission control area, fuel oil changeover to low-sulphur oil must be started. The time for starting depends on how much volume of fuel is used by the system. Logbook entries must be made accordingly, recording the volume of low-sulphur fuel in the tanks as well as the date, time, and position of the ship when the fuel oil changeover was completed. It is a mandatory requirement that changeover procedures are available on board in the proper written format.

Measuring and reporting distance travelled

Distances travelled are measured over ground between the point or port of departure and the point or port of arrival. These are recorded in the position and arrival reports. Distances travelled over ground may be taken from an electronic chart display and information system (ECDIS), global positioning system (GPS), or by manual measurement using the ship's charts. Distances travelled through water are also reported in the position and arrival reports and are to be taken from the (water) speed log. Distances that may be travelled between the arrival and departure reports (such as during transit from anchorage to berth or when shifting between terminals within a port) are not required to be reported in the voyage reports but should be noted in the ship's logbook.

Method to measure hours underway

'Hours underway' from the last berth at the port of departure to the first berth at the port of arrival is calculated from the departure and arrival times (Greenwich Mean Time (GMT)) and dates (GMT) recorded in the departure and arrival reports. Times and dates must be recorded in GMT as well as in the ship's mean time (SMT). Time spent between the first berth at the port of arrival to the last berth at the port of departure is considered time spent in port. This includes periods at berth and at anchor, and periods spent manoeuvring within the port. Fuel flow meters, fixed tank gauging devices, and temperature measuring devices and gauges are to be checked and calibrated for accuracy at intervals as recommended by the manufacturer or as stated in the ship's preventative maintenance schedule. Certificates of calibration are to be issued after the completion of these checks and thereafter retained on board. The validity of calibration certificates is inspected as part of the ship's annual marine assurance audit.

Emission factors

C_F is a non-dimensional conversion factor between fuel oil consumption and CO_2 emissions. It was first promulgated in the 2014 Guidelines on the Method of Calculation of the attained Energy Efficiency Design Index (EEDI) for New Ships. The annual total amount of CO_2 emitted by a vessel is calculated by multiplying the annual fuel oil consumption and C_F for the specific type of fuel. Table 24.5 provides an illustration of how the C_F is calculated for different types of fuels.

Information to be submitted to the IMO ship fuel oil consumption database

Since 2019, every ship above 5,000 gross tonnes must collect certain information about the ship and its fuel consumption properties, and submit these to the IMO. This information includes the ship's particulars, the period of the calendar year for which the data is being submitted, the fuel oil consumption in MT, fuel oil types, the methods used for collecting fuel oil consumption data, and the distance travelled and hours spent underway. When this data is provided by the ship, it enables the IMO to calculate fuel oil consumption across the entire global merchant fleet, which can then be used to research and establish

Table 24.5 Ship emission factor

Fuel Oil Type	C_F (t-CO_2/t-Fuel)
Diesel/gas oil (e.g., ISO8217 grades DMX through DMB)	3.206
Light fuel oil (LFO) (e.g., ISO 8217 grades RMA through RMD)	3.151
Heavy fuel oil (HFO) (e.g., ISO 8217 grades RME through RMK)	3.114
Liquefied petroleum gas (LPG) (Propane)	3.000
Liquefied petroleum gas (LPG) (Butane)	3.030
Liquefied natural gas (LNG)	2.750
Methanol	1.375
Ethanol	1.913
Other	

new and innovative ways for reducing emissions, marine pollution, and improving ship technologies.

In this chapter, we have examined the process of bunkering and some of the main hazards and problems marine engineers may experience during the bunkering process. We have also covered several of the nefarious efforts that some unscrupulous bunkerage agents use to undersell and overcharge unsuspecting vessels. In the next part of this book, we will turn our attention to the main engine room drills firefighting procedures and apparatus.

NOTES

1. The International Code of the Construction and Equipment of Ships Carrying Liquefied Gases in Bulk (IGC Code), adopted by resolution MSC.5(48), has been mandatory under SOLAS Chapter VII since 1 July 1986. The IGC Code applies to ships regardless of their size, including those of less than 500 gross tonnes, engaged in carriage of liquefied gases having a vapour pressure exceeding 2.8 bar absolute at a temperature of 37.8°C (100.04°F, and certain other substances listed in Chapter 19 of the Code. The aim of the Code is to provide an international standard for the safe carriage by sea in bulk of liquefied gases and the substances listed in Chapter 19 by prescribing the design and construction standards of ships involved in such carriage and the equipment they should carry to minimise the risk to the ship, to its crew, and to the environment, having regard for the nature of the products involved.
2. For example, the Singapore Bunkering Procedure SS 600 prohibits the use of compressed air from bottles or compressors during the pumping period or during stripping and line clearing. It should be confirmed with the barge master that they will follow the procedure (Reference SS600 paragraphs 1.12.10/11/12/13).
3. A lighter is a shallow-draught boat or barge, usually flat-bottomed, and used in unloading (lightening) or loading ships offshore. Use of lighters requires extra handling and thus extra time and expense and is largely confined to ports without enough traffic to justify construction of piers or wharves.

Part VI

Engine room fires and emergency response

Chapter 25

General emergency drills, alarms and emergency systems

GENERAL ALARMS AND EMERGENCIES

Usually, the first indication of an emergency on board is the sounding of the *general alarm*. In the event the general alarm is activated, make way to the muster station. Don a life jacket and if possible, take an immersion suit. Act according to the vessel's Muster List. At the muster station, the officer in charge will explain the nature of the emergency and issue orders accordingly. Obey the commands given and ensure any duties are conducted. In the event the *fire alarm* sounds, contact the officer of the watch on the bridge and establish whether the alarm is genuine or a false activation. If the alarm is genuine, make way to the muster station and wait for further instructions from the officer in charge. If a fire breaks out in a manned space or compartment, raise the fire alarm immediately. If safe to do so, attempt to tackle the fire using the portable fire extinguishers. If there is more than one person in the affected compartment, one person should attempt to fight the fire, and the other person should notify the bridge of the nature of the incident and the extent of the fire. If the fire is too far involved or is likely to impinge on safe escape from the compartment, exit the compartment immediately. Do not attempt to fight the fire. Close all doors or hatches as they are passed through to prevent the spread of fire gases and flames.

> Only attempt to fight a fire if it is safe and reasonable to do so.
> Never put yourself, or others at unnecessary risk.

Given the engine room is located well below the waterline, it may seem perverse to include guidance on the *man overboard alarm*. Even so, every member of the ship's staff must be fully acquainted with every alarm and emergency that may evolve. When off duty, engineering personnel may find themselves out on deck or assisting the deck crew in their duties. Whatever the case, man overboard is a serious condition that has ramifications for the entire ship and its crew. In the event the man overboard signal is sounded, make way to the weather deck, and try to locate the crew member who has allegedly fallen overboard. If located, throw the nearest lifebuoy. Visually and orally indicate the location of the crew member. Try to attract the attention of other crew members to notify the bridge. Do not take eyes off the fallen crew member. When speaking, remain looking directly at the crew member; even just a split second is sufficient time to lose the position of the fallen individual. The one alarm every seafarer fears the most is the abandon ship alarm. Abandoning ship is the act of evacuating the vessel when all other methods and techniques for saving the ship have been exhausted. It is the sole responsibility of the master to authorise the command to abandon ship. Any attempt to do so without the master's authorisation is illegal and counts as desertion and dereliction of duty. When the command to abandon ship is given, the alarm will be activated. In this situation, make way to the muster station as calmly as possible. On route, don as much warm and dry clothing as possible. If safe to do so, attempt to collect and carry freshwater and rations to the muster station. At the muster station, obey all commands issued by the officer in charge.

Remember, never put yourself or others unnecessarily in harm's way.

Attempting to do more than is necessary in an emergency only puts yourself and other crew members at risk.

In the event of any *oil spill or oil pollution*, immediate action should be taken in accordance with the ship's SOPEP. SOPEP equipment is usually located in the deck stores and should be used in the event of an *oil spill*. Sometimes the engine room personnel are the first to encounter *flooding in the cargo holds*. In the event of cargo hold flooding, inform the bridge immediately. Attempt to stem the inflow of seawater (if the hull is breached) or attempt to repair any damaged valves or piping (if the leak is internal). If safe to do so, close all local watertight doors and hatches and raise the general alarm.

ENGINEER'S CALL AND ALARMS

In addition to the aforementioned general alarms, the engineers have their own distinct alarms. These are often referred to as the *engineer's call*. In the event of an engineer's call, all members of the engine department should make way and assemble in the engine control room. The engineer of the watch and or chief engineer will explain the reason for the call and issue orders accordingly. Should the CO_2 *alarm* sound, leave the engine room immediately. In the event of *engine room flooding*, the chief engineer should be called immediately, and the *general alarm* raised. Conduct immediate actions to stem the flow of seawater into the engine room (if the hull is breached) or attempt to repair any damaged valves or piping (if the leak is internal). Activate the engine room bilge pumps and obey all commands issued by the chief engineer. Inform the bridge and keep the officer of the watch updated on progress.

ENGINE ROOM DRILLS AND TRAINING PROCEDURES

Drills on board ships play a significant role in preparing the crew for emergency situations. The ship's engine room is a hazardous place where any manner of accidents can take place. Engine room personnel are therefore required to perform drills and training procedures on a regular basis to ensure they can react competently in the event of an engine room emergency. There are ten main types of drills and training that engine room personnel must complete. These are engine room fire drills, flooding drills, enclosed space drills, scavenge fire drills, crankcase explosion drills, uptake fire drills, oil spill drills, bunkerage training, pollution prevention training, and blackout training. We will now briefly discuss each of these drills and training procedures.

Engine room fire drills. Accidents caused by fire are the most common in the ship's engine room. Fire drills, which must include firefighters from both the deck and engine departments, must be conducted frequently to ensure that the ship's crew members are well prepared for any such eventuality. Fire drills must be performed at various levels and on different machineries within the engine room, for example, on the boiler, generator, purifier, and main engine.

Engine room flooding drill. A delayed action during engine room flooding can lead to the loss of critical machinery including the generators, main engine, etc. This can have knock-on effects leading to a complete blackout of the ship. Engine room flooding response training and immediate repair actions must be taught to all engine crew and thereafter rigorously reinforced. The flooding training must include response actions to different emergency situations, such as groundings and collisions, which can lead to structural damage and flooding of the engine room.

Enclosed space drill. The engine room comprises several tanks and confined spaces which are unsafe to enter without proper preparation and authorisation. Enclosed space training, with a risk assessment and dedicated checklists, must be conducted for the entire ship's crew.

Scavenge fire drill. All engine room crew members must know the engine scavenge firefighting procedure. The crew must know about the systems to be employed for scavenge space firefighting, together with the precautions to be taken before implementing any firefighting method; for example, if steam is used to suppress the fire, the line should be drained before the steam insertion, as water in the line may lead to thermal cracks developing in the crankcase. We will cover scavenge space fires in more detail later.

Crankcase explosion drill. Crankcase explosions can lead to fatal conditions and heavy loss of the ship's machinery. The crew should be prepared to conduct the correct procedures whenever the engine's oil mist detector activates the alarm. We will cover crankcase explosions in more detail later.

Uptake fire drill. The engine crew must be well trained through frequent and rigorous drills on fighting boiler uptake fires. This includes recognising the various stages of uptake fire and the different procedures for responding to an uptake fire. We will cover boiler fires in more detail later.

Oil spill drill. The oil carried on the ship as cargo or for use by the ship's machinery is overseen by the engine department. It is important to know the correct oil transfer procedures, and how to respond to an oil spill. The correct actions and procedures are listed in the ship's SOPEP.

Bunker training. Bunkering is one of the most dangerous operations conducted on board, as there is the constant risk of oil spill and fire. At least 24 hours before every bunkering, the ship's crew must be called for a meeting where the bunkering operation is discussed. The crew should be trained in appropriate safety signals, oil spill reporting procedures, and emergency response procedures.

Pollution prevention appliances training. Port State Control and other government authorities are extremely strict when it comes to compliance with maritime pollution prevention. It is therefore important for the ship's crew to be trained in the ship's pollution-preventative measures. This includes the crew's knowledge of the pollution prevention equipment present onboard (for example, the oily water separator, incinerator, and sewage treatment plant). The ship's crew must be trained in the operation of this equipment together with the regulations governing the discharge of processed waste at sea.

Blackout training. Whenever a ship loses its power source, i.e., the generator, the ship becomes entirely dependent on the forces of the sea and wind. This means the ship is effectively dead in the water. It is the responsibility of the engine department to restore the ship to full power. Blackout emergency training must be given to all members of the engine room crew.

In this chapter, we have discussed the general emergency drills, alarms, and response arrangements on board. In the next chapter, we will look at the main types of explosions and fires to affect the ship's engine room.

Chapter 26

Engine room explosions and fires

On most ships, the engine room is the most dangerous compartment. High concentrations of fuels, oils, flammable chemicals, and machineries all make the engine room a potentially volatile and explosive area. In this chapter, we will explore some of the fundamental issues that can develop in the engine room when proper maintenance and poor housekeeping are allowed to happen. We will begin by looking at the causes and consequences of crankcase explosions, starter air-line explosions, purifier room fires, scavenge space fires, exhaust gas boiler fires, incinerator fires, electrical fires, and the most unusual of all ship fires, those caused by bacteria. Before we do, though, it is worth spending a few moments explaining the chemistry of fire. With this knowledge in hand, we are better prepared to understand the causes of fire, and from there, take appropriate actions to prevent them. Most people have heard of the fire triangle. Up until recently, the fire triangle was the standard mechanism for explaining the causation of fire. This has since been replaced by the fire tetrahedron. The reason for this is not particularly important for us but suffice it to say for any fire to become self-sufficient, it requires three core elements: heat, fuel, and oxygen. A fourth element, chemical reaction, is also needed to complete the process. Unfortunately, the ship's engine room has all three (or four) elements and in abundance. All fuels have a lower and higher flammability limit. This limit is the lowest and highest point at which the fuel will ignite. Different fuels have different ignition points. Diesel, for example, is notoriously difficult to ignite, but with the right ratios of heat and oxygen, diesel is highly flammable. Other materials often found in the engine room, such as acetylene (used in welding), have much lower ignition points, which makes acetylene one of the most dangerous elements contained on board ships. In any case, for a fire to become self-sufficient, it needs a constant source of heat, a source of fuel, and a constant source of oxygen to feed the fire. Before we start to look at the main types of explosions and fires that can affect the engine room, we should first look at the Fire Control Plan (FCP).

FCP

The FCP is a mandatory requirement of the SOLAS convention and is described in regulation 15 of Chapter II. The FCP provides information about the fire station on each deck of the ship, the location of each bulkhead, and whether the bulkheads are categorised as class 'A' divisions, 'B' class divisions, or 'C' class divisions. It also explains the type of fire detection and firefighting systems available throughout the ship such as the various fire alarm systems, sprinkler installations, extinguishing appliances, means of escape to different compartments and decks, and the ventilation system, including the particulars of the remote operation of dampers and fans. The FCP also states the position of each damper, their markings, and which fan is for which compartment or deck, and whether a damper can be closed in the event of fire. The graphical symbols used in the FCP should be as per the firefighting equipment symbols established by IMO Assembly Resolution A.654(16). It is the responsibility of each member of the ship's crew to know and recognise the meaning of the symbols used in the FCP. The FCP must be written in the working language of the ship, and in English. An integral part of the FCP is the ship's general arrangement plan, a copy of which must be permanently exhibited on the bridge, in the engine room, and in the ship's accommodation. At least one copy of the FCP should be available onshore at the offices of the shipowner and the vessel operator. On passenger vessels, copies of the FCP must be provided to each member of the fire patrol team and posted at each continuously manned central control station. For all ships, a copy of the FCP should be permanently stored in a prominently marked weathertight enclosure outside the bridge for the assistance of shoreside firefighting personnel. Also, with the permission of the ship's maritime administration, i.e., the ship's Class and Flag state authority, the details of the FCP can be summarised in a booklet format and supplied to each of the ship's officers. The renewal and update of the FCP fall under the responsibility of the ship's master, the ship's owner, and the shoreside ship management team. Any amendments or revisions must be recorded and all copies of the FCP updated to reflect the most current information.

With the FCP in place, and the crew fully aware of their responsibilities and duties in the event of a fire emergency, we can turn our attention to the first of our incidents, the crankcase explosion.

CRANKCASE EXPLOSIONS

The main cause of crankcase explosions is the development of hot spots in the crankcase. Hot spots are a prime example of how poor maintenance can lead to potentially lethal conditions. The hot spot is formed by two metal surfaces rubbing against each other, or through abnormal friction between two metallic parts such as the piston rod and gland, crosshead guides, or chain and gear drive. When oil meets the hot spot, these oil particles vapourise into much smaller particles. These particles are naturally drawn towards the cooler sections of the crankcase. When the particles contact these cooler surfaces, they form a fine white mist. Over time, the formation of mist increases to such an extent that the optimum air/fuel ratio is achieved – i.e., it is high enough to exceed the lower explosive limit (LEL) of the fuel. Once the fine mist contacts the hot spot again, the elevated temperature ignites the mist, causing an explosion. This explosion, or the primary explosion, produces a shock wave which propagates inside the crankcase with increasing speed and distance. This shock wave causes a breaking effect which further reduces the size of the oil droplets, in effect producing yet more fuel for ignition. Now the pressure front is followed by a low-pressure area which tries to suck in more air from outside the crankcase. This is usually through leaky piston glands or relief valves in the scavenge space below the crankcase. This new air and fuel, which is supplied by the first explosion, meets the hot spot, causing another explosion. This secondary explosion is far more potent than the first as the amount of fuel in the crankcase is high. The veracity of the explosion is dependent on the volume of oil mist inside the crankcase, but suffice it to say, even mild primary explosions are sufficient to incapacitate the crankcase. Large secondary explosions can completely disable the main engine, destroy the steering gear, and even punch holes into the ship's hull below the waterline.

STARTER AIR-LINE EXPLOSION

The second main type of engine room incident is the starter air-line explosion. In the starter air system, fuel is often present in the form of lube oil carried over from the air compressor. Moreover, oxygen is present in the system in abundance. The heat source may come from a leaking starting air valve fitted on the cylinder head or some other defective valve. The combination of these three elements in the correct ratio can lead to a devastating starter air-line explosion. To prevent starter air-line explosions from happening, there are several safety devices which can be installed on the starter air-line arrangement. These include the relief valve, the bursting disc, the non-return valve, and flame arrestors. *Relief valve.* The relief valve is fitted on the common air manifold which supplies air to the cylinder head. It is normally installed at the end of the manifold and works by lifting the valve in the event of excess pressure developing inside the manifold. The advantage of the relief valve is that it will return to its position after removing the excess pressure. This means a continuous supply of air is provided to the engine when manoeuvring or in traffic. *Bursting disc.* The bursting disc is fitted inside the starting air pipe and consists of a perforated disc protected by a sheet of material which bursts in the event excessive pressure develops with the potential to cause an air-line explosion. The bursting disc also consists of a protective cap that is designed in such a way that even if the engine is required to run after the disc has been ruptured, the cap will cover the holes when it is turned. This ensures that the ship can continue to manoeuvre without disrupting the ship's progress or causing a hazard to other marine traffic. *Non-return valve.* The non-return valve is positioned between the air manifold and the air receiver. It is designed in such a manner that it will not allow any back flow of air to reach the air bottle. This ensures the air bottle cannot be inadvertently ignited by back flow from the starter air line. *Flame arrestor.* The flame arrester is a small unit consisting of several tubes which work by arresting any flames emitted from the cylinder through a leaking starter air valve. There is a flame arrester fitted to every cylinder before the starter air valve. To prevent a starter air line from occurring, it is important to ensure that all safety devices are fitted and in working order. Always drain the air bottle during every watch and ensure the auto-drain is checked for proper functioning. Ensure the air compressor is well maintained to avoid any oil carryover. Regularly inspect the oil separator at the air compressor discharge point for signs of damage or defects. The starting air manifold pipe should be cleaned regularly and checked for paint deformation. This will indicate any overheating of the pipe. Last of all, overhaul the starting air valve at the intervals stated by the manufacturer to avoid leaks and poor operation.

We have now covered the two main causes of engine room explosions. Both are extremely dangerous. Fortunately, crankcase and starter air-line explosions are completely avoidable through effective maintenance and overhauling procedures. There is no legitimate reason either type of incident should occur other than through poor engine room housekeeping. In the next section of this chapter, we will look at the most common types of engine room fires. Again, these can be easily avoided by following proper engine room procedures.

PURIFIER ROOM FIRES

The purifier room is one of the most probable locations within the engine room to catch fire; indeed, purifier room fires have been the cause of several major incidents in the past. As we know, fires require three elements, all of which are readily available in the purifier room. To prevent an outbreak of fire in the purifier room, the following safety precautions are recommended: (1) All pipes leading to the separator should be double sheathed; the reason for this is that if the inner pipe leaks, it will not spray oil all over the place but instead will leak into the outer pipe. (2) Drip trays should be provided below the purifier or separator so that in the event of an oil spill the oil will not flow and spread throughout the purifier room. This will prevent inadvertent contact with hot surfaces which can lead to fires. (3) All pipes with flanges or connections should be covered with anti-spill tape, which can prevent spills from the flanges in the event of a leak. (4) Firefighting systems such as water mist and CO_2 should be installed and kept on standby. (5) Quick-closing valves and remote stopping of pumps and purifiers should be provided and regularly tested for operation. (6) Fire detection and alarm systems should be provided and tested regularly for operation. Bearing these safety precautions in mind, a small purifier fire can be easily stopped with the help of a portable fire extinguisher. In the event the fire becomes self-sufficient and established, the following steps should be taken: (1) as soon as the fire alarm is sounded, inform the chief engineer and locate the fire; (2) close the quick-closing valves from which the oil is leaking; (3) stop the transfer pump and quick-closing valves (these can be operated from a remote location, such as the engine control room, the bridge, or the ship's control centre); (4) stop all motors and electrical equipment; (5) if the fire can be extinguished locally, attempt to fight the fire with a portable fire extinguisher; (6) if the fire is far too gone, close the air supply pump and exhaust from the purifier room; (7) release the water mist system if installed on the ship; (8) enter the purifier room only after donning firefighter protective equipment, self-contained breathing apparatus, and manhandling the fire hose; (9) fight the fire by spraying cooling water along the boundaries of the compartment and at the seat of the fire; (10) if the fire cannot be extinguished with cooling water, inform the chief engineer. The chief engineer will discuss the situation with the master and authorise the use of the CO_2 system; (11) exit the purifier room and secure the entry point. No one should be permitted to enter the compartment after authorisation to activate the CO_2 is given. Once the fire is fully extinguished, begin the clean-up procedure in accordance with the FCP.

SCAVENGE FIRES

Of all the types of fires to involve the engine room, the scavenge fire is the deadliest of all. To understand why, we must first begin by covering the basic principles of scavenge fires. By now we are aware that all fires require heat, fuel, and oxygen. In the scavenge space, all three elements are readily available and in great abundance. Oxygen is omnipresent, as it is needed in many engine room operations. The heat source may come from the blowing by of gases between the piston rings and the liners, or because of rubbing between two metal surfaces. The fuel may be unburnt fuel, carbon, or cylinder lubricating oil which has leaked into the scavenge space. When all these elements are present in a proportional ratio and lie within the flammable limit inside the scavenge space, the latter becomes a prime location for combustion. The fire which results from this combustion is known as scavenge fire. There are many causes for scavenge fires, though the main causes include excessive wear of the liner, the piston rings becoming worn out or developing loose ring grooves, broken piston rings or rings seized in the grooves, a dirty scavenge space through poor housekeeping, poor combustion due to leaking fuel valves or improper timing, and insufficient or excessive cylinder lubrication. Fortunately, there are a few signs which may indicate a scavenge fire is developing. These include (1) the scavenge temperature is starting to increase; (2) the turbochargers are starting to surge; (3) there is an abnormally high exhaust temperature; (4) there is a loss of engine power and reduction in rpm, which happens because back pressure is created under the piston space due to the fire; (5) smoke is seen coming out of the scavenge drains; (6) the development of paint blisters forming on the scavenge doors due to excessively high temperatures.

The actions to be taken in the event of a scavenge fire depend on the type of fire, and whether it is incipient or established. Established fires are usually easier to detect as physical manifestations will begin to appear, such as the peeling or blistering of paint, large reductions in engine rpm, and surging of the turbocharger. Incipient, or smaller fires which have not yet become established, are much harder to detect. If a small scavenge fire is suspected, start by reducing the engine rpm to slow or dead slow. Increase the cylinder lubrication of the affected unit. Special attention must be given to ensure the lube oil does not feed the fire. If the fire is seen to increase in intensity, cease all lubrication. The cause of the fire may be due to leaky fuel valves, so lift the pump of the affected unit. Keep the scavenge drain fully closed. Keep monitoring the scavenge and exhaust temperatures and let the fire starve and burn itself out. After confirming that the fire is extinguished, start to slowly increase the rpm. Keep monitoring the scavenge temperature for any signs of re-ignition. For large fires, stop the engine immediately and engage the turning gear. Keep the engine rotating with the turning gear. Extinguish the fire with the fixed firefighting system for the scavenge space. This may be the CO_2 system or a steam connection provided for smothering the fire. In the event a fixed firefighting system is not available (for example on incredibly old ships) apply external cooling to prevent heat distortion. After confirming the fire is extinguished, allow the scavenge space to cool down before entering for inspection and cleaning.

EXHAUST GAS BOILER FIRES

Every system, which is operated at elevated temperatures, has a constant risk of fire. This also applies to the exhaust gas boiler (EGB), which is a type of heat recovery system installed on ships. It allows the exhaust heat of the main engine to produce steam whilst being exhausted out into the atmosphere. Most EGBs operate with an inlet temperature of between 300°C and 400°C (572°F –752°F). With the water tube type arrangement, which is the most popular arrangement, the water passes through a tube stack, which is arranged in the path of the exhaust gas inside the exhaust gas trunking of the main engine. The exhaust gas flows over the tube stacks heating the water. This produces steam. Because the main engine is not 100% efficient, soot deposits can build up. The main constituents of these soot deposits are particulates in addition to unburnt residues of fuel and lubricating oils. If these soot deposits and particulates are not regularly removed from the EGB, they can ignite and combust. The main reasons for soot and deposit accumulation are poor combustion of fuel in the main engine, prolonged slow steaming, long manoeuvring periods, frequent starting and stopping of the main engine, use of poor grade fuel oil and or cylinder oil, low exhaust gas velocity passing through the EGB, low water inlet velocity in the water tubes, and low circulation water flow ratios. To better understand the causes of EGB fires, it is useful to distinguish the fires in stages rather than as types. EGB fires can be categorised in two or three stages depending on the intensity of the fire:

- Stage 1: Normal soot fire

- Stage 2: Hydrogen fire
- Stage 3: Iron fire

Stage 1: Normal soot fire. Soot is deposited in the water tube of the EGB. When the ship is at slow speed, the exhaust temperature of the main engine may vary from 100°C to 200°C (212°F–392°F). This temperature is sufficient to ignite 'wet soot' whose ignition temperature is around 150°C (302°F). If the soot is 'dry', it will not get ignited at such low temperatures (i.e., 150°C (302°F)) but when the engine is running at higher speeds and the temperature of the gases reaches above 300°C (572°F), then in the presence of excess oxygen, the deposits of combustible materials will liberate sufficient vapour, which can be ignited by a spark or flame. These types of soot fires are called small or normal soot fires because the heat energy is carried away by the circulating boiler water and steam. Moreover, the sparks remain inside the funnel or diminish while passing through the flame arrestor at the funnel top.

Stage 2: Hydrogen fire. Hydrogen fires in an EGB occur when the chemical reaction of dissociated water takes place at a temperature above 1,000°C (1,832°F). This leads to the formation of hydrogen (H_2) and carbon monoxide (CO), both of which are highly combustible, as provided below:

$$2 H_2O = {}^2H^2 + O^2$$
(Dissociation of water leading to formation of hydrogen (H_2))

$$H^2O + C = H_2 + CO$$
(Reaction of water with carbon deposits leading to the formation of carbon monoxide (CO))

Stage 3: Iron fire. At this stage, the chain reaction of oxidation of iron metal starts at an elevated temperature of 1,100°C (2,012°F), which means at such elevated temperatures, the tube itself will start burning. This leads to a complete meltdown of the tube stacks:

$$^2Fe + O^2\, 2 = FeO + heat$$

It is strongly advised not to use water or steam at this stage to fight the fire, as the overheated iron will react with water to continue this reaction:

$$Fe + H_2O = FeO + H_2 + heat$$

To prevent EGB fires from starting in the first place, avoid slow steaming wherever possible. Ensure good fuel combustion in the main engine. Ensure fuel is treated and is of excellent quality. Conduct regular soot blowdown of the boiler tubes. Perform water washing in port at regular intervals. Ensure the design of the exhaust trunk is such as to provide uniform heat to the complete tube stack. Preheat circulating water to the boiler at the time of start-up. Avoid turning off the circulating pump whenever the main engine is running and allow to operate for a further two hours after the engine has stopped. Start the circulating pump at least two hours before starting the main engine. By following these simple precautions, it should be possible to minimise the opportunity for EGB fires. In the unfortunate event that an EGB fire does break out, the response for tackling the fire will depend on the stage the fire is at. When there is a Stage 1 fire, i.e., normal soot fire, stop the main engine, and cut the supply of oxygen to the fire. Continue operating the water-circulating pump – never stop the pump! Never use the soot blowers when firefighting – both steam and air will accelerate the fire. Ensure the exhaust valves in the main engine are in the closed position to reduce any air supply to the fire. Cover the filter of the turbocharger. If fitted, use the water-washing equipment to fight the fire. This is normally connected to the ship's firefighting water system. Perform external boundary cooling to reduce the effect of heat and prevent buckling and metal formation. For major fires, i.e., Stage 2 and Stage 3 fires, stop the main engine, if it is not stopped already. Stop the circulating water pump. Shut all the inlet and outlet valves on the water circulation line. Discharge the (remaining) water from the EGB sections by draining. Cool down with plenty of water to boundary cool the seat of the fire. Take care not to splash water onto other parts of the EGB assembly, as the water may accelerate the fire.

INCINERATOR FIRES

Incinerator fires are an unusual occurrence, as the incinerator is fitted with a bevy of safety devices. Proper operation of the incinerator, as covered earlier in this book, should be sufficient to prevent an outbreak of fire from occurring. Incinerator fires have happened in the past, as this example demonstrates. A vessel which was underway started its incinerator to burn off oily rags and sludge. About five hours after the incinerator was started, the incinerator was stopped. All parameters appeared normal, and the furnace temperature was recorded at 950°C (1,742°F). Following company procedures, the engine crew continued to monitor the incinerator during the cooling-off period. By 1900 hours, a further five hours after the incinerator had been turned off, the temperature of the furnace was recorded at 280°C (536°F). Oddly, the blower fan was still running. At 2032 hours, the duty engineer noticed smoke coming from the outer body of the incinerator. On closer inspection, he could see paint peeling off the incinerator's body. The temperature of the incinerator body was checked and recorded as between 250°C and 350°C (482°F –662°F). The duty engineer informed the chief engineer, who authorised the emergency response. The crew mustered and fire parties began boundary cooling. Boundary cooling was continued for about four hours until heat indications suggested that the fire was extinguished. During the investigation, it was found that the fire had started in the air-cooled incinerator chamber jacket. Later, it was found that the refractory and outside body plates were intact. Traces of oil were found between the sludge dosing door and the combustion chamber, which was an indication that oil had accumulated in the double-shell refractory lining. What this incident tells us is that even during the cool-off period, the incinerator must be addressed and regularly checked. Moreover, by initiating the boundary cooling, the chief engineer was able to prevent the fire from growing out of control. Hopefully, this real-life incident reinforces the need for diligence and for maintaining a cool head even in the face of a potentially lethal situation.

ELECTRICAL INSULATION FIRES

The insulation on electric cables is typically made from rubber or plastic. When this rubber or plastic burns, it gives off fire gases which differ according to the type of insulation material used, any additives in the manufacturing process, the intensity of the flame, and any ventilation arrangements. Most rubbers and plastics produce very dense fire gases when heated. In this section, we will cover the hazards associated with electric cable insulations when subjected to fire. Some plastics burn cleanly when subject to heat and flame, producing extraordinarily little in the way of smoke. Other types of insulation, such as urethane foam, however, produce very dense smoke meaning visibility in the affected compartment can be lost in as little as one minute. Some plastics contain polyvinylchloride or PVC, which produces hydrochloride (HCl) gas as a by-product of combustion. This is a very deadly gas and has a pungent, irritating odour. Alternatively, when rubber is used for insulation, and is subject to intense heat and flame, it produces a dense black, oily smoke which is also highly toxic. The most common gases produced during rubber combustion are hydrogen sulphide (H^2S) and sulphur dioxide (SO^2). These gases are both fatal.

BATTERY ROOM FIRES

The battery room of a ship is at risk of explosion as the ship's batteries constantly release hydrogen during charging. Hydrogen (H) is a highly explosive gas, and it is therefore important to take the following necessary steps while working inside the battery room during maintenance: (1) provide proper ventilation inside the compartment and (2) prevent any source of ignition inside the compartment. Ventilation is provided with the help of ventilation fans. The ventilation arrangement should be such that there is no accumulation of hydrogen in the space. Hydrogen is lighter than air and thus tends to accumulate at the top of the compartment. The fans should be of a non-sparking type and should not produce any static charge. The ventilation ducts should be below battery level, which helps force the gases out. The motor used should be of a standardised approved type so that there is no spark from the motor. The tools used for maintenance should be coated in a rubberised layer to prevent any chance of short circuiting by mistake. The coating will also prevent any kind of spark from occurring should the tools fall on the floor. The paint used in the battery room and the materials for ducting should be corrosion resistant. Metal jugs should not be used for filling distilled water inside the batteries. Additional precautions include preventing the usage of naked lamps and prohibiting smoking in the battery room. The battery must never be placed in the emergency switchboard room as there is a high chance of sparking due to circuit breakers arcing. To avoid malfunctioning, the batteries should be maintained in a fully charged condition. This is done by the charging circuit. The state of charge can be seen with the help of a hydrometer. A sample is taken using the hydrometer which checks the condition of specific gravity. For a fully charged lead acid battery, the specific gravity is 1.280 at 15°C (59°F). There is no change in specific gravity in alkaline batteries during charging and discharging; therefore, the hydrometer test is used only for lead acid batteries. Top the batteries up with distilled water to compensate for the loss of water during charging. Always keep the battery terminals clean. This can be achieved by smearing the terminals with petroleum jelly.

BACTERIA FIRES

Bacteria fires are one of the more obscure and least understood incidents to happen on board ships. That said, like most issues, bacteria fires can be easily avoided through good housekeeping and diligence. To put this into context, it was a fine calm morning on a cargo ship and the engineers were busy with their routines. The chief engineer was also on his regular rounds. On the cylinder head platform, as he was checking the parameters and engine components, he observed some hazy white patches of smoke emerging out of a partially covered waste bin. The chief engineer attempted to open the cover to check what was going on inside the bin. As he opened the bin, the in rush of air caused the fire to spike out from the bottom of the bin. The chief engineer immediately raised the fire alarm and, taking a nearby dry powder extinguisher, managed to safely extinguish the fire. After the commotion settled down, an investigation was made to analyse the cause of the fire. The chief engineer suspected that other engineers or engine crew might have dropped a lit cigarette butt into the bin. When interviewed, it transpired none of the engine crew were smokers! The chief engineer immediately informed the master about the incident and despite the initial investigation, no obvious cause was identified. What we do know from the incident report is that the chief engineer asked the oiler, who oversaw that platform, to clear out the garbage daily. The oiler explained that he followed standard procedure. The oiler dumped milk packs, oily rags, some waste bread, and other organic materials such as cotton waste into the bin. At the end of the day, the bin was emptied, and the contents incinerated the next day. The master informed the ship manager about the incident, who decided to send samples ashore for further laboratory investigation. After a month, the laboratory established what they considered was the most probable cause for the fire. The entire sample was divided into 25 to 30 equally sized smaller samples, with each sample kept at a certain temperature to observe their behaviour. The samples which were kept at a temperature of 20°C to 30°C (68°F– 86°F) did not show any considerable change in their behaviour. However, the samples which were kept at 40°C (104°F) and above showed unusual behaviour. Even after the heating element for the sample was turned off, their temperature continued to rise inexplicably to more than 9°C (48.2°F). Later upon testing, it was found that the steep increase in temperature was due to a peculiar kind of bacteria called 'thermophilic bacteria'. Thermophilic bacteria grow in warm and moist environments. Ship engine rooms are an ideal breeding ground for Thermophilic bacteria. When they multiply, the bacteria give off intense heat, which in the scenario was sufficient to ignite the oily rags disposed of in the bin. Further investigation found the auto-ignition temperature of oily cotton rags is 120°C (248°F) with oily cotton rags having a greater affinity for self-heating and spontaneous combustion. The outcome of the investigation was to demonstrate the importance of having a well-thought-out garbage management plan on board and to follow it. This incident led to the now mandatory practice of segregating garbage. For example, oily rags must be disposed of in a separate bin with food waste, metal scraps, ash, and batteries assigned to assorted colour-coded bins. Though not mandatory, it is good practice to clear out the bins before unmanning the engine room.

In this chapter, we have covered some of the most common fires to break out in the ship's engine room. We have also looked at the main ways these fires can be avoided. At the heart of fire prevention is good housekeeping and following processes and procedures. Unfortunately, sometimes it is impossible to prevent accidents from happening. But in most situations, fires are the result of carelessness, poor maintenance, and ill-guided attempts to cut corners and save money. Safety should always be the number one concern for everyone on board. Now that we know the most common types of engine room fires, in the following chapter, we will look at engine room drills, firefighting equipment, and firefighting procedures.

Chapter 27

Engine room drills, firefighting procedures and apparatus

The engine room is particularly at risk of incidents that can quickly overcome the engine department personnel, rendering the vessel unseaworthy. The purpose of any shipboard drill is to acquaint the officers and crew with the various procedures to be followed in an emergency. It is an opportunity to train and evaluate the ship's staff's ability to respond to potentially life-threatening incidents. The fire drill is one such drill which is of immense importance on ships. It helps the ship's crew to understand the basics of fire prevention, and helps to prepare the crew for dealing with an emergency situation that may arise because of a fire on board the ship; it enables every crew member to become familiar with the duties they are assigned in the event of an actual emergency; and it enables the crew to train and become proficient in the use of firefighting appliances such as breathing apparatus, different types of fire extinguishers, the CO_2 flooding system, the inert gas system, donning and doffing personal protective equipment in emergency scenarios, and using the Neil Robertson stretcher to rescue incapacitated or injured crew members. Fire drills are also an excellent way of training the crew to understand the procedures and operational limitations of specific firefighting systems and the precautions to be taken before operating those systems. The engine room, because of its location and position within the vessel, has unique firefighting systems that differ from other parts of the ship. These systems require specific and dedicated training to be effective in stressful conditions. Moreover, engine room personnel are more at risk of succumbing to smoke and flooding. Again, engine department personnel need to be well versed in the location of emergency escape routes should the conventional means of egress become inaccessible.

FIRE DRILLS

To be effective, fire drills need to be regular and as realistic as possible. Under SOLAS, all crew members are required to participate in fire prevention training, which is renewed, onshore, once every five years. In addition, to meet the requirements of SOLAS, additional fire drills must be conducted within 24 hours of leaving port if more than 25% of the current crew have not participated in fire drills within the preceding month. Muster lists for drills must be displayed prominently throughout the ship in locations where the list can be easily accessed. Copies of the muster list must also be displayed on the bridge, in the engine control room, and throughout the crew accommodation areas. A clear fire control plan should be prominently displayed in critical areas throughout the ship. Each crew member should be provided with clear instructions which they are required to follow in the event of an emergency. The duties of each crew member, together with their assigned lifeboat number, must be written on individual cards and made available inside and outside their cabins and their offices (where relevant). The timing of emergency drills should be amended each time to change the scenario and allow those crew members to participate who have not attended the previous drill. The location of the drills should also be changed to provide practice to the crew of different conditions and to train them to tackle diverse types of fires such as machinery space fires, accommodation area fires, storeroom fires, cargo hold fires, and so forth. The location of the muster station should be such that it is readily accessible from the accommodation areas and all workplaces. It should also be located close to the embarkation station. The muster point should have sufficient means of illumination provided from an emergency source.

Each area of the ship has a different method of approach to deal with emergency situations. Training with drills in different situations helps to prepare the crew members for all types of emergencies. It is the duty of every member of the ship's staff to acquaint themselves with the location of the emergency muster station upon joining the ship for the first time. Each crew member must also know their duties and responsibilities, as described in the muster list, and learn how to use the onboard firefighting appliances.

PREVENTATIVE MEASURES AND FIREFIGHTING APPLIANCES

Without a doubt, one of the main causes of accidents onboard ships is fire. This is because of the presence of elevated temperatures, copious quantities of flammable oils, and other combustible materials. Today, ships are only approved to operate within international waters if they are designed and built to the IMO's Fire Safety System Code (FSSC) and carry the mandatory firefighting appliances required for the type and class of vessel. It is worth noting that several types and classes of vessels are fitted with diverse types of firefighting appliances. For example, one system that is commonly installed on oil tankers is the inert gas system. Although inert gas is remarkably effective at suppressing fires, it would be inappropriate to use inert gas against a fire on a passenger ferry or cruise ship. There is a litany of fire suppressants and preventative measures available on ships today. In this section, we will briefly discuss some of the most common.

(1) *Fire retardant bulkheads*. To prevent the propagation of fire from one compartment to another, all watertight bulkheads are provided with fire-resistant panelling. Depending on the level of fire resistance, bulkheads are rated into one of three classes: Class A, Class B, or Class C. *Class-A panels*. All watertight bulkheads are Class-A type. Bulkheads of Class-A type must be constructed of steel or an equivalent material and must pass the standard fire test, preventing the passage of fire or smoke to the unaffected side for at least one hour. Where a Class-A bulkhead is installed, the average temperature on the unaffected side of the bulkhead must not exceed 120°C (248°F). In addition, there are three categories of Class-A bulkhead depending on the time up to which the temperature at any point on the bulkhead must not rise above 160°C (320°F), accordingly:

- A-60 panel: 60 minutes
- A-30 panel: 30 minutes
- A-15 panel: 15 minutes
- A-0 panel: 0 minutes

Class-B bulkheads. Bulkheads of Class B are constructed of materials that SOLAS and Class determine are incombustible materials. These must pass the standard fire test, preventing the passage of fire or smoke to the unaffected side of the bulkhead for a minimum of 30 minutes. Where Class B bulkheads are installed, the average temperature on the unaffected side must not exceed 120°C (248°F). There are two types of Class B bulkheads depending on the time up to which the temperature at any point on the bulkhead must not rise above 206°C (402.8°F). These are as follows:

- B-15 panel: 15 minutes
- B-0 panel: 0 minutes

1. *Class-C bulkheads.* Class C bulkheads and decks are constructed of materials that are approved by SOLAS and Class, and are determined to be incombustible, but they are not required to meet any requirements related to the rise in temperature or passage of smoke and flame to the unaffected side of the bulkhead.

   Class-A and -B bulkheads are most used adjacent to enclosed spaces within the ship, for example, the cargo holds, control stations, stairways, lifeboat embarkation stations, galleys, machinery spaces, tanks, public spaces, and accommodation areas. Class C panels are mostly used on open decks and promenades, where the fire safety requirement is minimum. They can also be used between two similar spaces if they are not separated by a watertight bulkhead, in which case Class-A bulkheads are mandatory.
2. *Fire doors.* Fire doors are installed in fire retardant bulkheads to provide access to and from the compartment. They are self-closing type doors with no holdback arrangement.
3. *Fire dampers.* Dampers are provided in the ventilation system of cargo holds, the engine room, and the ship's accommodation to prevent oxygen from supplying the fire.
4. *Fire pumps.* As per the SOLAS regulations, every ship must carry a main fire pump and an emergency power pump of an approved type and capacity. The location of the emergency fire pump must be outside the space where the main fire pump is located.
5. *Fire main piping and valves.* The fire main piping, which is connected to the main and emergency fire pump, must be of an approved type and capacity. Isolation and relief valves must be provided in the line to avoid the build-up of overpressure.
6. *Fire hose and nozzles.* Ships are mandated to carry fire hoses with a length of at least 10 m (32 ft). Class determines the number and diameter of these hoses. Nozzles with diameters of 12 m (39 ft), 16 metres (52 ft), and 19 m (62 ft) are used for dual-purpose types, providing both jet and spray modes.
7. *Fire hydrants.* Fire hoses are connected to fire hydrants from which the water supply is controlled. These are made from temperature-resistant materials to prevent damage from heat and sub-zero temperatures.
8. *Portable fire extinguishers.* Ships carry an array of portable fire extinguishers consisting of CO_2, foam, and dry chemical powder (DCP). These are typically located throughout the ship's accommodation, in the deck and machinery spaces, and in any other spaces and compartments stated in the appropriate regulations or as dictated by Class.
9. *Fixed fire extinguishing systems.* CO_2, foam, and seawater are all used as firefighting media by fixed fire extinguishing systems. Because they blanket the entire compartment, the system must be operated remotely from outside the protected compartment.

10. *Inert gas system.* The inert gas system is provided on the oil tankers of 20,000 gross tonnes deadweight and above, as well as those which are fitted with crude oil washing mechanism. The Inert gas system is designed to protect the cargo space from fire hazards by smothering the fire and replacing any oxygen within the affected compartment. Inert gas systems are extremely toxic to human health and must never be deployed whilst personnel are in the compartment.
11. *Fire detectors and alarms.* Fire detection and alarm systems are installed in the cargo area, accommodation, deck areas, and machinery spaces along with an alarm system to notify the bridge of any outbreak of fire.
12. *Remote shut and stop systems.* The remote station shutdown is provided to all fuel lines from the fuel oil and diesel oil tanks in the machinery space and is done by quick closing valves. A remote stop system is also provided to stop machinery such as the fuel pumps, purifiers, ventilation fans, and boiler, in the event of a fire in the engine room or before discharging a fixed firefighting system.
13. *Emergency Escape Breathing Device (EEBD).* The EEBD is used to escape from any compartment or space with an irrespirable atmosphere. The location of emergency escape breathing devices and their spares must be as per the requirements given in the FSSC.
14. *Firefighter's personal protective equipment.* The firefighter's personal protective equipment, colloquially known as the firefighter's outfit, is worn when fighting shipboard fires. The outfit is manufactured from fire retardant material of an approved type. For cargo ships, at least two outfits must be carried on board, and for passenger ships, at least four outfits must be carried on board.
15. *International shore connection (ISC).* The ISC is used to connect shore water to the ship's fire main system in the event the ship's fire pump is not operational. The ISC can only be used when the vessel is in port, on layoff, or in dry dock. The size and dimensions of the ISC are standard for all ships and at least one coupling with a gasket must be carried on board.
16. *Means of escape.* Escape routes and passages must be provided at various locations throughout the ship together with ladders and supports leading to safe locations.

SPRINKLER SYSTEM: AUTOMATIC FIRE DETECTION, ALARM, AND EXTINGUISHING SYSTEM

The sprinkler system is an automatic fire detecting, alarm, and extinguishing system which is constantly on standby to deal quickly and effectively with the outbreak of fire that may occur in the ship's accommodation and other spaces. This system consists of a pressure water tank with water pipes leading to various locations throughout the ship. The water pipes consist of a sprinkler head which comes into operation when there is an outbreak of fire. The pressurised water tank is half-filled with freshwater through the freshwater supply connection. Compressed air is delivered from an electrically operated compressor or from the air bottle, which increases the pressure to the predetermined level. The pressure in the tank is such that it can deliver pressure to the highest sprinkler head in the system. For this purpose, the pressure should never be less than 4.8 bar. The sprinkler heads are grouped into different sections with no more than 200 sprinkler heads in each section. Moreover, each section has its own alarm system which activates on operation. The sprinkler head consists of a quartzoid bulb which bursts when the temperature in the compartment or

Table 27.1 Sprinkler head temperature ratings

Colour	Temp. (°C)	Temp. (°F)
Red	68°C	154°F
Yellow	80°C	176°F
Green	93°C	177°F

space increases beyond the prescribed limit. Once the bulb bursts, water starts flowing from the sprinkler head. These quartzoid bulbs are colour coded in red, yellow, and green in accordance with their temperature rating (Table 27.1).

Each sprinkler head covers a deck area of 16 m square (172 sq. ft) and the flow of water in each head should be a minimum of 5 litres per minute (1.09 gal) as per SOLAS. When the sprinkler head bursts and comes into operation, the non-return valve in the line opens allowing the water to free flow. Due to this flow, there is a drop in pressure in the line, which activates the alarm for that section. This indicates a fire in the affected compartment or space. The sprinkler system is connected to the seawater pump and supplies water to the system in the event the freshwater in the pressure tank is depleted. Various alarms and pressure switches are provided in the system for the aid of maintenance. This includes checking the alarms and the activation of the seawater pump by isolating the system. The sprinkler system is found in the ship's accommodation, the paint room, and other sensitive locations around the ship.

CO_2 FIREFIGHTING SYSTEMS

The most used system for firefighting in the engine room and dry cargo holds (i.e., those on any vessel not designated as an oil tanker, gas carrier, or chemical tanker) is the CO_2 flooding system. The CO_2 system consists of a fire detection system (usually smoke detectors) and an alarm system, together with an assembly of CO_2 cylinders. When there is a likely indication of fire in the engine room or cargo hold, the assembly of CO_2 bottles is released depending on the cargo permeability (i.e., how much space is empty over the cargo for the CO_2 to settle). The CO_2 firefighting system typically consists of a 20-millimetre (0.06 ft) diameter sampling pipe which is fed into all the cargo holds. This is controlled via a cabinet on the ship's bridge or in the ship's control centre. Air is continuously drawn through these pipes to the cabinet by way of suction fans. When there is a fire in any of the cargo holds, fire gases are sucked into the sampling pipes and passed through the diverting valves' bridge-mounted cabinet. This means the bridge is warned almost immediately of any fire in the holds. At the same time, the sample from the pipes is passed over a smoke detector which senses the fire gases and activates the audio-visual alarm indicating the outbreak of fire. The main advantage of the audio-visual alarm is that even if the bridge is unattended, for instance when the ship is in port, the ship's staff will still be advised there is an outbreak of fire. In the cabinet, the sample is passed over small propellers made from nylon, and through a transparent tube approximately 13 mm (0.04 ft) in diameter, which indicates the direction and clarity of the airflow. If the propellers cease running, this indicates that the pipe is blocked. In the event of a fire, the sampling pipe will turn dark, which is demonstrative of fire gases flowing through the sampling pipe. The sampling pipe is connected to the assembly of CO_2 bottles by way of a changeover valve. The CO_2 is released by opening the appropriate valve for the hold. A crucial point to note before releasing the CO_2 is that the amount of CO_2 to be released must be calculated, accordingly:

Total volume of the hold – the volume of cargo = The permeable space

30% *of the total volume of the hold must be flooded with CO_2 to be effective*

One bottle of CO_2 provides approximately 45.2 kg (99.6 lbs)

Tragically, there have been several instances in the past where crew members have lost their lives in the engine room, not because of the fire, but due to suffocation caused by the release of CO_2. Subsequently, the CO_2 system must always be seen as the last resort for engine room firefighting and only after all other methods have been exhausted. Furthermore, it should be remembered that once discharged, the CO_2 system cannot be used again until the CO_2 cylinders are recharged. On most, if not all vessels, the CO_2 system is managed and controlled by the chief engineer or the second engineer in their absence. To operate the CO_2 system, the following protocols must be followed: (1) Following an outbreak of fire, the fire alarm will sound alerting the OOW. If the fire is small enough to be fought exclusively with portable extinguishers, the crew should be gathered at the muster station for a head count. If the fire is too far involved or the location of the fire makes it unacceptably dangerous to tackle in person, the chief engineer must discuss with the master the option of activating the CO_2 system. (2) If the decision is made to activate the CO_2 system, the emergency generator should be started as CO_2 flooding requires all engine room machinery, including the auxiliary power generator, to be stopped. (3) From the bridge, reduce the ship's speed and stop the main engine once the vessel has reached a safe position – ideally away from other marine traffic, stationary objects such as buoys and sea marks, or potential grounding areas such as sandbanks and submerged rock outlets. The master must inform the nearest coastal authority if the ship is within their coastal zone. (4) Using the key in the ship's fire station, open the CO_2 system cabinet. This will automatically activate the audible CO_2 alarm in the engine room. (6) Some systems and machinery, such as the engine room blowers and fans, will trip when the CO_2 cabinet is opened. Before proceeding, check that all systems have tripped. (7) Ensure there is no one inside the engine room by repeating the head count. (8) Operate the remote closing switches for the quick closing valve, funnel flaps, fire flaps, engine room pumps, and machinery, watertight doors, etc. (9) Stop and isolate the engine control room air-conditioning system. (10) Close all the entrance doors to the engine room and ensure the compartment is fully airtight. (11) Operate the control and master valve in the CO_2 cabinet. This will sound another alarm, after which the CO_2 will be released. This takes about 60 seconds. (12) If there is any need to enter the engine room after the CO_2 has been released (for example, to rescue a trapped member of the crew), self-contained breathing apparatus sets (SCBA) and lifelines must be donned. Under no circumstances should anyone enter the affected compartment until authorised by the chief engineer and the master.

ISC

The ISC is a universal hose connection that is to be provided on all ships as per SOLAS Chapter II-2, regulation 10.2.17, which states all ships over 500 gross tonnes deadweight or more must carry a minimum of one ISC. The purpose of the ISC is to keep a

Table 27.2 Dimensions of the ISC

Description	Dimension
Outside diameter (OD)	178 mm (7.0 in)
Inside diameter (ID)	64 mm (2.51 in)
Bolt circle diameter (PCD)	132 mm (5.19 in)
Slots in flange	× 4 holes, 19 mm (0.74 in) in diameter spaced equidistantly on a bolt circle of above diameter, slotted to the flange periphery.
Flange thickness	14.5 mm (0.57 in) minimum
Bolts and nuts	× 4, each of 16 mm (0.62 in) diameter and 50 mm (1.96 in) in length.

standby hose attachment to facilitate the coupling of shore-based fire hoses in the event of a total failure of the ship's fire pumps. When using the ISC, seawater is supplied at a predetermined pressure and is connected to the ship's fire main. This ISC flange is kept at a convenient and accessible location either on the bridge or in the fire locker so that in the event of an emergency it is readily available. The ISC flange has a standard size and is the same for all countries and ships. The dimensions of the ISC are as follows (Table 27.2)

By international agreement, all ships, jetties, and offshore platforms likely to require an emergency source of fire water or to provide such must have at least one international shore fire connection. The connection is typically formed from steel or another suitable material and is designed for 1.0 N/mm^2 services. The flange should have a flat surface on one side and on the other side a permanent connection or attachment to a coupling that can be easily fitted to the ship's hydrant and hose connection. The connection should be kept on board with a ready gasket of material that can manage a pressure of 1.0 N/mm^2 together with four 16 mm (0.62 in) bolts, 50 mm (1.96 in) in length, and eight washers so that the connection can be readily assembled in the event of an emergency. Where the internal shore connection is fixed to the vessel's fire main, the shore connection should be accessible from either side of the ship and be clearly marked. A notice should be posted near the ship's fire main close to the accommodation in English and the local language of the ship's Flag, indicating the connection's location and the maximum working pressure of the piping system which the port must acknowledge and supply. The velocity of water supplied by the port or another ship to the affected ship should not exceed 5 m per second (16 ft per second) in the firewater distribution network. The fitting and joints must be suitable for a working pressure of at least 10.5 bar. To ensure the ISC is kept in good operational condition, the connection flange should be included in the preventative maintenance schedule plan and not in the Condition Monitoring plan for the ship. Always ensure the gasket is kept together with the ISC flange. It is advisable to prepare more than one gasket and keep it with the shore connection, as any damage to the gasket may render it inoperable. When making a duty list for the firefighting drill, it is good practice to assign one member of the crew to bring the ISC to the open deck area from where the shore connection is to be taken. Turning this action into a habit will reduce the amount of time it takes to set up the short connection in a real emergency. Last of all, if the international shore coupling is permanently installed on the ship's fire main, always check its tightness before using the connection with water.

FIREFIGHTING ON OIL TANKERS

Not strictly related to the engine room per se, but useful to discuss all the same, oil tankers have a specific firefighting system which is unique to their class of vessel. Oil tankers carry oils of different grades and quality, which have the properties to produce flammable vapours and gases when loaded for transportation. Even when there is no cargo loaded onboard, there can be harmful flammable vapours present in the holds. When the vapour produced by the cargo mixes with a certain concentration of air (primarily containing oxygen) it can result in a considerable explosion. Because oil tankers carry highly flammable and explosive cargoes, the media used to tackle fires must not present a greater risk to the ship than the initial hazard posed by the cargo and the fire. For this reason, inert gas is used. The inert gas system is like the CO_2 system discussed earlier in the sense that it releases a gas which neutralises the fire by cutting off the supply of oxygen, thus starving the fire of one of the three basic elements. The inert gas can be supplied either through a separate inert gas plant or as flue gas produced by the ship's boiler. Inert gas is a gas that contains minimal oxygen (normally less than 8%) and is used to suppress the combustion of flammable hydrocarbon gases. The inert gas system spreads the inert gas over the oil cargo hydrocarbon mixture, which increases the lower explosive limit, which is the lowest concentration at which the vapours can be ignited, simultaneously decreasing the higher explosive limit, the highest concentration at which a vapour explodes. When the concentration reaches around 10%, an atmosphere is created inside the tank in which hydrocarbon vapours cannot burn. The concentration of inert gas is maintained at around 5%.

The typical inert gas system consists of the following components: (1) *Exhaust gases source*. The inert gas source is taken from exhaust uptakes of the boiler or main engine as contains flue gases in it. (2) *Inert gas isolating valve*. This serves as the supply valve from the uptake to the rest of the system, which isolates both systems when not in use. (3) *Scrubbing tower*. Flue gas enters the scrubbing tower from the bottom and passes through a series of water spray and baffle plates to cool, clean, and moisten the gases. The SO^2 level decreases by up to 90% and gas becomes clear of soot. (4) *Demister*. Normally made of polypropylene, the demister is used to absorb moisture and water from the treated flue gas. (5) *Gas blower*. Normally two types of fan blowers are used: a steam-driven turbine blower for inert gas operation and an electrically driven blower for topping-up purposes. (6) *Inert gas pressure regulating valve*. The pressure within the tanks varies according to the properties of the oil and the tank's atmospheric conditions. To control this variation and to avoid overheating the blower fan, a pressure regulator valve is attached after the blower discharge. This re-circulates the excess gas back to the scrubbing tower. (7) *Deck seal*. The purpose of the deck seal is to stop the gases from the cargo tanks returning to the blower. Normally wet type deck seals are used. A demister is fitted to absorb the moisture carried away by the gases. (8) *Mechanical non-return valve*. The mechanical non-return valve is an additional non-return mechanical device fitted in line with the deck seal. (9) *Deck isolating valve*. The engine room system can be isolated fully from the deck system by way of the deck isolating valve. (10) *Pressure vacuum (PV) breaker*. The PV breaker helps control the over or under pressurisation of the cargo tanks. The PV breaker vent is fitted with a flame trap to avoid the fire igniting when loading or discharging operations are going on when the vessel is in port. (11) *Cargo tank isolating valves*. Oil tankers have several cargo holds, and each hold is provided with an isolating valve. The valve controls the flow of inert gas to each hold. (12) *Mast riser*. The mast riser is used to maintain the positive pressure of the inert gas during the loading of cargo. During the loading time, the mast riser is kept open to avoid pressurisation of the cargo tank. (13) *The safety and alarm system*. The inert gas plant is provided with various safety features to safeguard the tank and its own machinery. These include (a) the high-pressure level in the scrubber alarm; when activated the alarm shuts down the blower and the scrubber tower; (b) low-pressure seawater supply (approx. 0.7 bar) to the scrubber tower leads to an alarm and shutdown of the blower; (c) low-pressure seawater supply (approx. 1.5 bar) to the deck seal leads to an alarm and the shutdown of the blower; (d) high inert gas temperature (approximately 70°C (158°F) leads to an alarm and shutdown of the blower; (e) low pressure in the line after the blower (approximately 250 mm wg)[1] leads to an alarm and shutdown of the blower; (f) oxygen content high (8%) leads to an alarm and shutdown of the gas delivery on deck; (g) low level in deck seal leads to an alarm and shutdown of the gas delivery on deck; (h) power failure leads to an alarm and shutdown of the blower and scrubber tower; and (i) emergency stop leads to an alarm and shutdown of the blower and scrubber tower.

The primary basis of inert gas production in the inert gas plant is the flue gas generated from the ship's boiler. The high-temperature gas mixture from the boiler uptake is treated in an inert gas plant which cleans, cools, and supplies the inert gas to the individual tanks via the PV valves and breakers to ensure the safety of the tank structure and atmosphere. The system can be divided into two basic sections: (1) the production plant, which produces the inert gas and delivers it under pressure, by means of blower(s), to the cargo tanks, and (2) a distribution system to control the passage of inert gas into the appropriate cargo tanks at the required time. The basic working principles of the inert gas production system start when the boiler uptake gases are drawn to the scrubber unit via the flue gas isolating valve(s) to the scrubber unit. In the scrubber unit, the gas is cooled, cleaned, and dried, before being supplied to the tanks. Motor-driven inert gas blowers supply the treated gas from the scrubber tower to the tanks. They are mounted on rubber vibration absorbers and isolated from the piping by rubber expansion bellows. The regulation of gas quantity delivered on deck is managed by the gas control valves and the deck pressure is managed by the pressure controller. If the deck pressure is lower than the set point, the output signal will be raised to open the valve more, and vice versa, if the deck pressure is lower than the set point. These valves then work in tandem to keep both the deck pressure/blower pressure at their respective setpoints without starving or overfeeding the circuit. Before entering the deck line, the gas passes through the deck water seal which also acts as a non-return valve automatically preventing the backflow of explosive gases from the cargo tanks. After the deck seal, the inert gas relief is mounted to balance built-up deck water seal pressure when the system is shut down. In the event of a failure of both the deck seal and the non-return valve, the relief valve will vent the gases flowing from the cargo tank into the atmosphere. The oxygen analyser, which is fitted after the blower, separates the "production" and "distribution" components of the plant and analyses the oxygen content of the gas. If the oxygen is more than 8%, the alarm will trigger the plant shutdown procedures.

With many thousands of tonnes of highly flammable and hazardous cargo and fuel oil on board, oil tankers are accident-prone zones, with a high probability of fire-related accidents. The frequent cargo-handling operations and loading and unloading procedures of hazardous materials make tankers highly vulnerable. Sometimes despite all the safety precautions, accidents happen. Tackling a fire on a tanker requires strategic planning and a systematic approach. It also requires the confidence that comes with a very cool head. In this chapter, we have discussed some of the main firefighting systems and appliances carried onboard ships. Although SOLAS and the FSSC provide the basic requirements, many shipping companies mandate their fleets carry systems which go over and beyond those listed in SOLAS and the FSSC. Unfortunately, not every shipowner and shipping company is as careful. The costs of installing additional firefighting systems and equipment can be prohibitively expensive. They also require initial and regular refresher training to maintain competence in their use. Irrespective of what systems and equipment are carried on board, it is up to every crew member to ensure they are fully qualified and competent. Ship fires are all too common, and the consequences can be, and are, severe.

In this chapter, we have looked at the main firefighting systems and apparatus in the engine room. In the next chapter, we will briefly look at engine room flooding, the causes and consequences of engine room flooding, and the means and methods of preventing flooding below the ship's waterline.

NOTES

1. MM WG, mm wg, millimetres water gauge.

Chapter 28

Engine room flooding

LEAKS FROM MACHINERY AND EQUIPMENT

Engine room flooding can take place due to leaks in the engine room space from machinery, or the seawater and freshwater systems. Leaks of this nature stem from the big seawater pump, from the seawater or freshwater cooler, or from cracks and fissures in the boiler feedwater system. Leaks may also occur along any of the fresh or seawater pipelines. Cracks in the ballast water tanks, leaks from manholes, or cracks in the water storage tanks can also lead to engine room flooding. Whatever the cause of the flooding, the following actions must be taken to reduce the risk of affecting the engine room machinery and the ship's stability: (1) call for maximum manpower to tackle the situation; (2) carry out immediate and deep depth troubleshooting to identify the cause of the flooding; (3) activate the other circulating systems and isolate the leaking pump, pipe, cooler, or machine; (4) close the inlet and outlet valves of the affected system to stem the ingress of water; (5) inform the chief engineer and the OOW on the bridge of the nature and extent of the flooding; action any commands as appropriate; (6) place notices or placards around the leaking equipment or system and trip the breaker until repairs have been completed; (7) in the event of tank leaks, start transferring excess content from the affected tank to another tank; and (8) do not use the defective tank, machine, or pipework until remedial welding, cement box, or repairs are complete.

LEAKS FROM THE OVERBOARD VALVE

If the source of the flooding is a leak after the overboard valve, and if the valve is holding, shut the valve if the system involved in that valve permits the normal operation of the ship with the valve closed. If the valve is not holding, then identify the source of the leak. This may be from the valve stem gland or flange joint; if possible, try to repair the leak. If the system for that valve can be isolated without disturbing the normal operation of the ship, install a blank into the valve. If the repair is temporary, ensure divers are called at the next port to install a blank into the valve opening from the outside until such time as permanent repairs can be made in dry dock.

FLOODING CAUSED BY CRACKS AND FISSURES IN THE SHIP'S HULL OR SIDEBOARD

When cracks, fissures, or holes are found in the ship's hull or sideboard, the ship is at very real risk. If the source of the following cannot be temporarily fixed, use all available means to stem the inflow of seawater. Inform the nearest coastal state and wait for assistance. If the source of the flooding is significant but not an immediate danger to the ship, try any means possible to stop the flooding, for example, using the cement box or welding a strip of plate against the location of the damage. When the vessel arrives in port, arrange for the necessary repairs to be conducted. In the event of an accident such as a collision or grounding, there is often little that can be done to stop the flooding. In these situations, the only course of action is to secure the engine room as best as possible and make way on deck. The master will decide whether the ship is safe to continue to a place of safety, or whether an emergency evacuation is needed. Abandoning the ship must always be the last resort as the ship provides the best safety and chance of rescue.

If for any reason water ingresses into the engine room, open the emergency bilge ejector valve and pump the water overboard. An entry must be made in the oil record book stating the date, time, and position of the ship and the reason for direct discharge. The duty engineer, the chief engineer, and the ship's master must sign this.

WATERTIGHT BULKHEADS AND FLOODING PREVENTION

The survival of a ship which is in a damaged condition is dependent on the strength and integrity of its watertight bulkheads. There are many factors that go into determining the location and position, design, and construction of watertight bulkheads. Watertight bulkheads are vertically designed watertight divisions or walls within the ship's structure. Their purpose is preventing the ingress of water into an unaffected compartment where the adjacent compartment is flooded. The position of bulkheads along the length of a ship is primarily decided through a process called *floodable length calculation*, which is performed during the damaged stability assessment of the ship. Once the position of each watertight bulkhead is fixed, various other factors must then be considered such as the type of watertight bulkhead, its strength and integrity requirements, height, width, depth, and so forth. The primary watertight bulkhead on any ship is the collision bulkhead (see Figure 28.1). The collision bulkhead is the forward-most bulkhead on a ship. There are two factors that determine the position of the forward collision bulkhead. The final position of the collision bulkhead is decided such that it takes into consideration three factors:

- Factor 1: position based on floodable length calculations
- Factor 2: position based on Class code books. Most Class rules have an allowable range of distance at which the collision bulkhead can be placed from the forward-most point of the ship's hull. This distance is usually a function of the length of the ship and any factors related to the shape of its bow
- Factor 3: position based on SOLAS rules, which state that the collision bulkhead should be located aft of the forward perpendicular at a distance not less than 5% of the ship's length of the ship or ten metres (32 ft) (whichever is less). The distance must also not exceed 8% of the ship's length

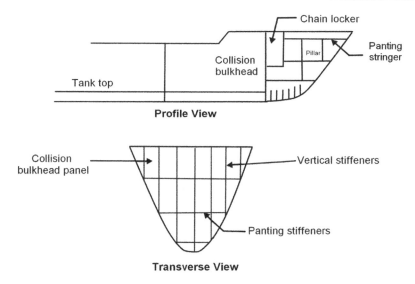

Figure 28.1 Collision bulkhead (profile and transverse view).

It is also recognised that the position of the collision bulkhead should be such that maximum cargo storage volume is achieved without compromising the safety integrity of the ship. The collision bulkhead is a heavily strengthened structure, with its main purpose being to limit the damage of a head-on collision to the part of the bow forward of the bulkhead. To limit the damage to its forward region also means that the collision bulkhead must be a watertight bulkhead. It is usually vertically stiffened with sections of scantlings higher than those on the surrounding structures. It is also stiffened by triangular stringers of higher scantling. These are called panting stringers. Panting stringers are usually provided at every two metres (6.5 ft) from the bottom, and forward of the collision bulkhead.

As per SOLAS regulations, the collision bulkhead must be watertight up to the bulkhead deck. The bulkhead deck is the deck level up to which all watertight bulkheads must extend. For providing access to the chain locker room and the forward part of the bulkhead, steps may be provided on the collision bulkhead. However, this must not violate Factor 3. There must be no doors, manholes, access hatches, ventilation ducts, or indeed any openings on the collision bulkhead below the bulkhead deck. However, the bulkhead may be allowed to have a maximum of one piercing below the bulkhead deck for the passage of one pipe to cater to fluid flow to the forepeak ballast tank. The passage of the pipe must be flanged and must be fitted with a screw-down valve which can be remotely operated from above the bulkhead deck. This valve is usually located forward of the collision bulkhead. However, the Class certifying the ship may authorise a valve aft of the bulkhead, provided it is easily serviceable at any condition and is not located within the cargo area. In cases where ships have superstructures positioned in the forward region (for example, refer to Figure 28.2), the collision bulkhead is not terminated at the bulkhead deck. Instead, it must be extended to the deck level below the weather deck. This ensures there is sufficient structural continuity and keeps the shear forces within safe limits. If the collision bulkhead is extended above the freeboard deck, the number of openings on the bulkhead should be restricted to a minimum to ensure sufficient buckling strength. All openings should be watertight.

Figure 28.2 Example of a vessel with a forward superstructure (*Mærsk Tracker*).

As we have noted, the primary function of the watertight bulkhead is to divide the ship into several watertight compartments. Though most watertight bulkheads are transverse in orientation, some ships may also have longitudinal watertight bulkheads within a compartment. This provides longitudinal compartmentalisation within a compartment. Other than watertightness, transverse bulkheads also add to the transverse strength of the ship, though we will investigate this aspect a little later. In small ships, a transverse bulkhead may be constructed from a single plate. However, for larger ships, the plating of a transverse bulkhead usually consists of a series of horizontal strakes welded together. What is particularly interesting about this is that the thickness of these strakes increases with depth, to strengthen the bulkhead against the maximum hydrostatic pressure in the event the compartment is fully flooded. Therefore, prior to construction, two-dimensional strakes are first cut out from plates of different thicknesses. The bulkhead plate itself is not resistant enough against large-scale transverse forces such as shear forces. Subsequently, they must be stiffened, either vertically or horizontally. Vertical stiffening (see Figure 28.3) is universally preferred, as horizontal stiffening in ships with high beams requires stiffeners of a long span, which would also increase the scantling and weight of the stiffener. This would have a deleterious effect on usable cargo volume. However, with vertical stiffening, the span (and thus, the scantling) of the stiffener can be kept low by introducing a stringer at mid-depth. The stringer acts as a fixed end, therein reducing the length of the span.

The sections used for stiffening the bulkheads are usually flat bars, angles, or bulb bars, depending on the required section modulus. An important aspect of the design of bulkhead stiffeners is meeting the end conditions. To meet the boundary conditions so that the stiffeners respond as per the naval architects' theoretical calculations, their end supports must be designed accordingly. At the upper end, they are attached to the underside of the

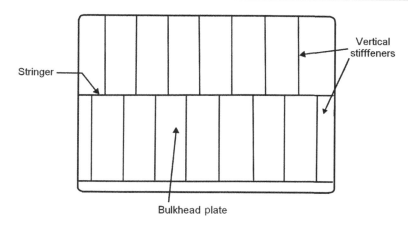

Figure 28.3 Vertical stiffening of a transverse watertight bulkhead.

deck plating with brackets, providing a hinged boundary condition. To achieve fixed ends, they are welded directly to the deck plate and the stringer. Most modern-day ships use advanced technology to achieve the required strength of bulkhead plates. They use corrugated bulkheads instead of stiffened ones. The corrugations are positioned in the vertical, except when the breadth of the bulkhead is significantly low. However, there is one trade-off that needs to be considered. Since the corrugations are provided on the bulkhead plate in the early fabrication stage, these corrugated bulkheads have uniform thick plates (that is, the thickness is equal to the lowermost strake in the case of conventional bulkheads). This increases the weight of the bulkhead when compared to a conventionally stiffened bulkhead. Despite this, the use of corrugated bulkheads has increased in popularity partly due to the ease with which they can be fabricated and the reduction of welded joints on the bulkhead, which provides additional strength and integrity (see Figure 28.4)

Figure 28.4 shows the elevation of a corrugated bulkhead from the side. With bulk carriers, to prevent the accumulation of cargo at the base of the corrugations, the lower end of the bulkhead is provided with angular plates called shredder plates. These help in shredding the dry cargo to the tank top. The bulkhead is connected to the tank top by a bulkhead stool, which is fillet welded to the tank top plate. The two forward and aft ends of the stool are in line with the transverse plate floors. This ensures proper stress flow from the bulkhead to the plate floors. The corners, where the bulkhead plate is welded to the side shell and the deck plate or the tank top, have separate corner plates which are welded to complete the joint after welding the remaining bulkhead plate to the hull. These corner plates are provided for the following reasons: (1) Fitting the entire bulkhead panel (with the corners) would be difficult from a production point of view since every structure is first fabricated with a certain amount of green material. Before final installation, the green material is removed, and structures as large as the bulkheads require repeated checks for proper dimensional adherence. Eliminating the corners from this stage would reduce the complexity of maintaining dimensional precision at the corners. (2) Stress concentration occurs at the corners due to the discontinuity of the structure. To prevent this, corner plates are provided with additional thickness when comparted to the adjacent bulkhead plating. After the installation of the bulkheads, they must be tested for their integrity and watertightness. Since it is not feasible to fill all the cargo holds or compartments with

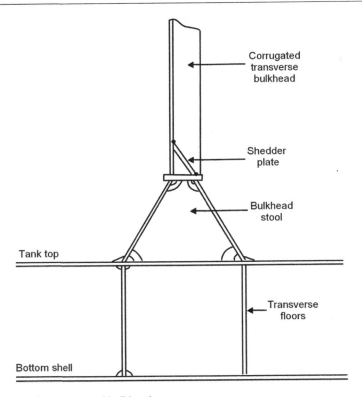

Figure 28.4 Elevation of a corrugated bulkhead.

water for this purpose, the test is done by a pressure hose. In this process, the bulkhead is subjected to predetermined water pressure from a hose for a fixed period, after which, the structural integrity of the bulkhead is inspected (checks are done for buckling and other deformations). Leak tests can also be done by pressurising the air in a compartment and checking for leaks.

WATERTIGHT DOORS

Intactness is the primary purpose of the watertight bulkhead. But, in most ships, there are unavoidable situations where access from one compartment to another is a necessity. This is particularly the case within the engine room and associated machinery spaces. For example, on most ships, access to the shaft tunnel is necessary to monitor the shaft oil temperatures or for conducting repairs. This issue is solved using watertight doors. The bulkhead panel is usually cut out in a rectangular shape to accommodate the watertight door. However, because the introduction of an opening will adversely affect the integrity of the watertight bulkhead, special considerations must be made to the structural design of the region around the watertight door opening. These special considerations usually include keeping the dimensions of the opening to the absolute minimum; as any opening results in a major structural discontinuity, resulting in stress concentrations around the opening, it is vital to maintain stress levels below acceptable limits. This is achieved by strengthened

Engine room flooding 367

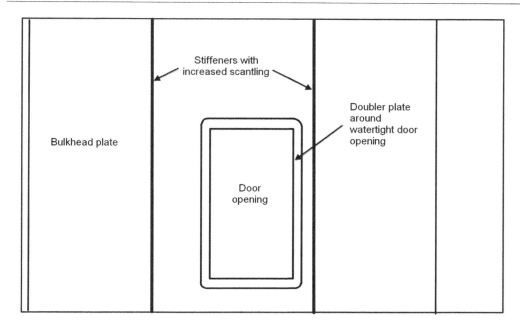

Figure 28.5 Opening for watertight door on a bulkhead plate.

the opening with doubler plates to increase the thickness of the bulkhead plate around the opening. If a vertical bulkhead stiffener is positioned in the way of the opening, the stiffener must be terminated at the upper and lower edges of the opening. However, designers might choose to increase the stiffener spacing to avoid this. In these cases, the scantling of the stiffeners adjacent to the opening are usually increased in relation to the other stiffeners (Figure 28.5).

Watertight doors are usually hydraulically or electrically operated and are either horizontally or vertically sliding. The reason swinging doors are not provided as watertight bulkheads is because it would be impossible to close a swinging door in the event of flooding. The watertight door must be easily operable even when the ship has listed to 15° to either side, and the control system should be designed such that the door can be operated from the local vicinity as well as remotely, i.e., from a position above the bulkhead deck. On all ships, visual indicators are provided at the remote-control location to advise whether the door is open or shut. Watertight doors are also subject to pressure tests after installation to check for their structural integrity at predetermined hydrostatic pressures in case of complete flooding up to the bulkhead deck.

SOLAS rules relating to watertight bulkheads

One of the most important regulations that apply to the design and operation of watertight bulkheads and watertight doors are the SOLAS regulations. SOLAS sets out the minimum standards that watertight bulkheads and apertures must comply with for the ship to be classified as seaworthy. These minimum standards pertain to (1) the number of openings for pipes and accesses, which should be kept to a minimum to retain the strength of the bulkhead. Where such openings are provided, proper reinforcements must be provided to prevent stress concentrations and to retain the watertightness of the structure. Moreover, proper flanging must be incorporated into the openings for pipelines and cables. (2) Not more than one watertight door is permitted per watertight bulkhead. However, in the case of ships having twin shafts, two watertight doors, each providing access to the two shaft tunnels on either side, may be allowed. The mechanical gears required for the manual operation of these doors must be located outside the machinery spaces. (3) The time required to close or open any watertight door when activated from the control room or bridge must be 60 seconds or less when the ship is in stable upright condition. (4) The transverse location of the watertight doors should be such that they must be easily operated even when the ship is damaged to within one fifth of the ship's breadth from its side shell. (5) Every watertight door must be equipped with an audible alarm which is distinct from all other alarms in the vicinity. In the event the door is operated remotely, the alarm should start sounding no less than five seconds before the door begins to slide either way and must continue until the door has completely opened or closed. However, if operated in situ, the alarm must sound only when the door is sliding. On passenger ships, the audible alarm must be accompanied by a clear and visual alarm. (6) All watertight doors that are accessible when the ship is at sea must be locked. (7) Any access doors and hatches on watertight bulkheads must remain closed when the ship is at sea. Visual indicators must be provided for every access hatch to indicate their status on location and on the bridge.

In this chapter, we looked at the causes and consequences of engine room flooding, and the design and function of watertight doors. In the next chapter, we will turn our attention to typical engine room watch procedures.

Part VII

Engine room watch procedures

Chapter 29

Engine room watch procedures

Watchkeeping is an integral part of the marine engineer's duties on board ship. A lot of maintenance work can be reduced by following an efficient watchkeeping routine in the ship's engine room. Moreover, it can also avoid serious accidents from taking place. A smooth-running ship is a product of efficient handling on the bridge and effective management of the ship's engine room under any seagoing condition. When a marine engineer is approved to be put in charge of the engine room, he is eligible and officially authorised to handle the ship's 'unlimited power'. It is therefore important that the watchkeeping procedures, which are a daily routine, are conducted systematically to prevent any kind of malperformance. Though there is no official yardstick against which we can measure the efficiency of a watchkeeping procedure, there are various established methods which, when followed, will ensure the vessel is kept safe and operating at its optimum efficiency.

Knowledge. Every good watchkeeper knows the mantra 'knowledge is the foundation of all good watchkeeping'. The first and most crucial step to good watchkeeping is to have an extraordinarily strong knowledge base of the machinery, equipment, and systems under supervision. The watchkeeper must know the fundamental basics of the engine room and its operations, as well as new and emerging trends and technologies. It is equally important to keep abreast of the main maritime regulations and their amendments. Engineers should also remain aware that engine room operations also require inputs from other engineering domains such as mechanical, hydraulic, pneumatic, refrigeration, and electrical and electronic systems. Knowing these fundamentals makes an engineer's foundation stronger. *Follow your instincts.* It is commonly said on ships that to become a good watchkeeper, an engineer must use all five senses: touch, hearing, smell, vision, and taste. It is also important to use our kinaesthetic sense (often referred to as the 'sixth sense'). All these senses when applied correctly can better improve the engineer's understanding of the condition of their machineries. Consider the following, for instance: (1) *Touch*: feeling a machine for its temperature can indicate the condition inside the machinery. In most cases, when a piece of equipment is abnormally hot to the touch, this usually means there is some malfunction or fault. (2) *Hearing*. It is always advisable to keep a track of sounds coming from different machineries in the engine room, as any abnormalities usually manifest in some form of sound (i.e., banging, knocking, humming, squealing, etc.). (3) *Smell*. One of the most powerful senses humans have is the sense of smell. A trained and experienced engineer can easily determine abnormal smells in the engine room. Burning, smoke, oil leaks, and sewage leaks all have unique and characteristic odours that can be distinguished over and beyond the usual smells of the ship's engine room. (4) *Taste*. The human tongue can identify many different tastes. Perverse as it may sound, this sense can be applied to everyday watchkeeping routines. For instance, ascertaining the difference between freshwater and saltwater, both of which are used as cooling media throughout the engine room. (5) *Vision*. Our eyes are our greatest asset, and without them, we simply cannot function as marine engineers. (6) *Kinaesthetic senses or the sixth sense*. We all experience gut feelings from time to time; these may not be based on any sound logic or engineering experience, but even so, we should always take note when our 'inner feeling' tells us something is amiss.

Follow the book. Every engine room is provided with a library of hundreds of documents such as manuals, operating instructions, and safety and pollution prevention instructions; logbooks; records; and checklists. These documents are there for a reason, either to advise or guide, or to record legal information that may be needed in the future. *Interpret the logbook*. Experienced engineers recognise the importance of the engine room logbook and know how to interpret previous entries to the best effect. This is a skill that comes with time, but nevertheless, it is critically important to read the logbook at the start of each watch and to record all relevant events during and at the end of the watch. Remember, the engine room logbook, like the ship's logbook, is a legal document and may be used during legal proceedings to exonerate or prosecute. *Use clear communication*. Efficient communication between maritime professionals is a principal factor in ensuring the safety of the ship. Keep commands simple and easy to understand; avoid using large or complicated words when simpler and smaller words would do. Be authoritative when required, and sympathetic when appropriate. Always maintain clear communications on radio and try to avoid words or commands that could be misinterpreted or misheard. If unsure, ask. It is always better to explain something again than to assume a command has been understood, when in fact, it has not. *Always conduct complete rounds*. There is always the tendency to slacken off around the end of the watch, especially when the watch has been busy or challenging. Even so, always maintain the highest professional standards of conduct. Never skip compartments, machineries, or systems when conducting rounds; this is how leaks and faults are most easily missed. Furthermore, always record any abnormalities in the engine room logbook, no matter how insignificant they may seem at the time. A slight knocking in the crankcase during your watch could escalate into a full crankcase explosion if left unreported. *Never ignore or neglect an alarm*. The alarm systems in the engine room provide an indication or pre-warning of any abnormality. Sometimes alarms can be triggered by technical faults, in which case the fault should be fixed and the alarm reset. If an alarm activates for any reason whatsoever, never ignore it and never silence it without first establishing why the alarm was activated. Even small faults or malfunctions can quickly escalate into serious problems that can affect the engine room or even the whole ship. No engineer wants to be 'that person' who silenced the alarm instead of checking what was wrong.

Never hide faults. If any faults are identified whilst conducting rounds, these should be reported and rectified as soon as practicably possible. If conducting end-of-watch rounds, ensure the oncoming engineer of the watch (EOW) is made aware of the fault, the location, and any other relevant information which may aid its rectification. It is perfectly natural to make mistakes. When we are new in our roles and lack the experience that comes with years spent at sea, mistakes and errors are bound to happen. There is nothing wrong with this if we own up to our mistakes and learn from the experience. Hiding our errors, however, is unprofessional and extremely dangerous. *If in doubt, ask!* The engine room is the beating heart of the ship. Without the engines and all the associated machinery that ensures the ship can operate, the ship is dead in the water. A dead ship is an extremely dangerous ship. If we are ever unsure of our duties, responsibilities, what actions should be taken, or how to respond to a particular incident or set of circumstances, always ask a senior officer. There is no shame in not knowing the correct procedure; even chief engineers were once trainees! Being on a ship is all about teamwork; remember, there is no 'I' in team for a particularly good reason. *Obey orders*. The watchkeeper must always follow orders and commands given by their senior officers. Unless an order is patently illegal or is likely to place the ship, its crew, or its cargo in jeopardy, there is no good reason to disobey a direct command. Even so, always exercise good judgement, and if unsure, ask for the order to be repeated or explained differently. *Abide by the ship's drug and alcohol policy*. Never, under any circumstances, is the EOW to stand duty under the influence of drugs or alcohol. This includes prescription medication where cognitive or agile ability is adversely affected (for example, drowsiness). Last of all, *try to avoid fatigue*. Fatigue is one of the main causes of human mistakes on ships. Wherever possible, take advantage of off-duty periods to catch up with sleep, eat healthily, and exercise regularly.

Hopefully, these points should provide some guidance on how to conduct an engine room watch safely and effectively. In the next section, we will look at the actual procedures for handing over the engine room watch when the ship is underway at sea.

ENGINE ROOM WATCHKEEPING

As responsible marine engineers working in the ship's engine room, it is our duty to do whatever it takes to safeguard the safety of the ship, her crew, and her cargo. Following the correct watchkeeping procedures is critical not only when standing the engine room watch but also during handovers. As such, the EOW must follow the instructions as stated by the chief engineer and company procedures when handing over the watch to the relieving officer. The following are some of the main points the oncoming EOW should note when preparing to take over the watch of a ship when in port. (1) *Take account of port regulations and local laws*. Every port has its own set of regulations and local requirements regarding pollution, the discharge of ship's effluence, ship readiness, and so forth. Failure to follow these regulations and laws can lead to prosecution and the imposition of fines, detention of the vessel and its master and crew, and in the most egregious circumstances, even imprisonment. When preparing to hand over the watch to the relieving EOW, ensure that all essential information is passed on. (2) *Maintain effective communication*. The EOW must inform the chief engineer and the bridge of any abnormal occurrences during their watch, including but not limited to accidents, near misses, injuries sustained, equipment or machinery faults, malfunctions, and maloperation. The EOW must also ensure that a full record of these is kept in the engine room log, and the relieving EOW is fully informed at handover. (3) *Standing orders and other important commands*. All standing orders of the day together with other important commands related to the ship's operation, maintenance, and repairs must be passed to the relieving officer. In the event bunkering operations are being conducted, the EOW should inform the oncoming officer of the quantity already taken, the level of the tanks, and the time remaining till completion. (4) *Condition of engine room machinery*. Ensure any essential information regarding the ship's machinery (such as the boiler, auxiliary engine, main engine) is reported to the relieving officer. These should also be recorded in the engine room logbook. (5) *Status of repair and maintenance work*. In the event any repair or maintenance work is going on in the engine room, the EOW must record this in the engine room logbook. It should also be reported to the relieving officer together with a full status update (time started, estimated duration, estimated end time, details of the job, any unusual or specific hazards and risks, etc.). It is always advisable, whenever possible, to commence and finish any repair and maintenance work within one watch period. (6) *Action any special orders of operation*. It might be possible that the ship is required to depart during the watch of the relieving officer. It is, therefore, necessary that they are informed about any conditions that may impact or adversely affect the safe and efficient operation of the vessel, including any known contamination, weather, visibility, ice, shallow water, hull damage, etc. (7) *Nature of work and number of personnel in the engine room*. The relieving EOW must be aware of all the jobs that are being always conducted in the engine room, together with the number of personnel, and their location. All personnel must be accounted for at the beginning of each watch, at the end of their duties in the engine room, or where their duties extend past the watch handover at the end of the watch. (8) *Information on power and other sources*. It is important for the EOW to inform the relieving officer of any anticipated demands on the existing and or potential power requirements, including propulsion power, heating, lighting, and their distribution. (9) *Level of tanks and amount of fuel*. The EOW must inform the relieving officer of the level and condition of water or residues in the engine room tanks, including the bilge tanks, slop tanks, reserve tanks, and so forth. The relieving officer must also be made aware of the availability and condition of the ship's fuel, lubricant, and water supplies. (10) *All information must be accurately recorded in the engine room logbook*. We have already covered this point, but it is worth reminding ourselves of the importance of recording accurate and factual notations in the engine room logbook.

The procedures for watchkeeping at sea are not so vastly different from those when the ship is alongside. When the ship is at sea, the engineering officers perform their duties in rotational shifts, each having a fixed and equal number of hours. This work shift, referred to as the watch, needs to be conducted in an efficient manner to ensure the safety of life and property at sea. Most of the points listed in the previous section apply even when the vessel is underway, the main difference being that when in port, the engineering officers do not usually stand watch in the same manner as when at sea. As a final note, it is the responsibility of each EOW to hand over the engine room in the most complete and safest manner possible. If the engine room watch cannot be handed over safely, for any reason whatsoever, do not do so. Even if the relieving officer is made to wait, never hand over the watch unless it is safe and appropriate to do so. Finally, when standing watch, the engine room is the sole responsibility of the EOW. Although the chief engineer retains ultimate responsibility (and authority), it is the EOW who will be held responsible in the event anything goes wrong.

Chapter 30

Engine room logbook entries and checklists

A ship, together with its engine room and experienced engineers, requires a set of important documents to sail safely and legally. A vessel can only travel from one foreign port to another with valid certificates and up-to-date records. In this chapter, we will briefly discuss the main logbooks and checklists used in the engine room.

ENGINE ROOM LOGBOOKS AND RECORDS

Engine room logbook. The first, and the most important, is the engine room logbook. Like the ship's logbook which is completed by the deck officers, the engine room logbook records everything of note in the ship's engine room. It provides a record of all machinery parameters, including the main propulsion plant, power generation system, boiler, purifier, refrigeration, and air-conditioning plant. Any abnormal findings or occurrences during each watch must be recorded in the logbook. Major maintenance activities, overhauling, replacements, and repairs must be logged together with a summary of why these activities were conducted. The logbook also records each voyage number, the ports visited, the running hours of each machinery set, the quantity and volume of fuel, diesel and lube oil used for each passage, and any special operations such as bunkerage. In short, the engine room logbook records each activity that occurs in the engine room. Each entry must be initialled by the officer making the entry, which is then countersigned by the chief engineer. In doing so, the logbook provides an official – and legal – record in the event of any accidents or disputes that may find their way into a court of law.

Oil record book. The oil record book is one of the most important documents on board, as it provides a written record for compliance with annex I of the MARPOL 1973/78 regulations. When operating the oily water separator, the operation is recorded with the time, position of the ship, the quantity discharged, and any retention. Any maintenance operations on MARPOL-regulated equipment (such as the oily water separator, sewage treatment plant, and incinerator) must also be recorded in the oil record book together with annotations summarising the type of maintenance performed and the date and time. Bunkering operations must be recorded including the date, time, bunkering grade, the quantity bunkered, the port of bunkering, and the retention of tanks used in the bunkering operation. Moreover, the weekly retention of wastewater, including bilge and sludge, must be recorded. Any internal bilge or sludge transfers must be recorded with the date, time, and quantity transferred. The oil record book should always be accompanied by the IOPP certificates. All bunkering delivery notes

(BDN) and bilge/sludge transfer notes must be retained for the statutory period. All operations and records are initialled by the officer overseeing the activity, which is then countersigned by the chief engineer. At the end of each page, the ship's master must sign the oil record book confirming the accuracy of the entries logged.

Engine room tank sounding log. The engine room tank sounding log is a written record of the soundings taken for all engine room tanks, including the wastewater tank, fuel oil and diesel oil service settling tanks, and the bunker tanks. On most vessels, the responsibility for maintaining the sounding log rests with the fourth engineer, though different ships may assign responsibilities and duties differently. The frequency of the tank sounding is done twice daily: once in the morning and once in the evening. A record must be kept of each sounding, which is initialled by the engineer, then countersigned by the chief engineer.

Sewage management log. The sewage management log consists of the ISPP certificate, the operating procedures of the sewage plant, and the maintenance procedures for the sewage plant. On most vessels, the second engineer is responsible for maintaining the sewage management plan log. Any discharge of sewage overboard must be accurately recorded with the date, time, position of the ship, and quantity discharged. Any maintenance conducted on the sewage treatment plant, including chlorine tablet dosing, must be recorded. Each entry is then initialled by the responsible engineer and countersigned by the chief engineer.

Oil-to-sea interface log. The oil-to-sea interface log is a record of those systems which facilitate the direct interface of oil with seawater. Given the importance of these systems, responsibility sits with the chief engineer to maintain the log. This usually includes the stern tube systems and lube oil coolers, where they are cooled by seawater. Ordinarily, the level or quantity of oil in the system is recorded to check for any indications of leaks. These entries must be made daily with all readings recorded and signed by the chief engineer.

Seal log. The seal log is kept by most shipping companies as a way of demonstrating compliance with the MARPOL regulations. The log acts as a record of all checks and inspections conducted on the oily water separator overboard line seals, sewage treatment plant seals, bilge system seals, etc. Every seal on board has a unique identification number assigned; these are individually inspected, with the date and time of each inspection being recorded by the chief engineer. If any seals are removed, repaired, or replaced, the details of which must also be recorded in the book.

Saturday/Monday routine log. All emergency equipment such as lifesaving appliances (LSA) and firefighting appliances (FFA), and their associated systems, must be evaluated either on a weekly, monthly, or annual basis depending on the equipment and the preventative maintenance schedule. All checks, tests, and inspections on the emergency generator; emergency fire pump; emergency compressor; lifeboat engine; emergency pump and fan stop; fire dampers; and indeed any other emergency response equipment must be recorded. Where individual officers are assigned specific responsibilities or duties, these must also be clearly recorded.

Chief engineer's night order book. The chief engineer is solely responsible for maintaining the engine room night order book. Like the master's night order book for the bridge team, the chief engineer lists their instructions for the night watch. Every engineering officer and trainee engineering officer is required to read and acknowledge the orders issued by the chief engineer. Once the order(s) have been actioned, these are initialled accordingly.

MANNED ENGINE ROOM CHECKLISTS

There are many important tasks that need to be conducted when a ship is due to arrive or depart a port of call. Failure to perform these tasks can lead to accidents and incidents which would otherwise have been entirely avoidable. To aid the engine room team, it is common for checklists to be made which set out in linear fashion each of the duties, responsibilities, and tasks that need to be conducted to ensure the ship is ready to enter port or safe to set sail. In this section, we will discuss some of the main checklists that marine engineers are well advised to follow in the course of their duties.

Engine department departure checklist

Note: This checklist assumes the vessel is leaving port, on a cold start, with the boiler and generator in running condition.

24 hours prior to departure

At 24 hours to departure:

- Check the oil level or sound the bunker tanks to measure the quantity of oil available. Ensure the temperature is maintained at around 40°C (104°F) or as per the analysis report. This ensures the smooth transfer of oil from the bunker tank to the settling tank. If the oil is cold, the viscosity will be thicker, which can damage the pump.
- Check the jacket water header tank for the correct level and refill accordingly. Do not fill beyond the maximum level, as the water will expand as the engine heats up, causing the water to overflow.
- After checking the level, start the jacket water circulating pumps (if separate from the main engine starting system).
- Check the jacket water temperature of the main engine and maintain to about 60°C (140°F). Any temperature below this may lead to water leaking into the scavenge space.

Six hours prior to departure

At six hours to departure:

- Check the oil level in the main engine sump and the turbocharger tank.
- The duty engineer should start the lubricating oil pump and the crosshead pump (in Sulzer engines) and the turbocharger pump.
- Check the oil flow through the sight glass of turbocharger outlet.
- Check the pressure of the lube oil pump, the turbocharger pump, and the crosshead pumps.
- Start the shaft bearing pumps and check the level of the header tank.

With unmanned machinery spaces (UMS) there is a programme for starting all these pumps in sequence. This programme should be activated only after completing the engine room rounds and once the levels are checked.

One hour prior to departure

At one hour to departure, the chief engineer will usually conduct a quick round of the engine room to ensure all machinery and systems are ready; this is then reported to the master on the bridge. On completion of the chief engineer's rounds:

- Check the oil level, header tank level, and the cylinder oil daily tank level.
- Check the pressure of the fuel oil pump, booster pump, and lube oil pump.
- Check the sump oil level in the air compressor.
- Drain the air bottles of any water.
- Check that the turning gear is set out.
- Check the parameters of the running machinery.
- Start an additional generator to satisfy the additional demand for power from the winches and thrusters.
- Start the exhaust gas boiler water circulating pump.
- Check the functioning of the ship's telegraph with the bridge OOW at both local and remote stations.
- Check the operation of the emergency telephones.
- Once the chief engineer has returned to the engine control room, make way to the steering gear and check for any leaks and port to starboard movement.
- Check the functioning of the limit switches in the steering gear.
- Check the gyro readings of the steering gear and cross-check with the bridge.

15 minutes prior to departure

At 15 minutes to departure:

- Open the main air starting valve for the main engine.
- Ensure all air compressors are in auto-mode and the air bottles are full.
- Ensure the main engine is blown with open indicator cocks with air to check for any water ingress.
- If no leaks or water ingress is found, report to the chief engineer and await instructions to close the indicator cocks.
- Close the turbocharger drain valves.
- Flow metre counts for the main engine, generator, and boiler are taken, calculated, and recorded.

Once these checks are complete, the chief engineer will report to the master on the bridge that the engine room is ready. On authorisation from the master, the controls are transferred from the engine room to the bridge.

Checks to be made when the main engine is running

Once the main engine is in full running mode:

- Conduct full engine room rounds.
- Check for any abnormalities.
- Check the parameters of the main engine, such as the temperature of the exhaust valve, jacket water, etc.

Engine room logbook entries and checklists

- Close the steam heating valve for the jacket if not auto-controlled.
- Once the ship is out from port, and the pilot has disembarked, open the sea suction valve.

Checks to be made when the vessel is full away

When the vessel is full away:

- Stop the additional generator.
- Stop the boiler if an exhaust gas boiler is installed.
- Start the freshwater generator.
- Open the sewage overboard valve.
- Start the turbine and shaft generator (if installed).
- Take flowmeter counts again for the calculation of fuel consumed from port to full away.
- If the engine room watches have changed, hand over to the next duty engineer.
- If the vessel has an UMS engine room, conduct final rounds and inform the bridge when complete.
- Resume normal engine room operations and procedures.

Engine department arrival checklist

The following tasks must be conducted by the duty engineer on receiving notice of arrival from the bridge.

One hour before the estimated time of arrival

- Inform the chief engineer of the estimated time of arrival.
- Prepare the engine room for port arrival as the bridge will start to reduce speed to manoeuvring speed.
- Start the additional generator in parallel with the running generator.
- Stop the steam turbine and shaft generator (if installed).
- Ensure sufficient power is available for the deck machinery and bow thrusters.
- Be prepared for the bridge to activate the additional steering gear unit.
- Conduct rounds of the steering gear room and check the oil level linkages and motor current.
- Start the auxiliary boiler and raise the pressure (if the exhaust gas boiler was in operation).
- Close the dampers for the exhaust gas boiler and open the bypass.
- Stop the freshwater generator.
- Shut off and lock the sewage direct overboard discharge and open the valve for the sewage holding tank.
- Close the sea chest valve on the ship's side facing the berth.
- Changeover from low sea suction to high sea suction.
- Drain the air receivers.
- Consult with the bridge to evaluate the engine for ahead and astern.
- Assess the steering gear for full rudder angle.

Once the manoeuvring is finished, the sea passage flowmeter count is taken for calculating the oil consumption from full away to the end of the sea passage.

On arrival at port

Once the vessel has berthed, control of the engine is returned to the engine control room:

- Stop the additional running generator, however, ensure the online generator can take the full load.
- Close the main air starting valve.
- Open the indicator cock and turbocharger drain valve.
- Engage the turning gear and turn the engine for at least ten minutes.
- Open the vent for the exhaust gas boiler.
- For UMS vessels: activate the arrival programme for the UMS to stop the lube oil pump, crosshead pump, shaft bearing and stern tube bearing pumps.
- For manned engine control rooms: follow the manual procedures for stopping the lube oil pump, crosshead pump, shaft bearing, and stern tube bearing pumps.
- Open the jacket water heating for the main engine.
- Conduct rounds of the engine room.
- For UMS vessels: inform the bridge or the ship's control centre that the engine room port procedures are complete, and the engine control room is unmanned.
- For manned engine control rooms: resume normal engine room duties or hand over to the next watch, whichever is applicable.

UMS CHECKLISTS

Unmanned machinery spaces, unattended machinery spaces, or UMS engine rooms are recent development on merchant vessels. Unlike the conventional watch system on normal ships, with unmanned machinery space class vessels, there are usually no engineering officers on watch in the engine room between 1700 hours and 0700 hours. If there is a malfunction of any machinery, an alarm will sound in the engine room as well as in the watch duty engineer's cabin. The watch engineer must then attend to the engine room to investigate the cause of the alarm. Before the duty engineer switches the engine room to unmanned machinery space mode at the end of the working day, they must conduct thorough engine room rounds to check the condition of all running machinery systems. They should also try to anticipate any probable alarms that may trigger during the night. Most shipping companies which have unmanned machinery space class vessels provide some form of checklist for the duty engineer to complete prior to signing off watch.[1]

Individual ships will have specific unmanned machinery space checklists tailored to the system installed, but as a generic illustration, the following are virtually universal:

- When the engine control room can be put into unmanned machinery space mode will depend on the company's operating procedures and the chief engineer's standing orders.
- When preparing to leave the engine control room, the duty engineer must consult with the bridge and inform the officer of the watch of their intention to head to the

mess, their cabin, or the ship's gym. It is vital the officer of the watch is always aware of the duty engineer's whereabouts.
- Before leaving the engine control room, the *Dead Man Alarm* must be activated. The Dead Man Alarm system provides a safeguard for the duty engineer who enters the machinery space during periods of unmanned machinery space operation in response to an alarm or for any other reason. Should the engineer get into difficulty whilst alone in the machinery space, the Dead Man Alarm system provides an indication to the other engineers and the bridge that the engineer is present in the machinery space but is unable to respond by activating the response switch. The Dead Man Alarm system can be started at the engine control room control console or at the entrance to the engine room. The engineer must reset the timer by acknowledging their presence in the engine room at intervals not exceeding 24 minutes. Timer reset buttons are located on each of the engine room signal light columns. If the countdown timer is not reset within a 15-minute period, an alarm is activated on the bridge. The timer may be reset by turning the switch to the OFF position and then to the PAT position. Before the duty engineer leaves the engine room, the switch must be returned to the OFF position.
- All fuel oil, lube oil, and freshwater tanks and sumps must be filled. This includes all engine room tanks such as the freshwater expansion tank, the main engine cylinder oil day tank, the generator sump tank, freshwater tanks, cascade tank, fuel oil tanks, marine gas oil tanks, main engine sump tank, and stern tube lube oil tanks.
- The bilges must be checked and confirmed dry, and the high-level alarms are made operational.
- Ensure the sludge and bilge tanks are at a minimum safe level. The sludge tanks will continue to fill up when the purifiers are running. Similarly, the oily bilge tanks will fill whenever the main engine is running. Always ensure they are below the minimum value to avoid them from overflowing and activating the overflow alarm.
- Ensure the smoke and fire alarms are set and no zone or detector is isolated. Often when conducting hot work in the engine room, the fire detectors of a particular zone are isolated. It is common to forget to reactivate the detectors after the hot work is complete. Therefore, it is necessary to check the detectors each time the engine room is left unattended.
- Ensure all alarms and safety cut-outs are operational. Ensure no alarms are in repose condition or in manual mode. Also, ensure all alarm set values are correct and not manipulated to avoid activation.
- Ensure the compressed air receivers are fully charged and one main air receiver is isolated for standby. The air bottles should be fully charged up to 30 bar. The pressure switches of the main air compressors ensure the air bottles have sufficient pressure all the time.
- Ensure the fuel oil separator feed inlets are suitably adjusted; the separating temperature is set to 98°C (208.4°F); the automatic temperature controller is operational. It is necessary to ensure maximum purification of the fuel oil is achieved at a minimum feed rate and separating temperature of 98°C (208.4°F). It should be noted that it may not be possible to always maintain the minimum feed rate, depending on fuel oil consumption and requirements.
- Ensure the fuel oil and diesel oil overflow tanks are empty.
- Check that the emergency diesel generator is on standby mode. The emergency diesel generator should always be on standby. In the event of a blackout, the emergency

generator should start automatically and supply the emergency switchboard within 45 seconds.
- Ensure the CO_2 and water mist systems are operational. In the event of a fire in the engine room, the water mist system should operate automatically. Hyper mist or water mist is provided for the main engines, generators, auxiliary boiler, purifiers, and the incinerator.
- Check that all sounding self-closing cocks are in the closed position and the sounding caps are closed. After taking soundings, confirm the self-closing cocks are free in their movement and the sounding caps are properly shut.
- Ensure the stopped diesel generators are on standby. All standby generators should be on standby and auto-start and synchronised in case of on-load generator failure.
- Ensure all standby pumps and machinery systems are on standby and auto-start. In case of any failure, the standby pumps or machinery should auto-start. This is referred to as *redundancy*.
- Ensure the auxiliary boiler is on standby and auto-start, and the auto-controllers are all functioning properly. When the main engine is running and steam pressure is supplied by the economiser or exhaust gas boiler, the auxiliary boiler should be put into auto-start mode and standby. If due to any malfunction, the economiser fails to provide sufficient steam pressure, the auxiliary boiler should fire to ensure the steam pressure does not fall.
- Ensure all ventilation fans are operational and running. The engine room should always be under positive pressure to assist the main engine and the generator's turbocharger. It is also necessary to maintain the ship's engine room temperature at a comfortable level.
- Ensure all combustible materials are stowed safely. To avoid any fire hazards when the engine room is unattended, make sure all combustible materials are locked away and secured properly.
- The pressure difference between the diesel particulate filters is set to minimum or not beyond the specified limit. The differential pressure filters are auto-backwash filters which have a self-cleaning mechanism. If the differential pressure is greater than the specified limit, the filter candles are choked, resulting in a decrease in pressure at the outlet. If left unresolved, this can cause other problems.
- Ensure the engine room and steering gear room watertight doors are shut and confirm the funnel flaps are locked. The ship must be left in a secure and seaworthy condition, and under no circumstances should there be scope for any water ingress into the engine room.
- Check that all operating parameters are within normal range. All working parameters, such as the freshwater cooling temperatures, fuel oil, and lube oil temperatures and pressures, and the exhaust temperature, should be within normal range.
- Ensure the engine control room air conditioning is operating correctly. It is important that the temperature inside the engine control room is appropriate for the electronics.
- Ensure any loose items are secured, including tools, manuals, and folders. Rolling and pitching movements can cause items to dislodge. These present a hazard for the engine room machinery.
- Ensure any domestic electrical appliances are unplugged. This includes coffee machines and kettles.
- Ensure the workshop welding plug is removed. Whenever the welding rectifier machine is left unattended, the plug should be removed to avoid fire hazards.

Engine room logbook entries and checklists 383

Vessel: _____   Date: _____

A tick indicates the check has been performed and appropriate action taken.
N/A indicates the check is not applicable to the vessel or the prevailing conditions.

Check		Check	
Has any part of the propulsion machinery been dismantled during the port stay?		Auxiliary engine: oil level, pressure, and temperature in order.	
If YES, the engine must be turned by the turning gear at least ONE revolution after permission is obtained by the OOW. The engine must be started TWO times: ahead and astern.		Fuel oil service tanks: level sounded and recorded; temperature in order; water drained off	
Lube oil sump main engines: water free; oil level in order, sounded and recorded.		Cylinder oil lubricators: level and temperature in order.	
Capacity of running generators is sufficient.		Whistle: steam, air open, electric motor ready.	
Emergency generator: ready and on auto start.		Lube oil: level in the reduction gear in order.	
Retractable thruster: secured in the housed position.		FW cooling system: valves correctly set; all inlet and discharge valves open.	
Bow and or stern thruster(s): ready.		SW cooling system: valves correctly set; all inlet / discharge valves open.	
Alarms: all connected, tested and in order.		Turning gear: disconnected and secured.	
Lube oil system for the main engines: valves are correctly set.		Level engine: pit / tunnel, well acceptable; bilge alarms tested and in order.	
Lube oil system, pumps for the main engines: pressure and temperature in order.		Engine room hoist, tools and heavy spare parts stowed and secured.	
Stern tube and seal: lube oil arrangement in order and ready for start.		Double bottom access openings secured in the closed position.	
Cooling water systems and pumps: pressure and temperature in order.		Telephone to the bridge: tested and in order.	
Main engine: preheated and temperature in order.		Engine room telegraph and emergency telegraph: tested from the bridge / ECR, and in order.	
Fuel oil system: viscosity in order and valves correctly set.		Pitch propeller: moved full ahead / astern before clutching in tail shaft.	
Fuel oil booster priming pumps and fuel valve cooling pumps (if appropriate): pressure and temperature in order.		Steering gear: tested in conjunction with OOW.	
Fuel oil system main engine: air bleed completed if appropriate).		Starting air: admission to main engine in order.	
Boiler plant: level and pressure in order; safety devices connected.		Engine staff: required number on duty.	
Starting air compressors: auto start / stop in order.		Auxiliary blowers: running.	
FW expansion tank: water level in order.		Engines: turned with starting air and open indicator cocks.	
Starting air and control air: pressure in order; water drained off.		Stand-by reported to the bridge for testing main engine.	
Starting air compressors: ready for service.		Testing of engines to be conducted according to agreement with the master.	
Pressure in the starting air vessels: in order.		If this is NOT possible, first manoeuvre is considered the test during which final control is to be conducted.	
Indicator cocks (main engine): in order.		Definitive stand-by for departure reported to the master by the chief engineer.	
Lube oil system (turbo chargers): expansion tank level in order.		Manoeuvring: agreed with the master to be conducted from bridge.	
Piston cooling (main engines): outlets in order.		Manoeuvring: agreed with the master to be conducted from the ECR.	
Upon completion of check, entry must be made in vessel's Engine Logbook as follows: - "BEFORE DEPARTURE CHECKLIST COMPLETED." Date: .... Time:			

Port / Installation: -   Date: _____   Time: _____

Completed By: -   _____   _____
                    Name (Engineer on Duty)   Signature (Engineer on Duty)

Reviewed By: -   _____   _____
                    Name (Chief Engineer)   Signature (Chief Duty)

Figure 30.1 Engine room departure checklist.

- Acetylene and oxygen cylinder line valves must be closed and secured.
- When preparing to leave the engine control room, advise the bridge that the main engine is on bridge control. Check that the main engine control on the bridge is activated. Also, make sure that any changes to rpm or other operational parameters are actioned before securing the engine control room.

- Ensure the data logger and alarm printer are working satisfactorily. In the engine room, all logs and alarms are recorded by two separate printers for records. Make sure they are functioning properly.
- Advise the chief engineer and second engineer that the engine control room is going into unmanned machinery space mode.
- Conduct final rounds then transfer control of the engine room to the bridge. Activate the remote unmanned machinery space cabin alarm. Record the date and time in the engine room logbook.

In summary, then, there are many things that the ship's duty engineers need to check depending on the condition of the engine room, operational requirements, the type of ship, and the chief engineer's standing orders. unmanned machinery space class engine rooms have the benefit of needing to be manned 24 hours a day; however, the responsibility that is placed on the duty officer before leaving the engine room is heavy. It should also be recognised that although the duty engineer may not formally be on watch, they remain responsible nonetheless for the same and smooth operation of the ship's machinery and systems. Any alarms and faults must be investigated and rectified accordingly. Conventional engine control rooms need to be manned continuously, and therefore follow standard watches. Whilst no one particularly likes standing watch during the early hours of the morning, the continuous presence of the engineer of the watch does provide a certain degree of confidence for the other crew members. Nevertheless, as technology continues to develop and improve, it is likely shipbuilders will continue to invest in UMS (Figure 30.1).

NOTES

1. It should be noted that although the duty engineer is not required to be physically present in the engine room, they remain fully responsible for the duration of their watch period. The OOW on the bridge must be always informed of the duty engineer's whereabouts and when the duty engineer intends to sleep or remain active.

Appendix
Recommended reading for marine engineers

Marine engineering is a specialised subject, which requires equally specialised books. Marine engineers, both students and seagoing professionals, must continuously keep themselves abreast with changing technologies and policies in the maritime field. Though there are many books available to read and learn about marine engineering and its affiliated areas, there are a few that are extremely comprehensive and useful for marine engineers around the world. The following books are some of those which the author has found especially useful when writing this publication.

1. *Introduction to Marine Engineering*
   One of the most-read books on marine engineering, *Introduction to Marine Engineering* by D. A. Taylor provides information on every aspect of the ship's machinery systems. An extremely valuable guide for maritime professionals, this is a "must-have" book for those studying marine engineering or those preparing for their competency examinations.
   - Author: D. A. Taylor
   - Language: English
   - Page count: 372
   - Publisher: Butterworth-Heinemann
   - Date published: 1996
2. *General Engineering Knowledge* (Marine Engineering)
   General Engineering Knowledge by H. D. McGeorge is the perfect textbook for those seeking a one-stop resource for marine engineering subjects. Though it does not provide in-depth insights into the working of all machinery, the book is quite useful for developing a general overview of the important aspects of every machine on modern ships.
   - Author: H. D. McGeorge
   - Language: English
   - Page count: 168
   - Publisher: Routledge
   - Date published: 2015
3. *Marine Auxiliary Machinery*
   Yet another masterpiece by H. D. McGeorge, *Marine Auxiliary Machinery*, is a highly respected book for both students and seagoing marine engineers. This book is the ultimate guide for marine auxiliary machinery and provides comprehensive

information on the operation, care, and maintenance of the ship's machinery and apparatus.
  - Author: H. D. McGeorge
  - Language: English
  - Page count: 528
  - Publisher: Butterworth-Heinemann
  - Date published: 1999

4. Lamb's *Question and Answers on Marine Diesel Engines*
Lamb's *Question and Answers on Marine Diesel Engines* is a one-stop resource for knowledge on marine engines. From the most basic concepts to the most advanced principles, this book covers everything in an easy-to-understand format.
  - Author: S. G. Christiansen
  - Language: English
  - Page count: 544
  - Publisher: Butterworth-Heinemann
  - Date published: 1990

5. Pounder's *Marine Diesel Engines and Gas Turbines*
Pounder's *Marine Diesel Engines and Gas Turbines* is one of the oldest books on marine diesel engines and gas turbines. Now in its ninth edition, the book has extensively helped both students and seagoing engineers to understand the technology and concepts behind marine diesel engines and gas turbines.
  - Author: D. Woodyard
  - Language: English
  - Page count: 928
  - Publisher: Butterworth-Heinemann
  - Date published: 2009

6. *Marine Electrical Equipment and Practice*
An important book in H. D. McGeorge's marine engineering series of textbooks, *Marine Electrical Equipment and Practice* is an important book for marine engineers who seek to understand ships' electrical equipment and operating principles.
  - Author: H. D McGeorge
  - Language: English
  - Page count: 168
  - Publisher: Butterworth-Heinemann
  - Date published: 1993

7. *Marine Boilers*
An extremely comprehensive and important book on marine boilers used on ships, this book on marine boilers provides information on boilers and associated equipment as used at sea. A must have for all marine engineers sitting their competency exams, it covers every aspect of marine boilers and their types used on ships.
  - Author: G.T.H. Flanagan
  - Language: English
  - Page count: 128
  - Publisher: Butterworth-Heinemann
  - Date published: 1990

8. Reed's *Basic Electrotechnology* – Marine Engineering Series
Reed's sixth volume of the marine engineering series deals with the electrotechnology and electrical engineering principles of the ship. A step-by-step solution to a variety

of electrical systems is provided in an easy-to-understand format. This book should be with every marine engineer going to the sea.
- Author: E.G.R. Kraal and S. Buyers
- Language: English
- Page count: 640
- Publisher: Adlard Coles Nautical
- Date published: 2004

9. *Practical Marine Electrical Knowledge*

Practical Marine Electrical Knowledge is an excellent book by D.T. Hall. The book provides knowledge of the ship's electrical systems, including the generation plant, switchboards, and distribution networks. It also focuses on the electrical safety requirements and safe working practices required on ships.
- Author: D.T. Hall
- Language: English
- Page count: 132
- Publisher: Witherby
- Date published: 1999

10. *Ship Construction*

Ship Construction by D. J. Eyres is a comprehensive book on shipbuilding and shipyard practices for marine engineers and naval architects. It also provides the latest developments in the construction of different types of ships and safety practices. This is an essential book for both maritime students and professionals working in the field of ship construction and maintenance.
- Author: D. J. Eyres
- Language: English
- Page count: 376
- Publisher: Butterworth-Heinemann
- Date published: 2007

Index

Actuator, 13–14
Air
    Circuit Breaker, 173
    Handling Unit, 181
Apparatus
    Self-contained breathing, 344, 356
Atmosphere, 66, 91, 94, 95, 128, 182, 223, 238, 239, 244, 299, 315, 345, 354, 358, 359
    Controlled, 223, 239
    Corrosive, 141
    Modified, 223
    Monitoring, 238
American
    Bureau of Shipping, 38
    North, Comb jellyfish, 255
    Wire Gauge, 134
Automatic Fresh Air Management Container, 223

Ballast Water
    BWM, 229–231, 235, 242, 254
    Certificate, 235
    Convention, 233
    Management, 225, 233, 234
    Officer, 232–234, 238, 254
    Operations, 234
    Plan, 230, 232, 233, 234, 235, 242, 254
    Procedures, 233
    Records, 234
Bearing
    Temperature Monitor, 117, 118
    Wear Monitor, 58, 117
Bellows, 94, 359
Biological Oxygen Demand, 276, 278, 280
Blower, 81, 82, 83, 128, 163, 172, 182, 183, 190, 262, 276, 277, 347, 356, 358, 359
    Aeration, 182, 183
    Air, 276
    Auxiliary, 81, 82, 83, 84, 86

    Cut-off, 81
    Soot free, 47
Boiling Liquid Expanding Vapour Explosion BLEVE, 315
Boil-Off, 266, 316
    Gas Management System, 316
    Hydrogen, 316
Bottom Dead Centre, 117
    BDC, 118
Bunker Delivery Note BDN, 314

Chemical Oxygen Demand, 276, 280
Chief Engineer, xxix, xxx, 17, 74, 81, 82, 86, 106, 114, 116, 121, 133, 134, 150, 157, 175, 232, 234, 252, 259, 261, 266, 273, 274, 312–314, 317, 319, 322, 328, 331, 338, 344, 347, 349, 356, 361, 362, 373–376, 378–380, 384
Clean Ballast Tank, 237, 238
Consumption
    Air, 61–63
    Fuel, 3, 19, 80, 82, 83, 85, 86, 88, 177, 210, 267, 328, 329, 333
    Specific fuel oil, 74, 85, 86, 88, 90
Colony-Forming Unit cfu, 231
Condition Assessment Scheme, 249
Controllable Pitch Propeller, 30, 36
    CPP, 30, 35, 36
Crankcase, 68, 104, 106, 107, 115, 116, 117, 118, 119, 128, 157, 182, 190, 192, 193, 194, 195, 225, 304, 339, 341, 327
    Chamber, 117, 119, 128
    Explosion, 52, 115, 338, 339, 341, 342, 372
    Inspection, 115, 116, 117, 130, 152, 157
    Lubrication, 104, 106, 107, 115, 157, 222
    Space, 116, 117
Crude Oil Washing COW, 247, 250, 306, 308, 354

Current
   Alternating, 144
   Direct, xxi

Diesel, xxvii, 40, 332, 341, 375
   Engine, xxvii, 45, 46, 48, 61, 129, 145, 199, 385
      Generator, 74, 137, 144, 147, 148, 160, 161, 175, 177, 212, 223, 328, 381, 382
      Oil, xxvii, 47, 53, 109, 113, 120, 122, 134, 193, 203, 266, 271, 303, 304, 311, 325, 327, 328, 376, 381
   Particulate, 94, 95, 262, 269, 345, 382
Direct
   Communication, 21
   Contact heat exchanger, 216
   Drive, 50
   Expansion R134a, 180, 181
   Online Motor, 163
   Thrust, xxviii
Double Bottom, 105, 183, 237, 304
Dry Chemical Powder, 183, 353

Effective Power, 145–147, 178
Electromagnetic
   Coil, 174
   Field, 168
   Force, 144
   Pick-up, 170
Electronic Chart Display and Information System ECDIS, 331
Electrotechnical Officer, xxviii, xxix
Emergency, 20, 41, 114, 122, 143, 145, 153, 154, 155, 159, 186, 208, 233, 273, 337, 338, 339, 342, 351, 352, 357, 362, 376, 378
   Air Bottle, 188, 190
   Air compressor, 186, 190
   Communication, 316
   Drill, 18, 300, 340, 352
   Escape, 351
   Escape Breathing Device, 354
   Evacuation, 362
   Generator, 12, 21, 114, 142, 144, 145, 158, 159, 160, 166, 175, 187, 304, 356, 376, 381
   Muster, 273, 352
   Power, 12, 18, 21, 40, 144, 158, 186
   Power pump, 353
   Lamps, 154
   Lights, 142, 145
   Procedures, 273, 313
   Response, xvii, 280, 339, 347
   Shutdown, 190, 315, 316, 317
   Standby, 149
   Steering, 18, 20, 21

   Stop, 152, 183, 205, 313, 317, 359
   System, 300
   Switchboard, 21, 141, 142, 158, 161, 164, 165, 166, 190, 348, 382
   Training, 340
Emission Control Area, 94, 100, 120, 304, 331
   ECA, 121, 130
Energy Efficiency Design Index, 145, 332
Engine Control Room, 21, 40, 46, 74, 150, 151, 232, 241, 268, 314, 338, 344, 351, 356, 378, 380, 381, 382, 383, 384
Engineer of the Watch, 274
   EOW, 372–374
Estimated Time
   of Arrival, 379
   of Departure, 238
Exhaust
   Gas, 61, 81, 83, 85, 88, 90, 91
   Gas Boiler, 83, 84, 91, 93, 113, 203, 206, 341, 345, 378, 379, 380, 382
   Gas Cleaning Systems, 94, 102
   Gas flow, 85, 91, 345
   Gas pipes, 91, 93, 94
   Gas receiver, 91
   Gas system, xvii, 90, 91, 93, 94, 95, 203
   Temperature, 84, 88, 91, 93

Fire, xvii, 66, 69, 84, 107, 152, 165, 183, 184, 190, 203, 205, 206, 268, 274, 280, 305, 337, 338, 339, 340, 341, 342, 343, 344, 345, 346, 347, 348, 349, 351, 352, 353, 354, 355, 356, 357, 358, 359, 360, 382
   Alarm, xxviii, 337, 341, 344, 349, 356, 381
   Bacteria, 348
   Boiler, 82, 206, 339, 341, 345
   Connection, 357
   Control Plan, 341, 342, 344, 351
   Control station, 184
   Crankcase, 115
   Dampers, 183, 353, 376
   Detection, 341, 344, 354, 355
   Drill, 338, 339, 351
   Electrical, 166, 222, 341
   Engine room, xvii, 52, 188, 190, 338, 343, 349
   Exhaust gas boiler, 206, 341, 345
   Extinguisher, 183, 309, 337, 344, 351, 353
   Flaps, 356
   Galley, 183
   Gas, 337, 347, 348, 355
   Hazard, xxviii, 354, 382
   Hose, 183, 344, 353, 357
   Hydrant, 183, 353
   Hydrogen, 346
   Incinerator, 341, 347

Iron, 206, 346
Locker, 357
LNG, 315
Main, 353, 354, 357
Patrol team, 342
Pool, 315
Pump, 353, 354, 357, 376
Purifier room, 341, 343
Resistant-panelling, 352
Safety System Code, 352, 354, 359, 360
Scavenge space, 53, 69, 341, 339
Soot, 206, 345, 346, 347
Station, 341, 356
Test, 352
Tetrahedron, 341
Triangle, 341
Tube, 203
Water, 145, 357
Firefighter, 338
   Outfit, 354
   Personal protective equipment, 344, 354
Firefighting, 183, 339, 347, 349, 355, 356
   Appliances, 114, 222, 351, 352, 376
   Drill, 357
   Equipment, 274, 316, 342, 349
   Media, 353
   Personnel, 342
   Procedures, 333, 339
   Systems, 145, 150, 341, 344, 345, 349, 354, 358, 359, 360
Fixed Pitch Propeller, 30
Forced Draft Fans, 204
Free Air Delivery, 187
Freshwater, xvii, xxvii, xxviii, 95, 99, 100, 102, 104, 134, 178–180, 187, 209, 211–215, 236, 258, 259, 267, 269, 281, 282, 283, 303, 338, 354, 355, 361, 372
   Cooling, xxviii, 129, 178, 182, 212, 213, 361, 382
   Heating, 212
   Generator, 84, 100, 179, 206, 207, 212, 281, 282, 379
   Inlet, 259
   Pump, 265, 281, 282
   System, 104, 157, 179, 180, 212, 213, 361
   Tank, xxviii, 128, 179, 180, 206, 281, 381
Fuel, 303, 304
   Bunker, 46, 47, 51, 311, 312, 313, 314, 315, 317, 319, 322, 323, 329
   Consumption, 3, 19, 80, 81, 82, 83, 85, 86, 88, 177, 210, 267, 328, 329, 330, 332, 379
   Injection system, 324
   Leak, 113, 310
   LNG, 45, 49, 311, 314, 315, 316, 317
   Low service, 304
   Marine, 81, 311, 320, 327
   Missing, 323
   Oil calculator, 121
   Oil deposits, 120
   Oil separator, 121, 381
   Oil, xxvii, xxix, 47, 49, 51, 53, 57, 59, 60, 63, 73, 74, 83–85, 88, 101, 105–107, 109, 113, 120–123, 134, 145, 153, 157, 190, 199, 200, 201, 247, 250, 251, 265, 266, 270, 274, 303–305, 308, 328, 329, 331, 332, 341, 345, 354, 359, 376, 378, 381, 382
   Oil, heavy, xxvii, 47, 49, 51, 53, 57, 101, 105, 106, 109, 120, 134, 247, 266, 308, 311, 314, 332
   Oil, high sulphur, 105, 121, 122, 199, 201, 304
   Oil, Light, 304, 332
   Pilferage, 317
   Quality, 327
   Residual, marine, 323
   Storage, 312
   Tank, 22, 121, 204, 237, 238, 248, 250, 251, 308, 319, 321, 327, 329–331
   Temperature, 113, 320, 329
   Testing, independent, 320, 324
   Undeclared, 326

Garbage, 183, 271, 279, 349
   Disposal, 183
   Management Plan, 183, 349
   Patch, Pacific, 279
Global Positioning System, 331

Heating, Ventilation and Air Conditioning, xvii, xxviii
   HVAC, 147, 196, 210
High
   Sulphur Fuel Oil, 105, 121, 122, 199, 201, 304
   Voltage, xxviii, 141, 142, 144, 161
Human Machine Interface HMI, 122
Hydrochloride, 348

Injection Control Unit, 60
International
   Code of Safety for Ship Using Gases or Other Low flashpoints, 315
   Code of the Construction and Equipment of Ships Carrying Liquefied Gases in Bulk, 333
   Convention for the Prevention of Pollution from Ships (1973/1978) (MARPOL), 49, 94, 99, 120, 121, 183, 237, 247–252, 257–259, 261, 264, 265, 268, 270, 271, 273, 275, 276–279, 375, 376

Convention for the Safety of Life at Sea (1965, 1974, 1980) (SOLAS), 17, 20, 21, 158, 188, 190, 274, 304, 333, 341, 351, 352, 353, 355, 356, 359, 360, 362, 363, 367
Maritime Organisation, 17, 49, 94, 229, 230, 232, 235, 241, 242, 250, 253, 270, 280, 332, 342, 352
Oil Pollution Prevention Certificate (IOPP), 235, 249, 252, 266, 310, 375
Shore Connection, 354, 357
Towing Tank Conference, 145

Lifesaving Appliances, 376
Light Diesel Oil, 201
Liquified Natural Gas (LNG), 45, 46, 49, 311, 314–317, 327, 332
   Engines, 49
   Industry, 49
Low
   Sulphur Marine Gas Oil, 105
   Voltage, 161
Lower Explosive Limit, 342, 358
Light
   Ballast, 34, 236
   Diesel oil, 201
   Lighting, xxix, 50, 143, 144, 147, 166, 210, 306, 307, 313, 374

Main
   Engine, xvii, xxvii, xxviii, xxix, 23, 27, 29, 31, 36, 42, 51–53, 55, 57, 59, 60, 63–66, 68, 69, 71, 73–75, 80–88, 90, 94, 103–106, 112–115, 118, 119, 121, 123, 128, 130, 133, 135, 142, 175, 186–188, 194, 201, 206, 212, 213, 265–267, 269, 281, 304, 328, 333, 339, 342, 345–347, 356, 358, 373, 377, 378, 380, 381–383
   Switchboard, 141, 142, 144, 157, 163–166, 172
Marine Environment Protection
   Committee MEPC, 253, 258, 270, 273, 278
Maritime Labour Convention MLC, 93
Maximum
   Allowable Working Pressure, 298
   Continuous Rating, 12, 82, 91
   Rated Power, 147
Metacentric Height, 232

Nitrogen, 101, 243, 245, 278
   Dioxide, 277
   Gas, 158
   Oxide, 94
Non-destructive Testing NDT, 38

Officer of the Watch, 79, 175, 274, 337, 338, 373, 380
Oil
   Discharge, 247, 248, 250, 259, 264, 273
   Discharge Monitor, 247, 249, 250, 257, 264
   Discharge Monitoring and Control System, 247, 249, 257, 264
   Record Book, 121, 210, 250, 252, 261, 266–268, 314, 362, 375, 376
Oily Water, 258
   Emulsion, 258
   Mixture, 195, 248, 257, 258, 261, 263, 265
   Separator, 180, 247, 251, 257, 323, 339, 375, 376
OWS, 258–264, 267–269

Permit to Work PTW, 254, 305, 309
Piston, 14, 15, 36, 46–48, 52, 53, 65, 67, 68, 70, 75, 76, 78, 79, 83, 88, 104–107, 114, 115, 122, 123, 133, 135, 158, 169, 174, 186, 190, 192, 194–196, 225, 266, 267, 304, 310, 342, 344, 345
   Ring, 52, 53, 68, 78, 83, 88, 105, 133, 190, 192, 194–196, 267, 344
   Rod, 67, 68, 104, 106, 133, 190, 266, 342
   Skirt, 67, 68
Polychlorinated Biphenyls, 271
Polyvinylchloride, 348
Port State, 230, 233, 253
   Authorities, 229, 233, 250, 309
   Control, 17, 114, 121, 123, 124, 153, 229, 230, 232, 234, 252, 264, 268, 274, 314, 317, 339
   Inspection, 268
Power
   Management, 117, 217
   Management system, 117, 150
   Management plan, 178
Preventative
   Maintenance, 114
   Maintenance Schedule/System, 42, 108, 113, 114, 117, 120, 123, 129, 153, 155, 166, 170, 188, 189, 208, 209, 294, 327, 332, 357, 376

Quick
   Acting valve, 190
   Closing valve, 65, 190, 344, 354, 356
   Connect, 316

Rapid
   Biological breakdown, 182
   Phase Transition, 315

Safety Management System SMS, 17
Scrubber, xvii, 90, 94–102, 359

Seawater, xxvii, xxviii, 23, 29, 33, 34, 47, 57, 95, 99, 100, 135, 157, 178, 206, 207, 209, 211–215, 229, 236, 238, 239, 246, 259, 262, 265, 267, 269, 275, 279, 281, 282, 306, 338, 353, 355, 357, 359, 361, 362, 376

Spaces
  Machinery, xxviii, 177, 248, 250, 353, 354, 366, 368, 377, 380
  Periodically unattended machinery, 177
  Unmanned machinery, xxviii, 213, 377, 380
Statement of Compliance, 249
Sulphur
  Dioxide, 95, 101, 348
  Fuel Oil, 105, 107, 121, 122, 145, 199, 201, 304, 314, 327
  Oxide, 94, 101

Time
  Ship's mean, 330, 332
Top Dead Centre, 60, 75, 115, 192
Total
  Base Number, 53, 105, 116
  Dissolved Solids, 207, 209
Turbocharger, 48, 58, 61–64, 68, 79, 81–83, 85, 88, 91, 93, 94, 104, 105, 128, 150, 151, 153, 180, 187, 216, 267, 304, 310, 345, 347, 377, 378, 380, 382

Ultra-Violet, 243

Variable
  Geometry Turbocharger, 64
  Pitch Propeller, 36
Very High Frequency VHF, 21
Vessel General Permit, 209

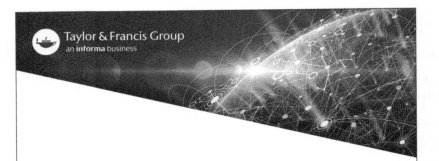

For Product Safety Concerns and Information please contact our EU representative GPSR@taylorandfrancis.com Taylor & Francis Verlag GmbH, Kaufingerstraße 24, 80331 München, Germany

Printed and bound by CPI Group (UK) Ltd, Croydon, CR0 4YY